AF614723

NEUROMETHODS

Series Editor
Wolfgang Walz
University of Saskatchewan
Saskatoon, Canada

For further volumes:
http://www.springer.com/series/7657

Muscarinic Receptor: From Structure to Animal Models

Edited by

Jaromir Myslivecek

Institute of Physiology, Charles University, Prague, Czech Republic

Jan Jakubik

Department of Neurochemistry, Acedemy of Sciences of the Czech Republic, Institute of Physiology, Prague, Czech Republic

Editors
Jaromir Myslivecek
Institute of Physiology
Charles University
Prague, Czech Republic

Jan Jakubik
Department of Neurochemistry
Academy of Sciences of the Czech Republic
Institute of Physiology
Prague, Czech Republic

ISSN 0893-2336 ISSN 1940-6045 (electronic)
Neuromethods
ISBN 978-1-4939-2857-6 ISBN 978-1-4939-2858-3 (eBook)
DOI 10.1007/978-1-4939-2858-3

Library of Congress Control Number: 2015945343

Springer New York Heidelberg Dordrecht London

Printed on acid-free paper

Humana Press is a brand of Springer
Springer Science+Business Media LLC New York is part of Springer Science+Business Media (www.springer.com)

Preface

Muscarinic receptors are involved in a number of physiological events like cognitive processes, motor coordination, attention, circadian rhythms, food reinforcement, drug addiction, and synaptic plasticity. This book provides methodology for the study of muscarinic receptors at the structural to systemic level. The chapters are primarily intended as a resource for scientists who want to newly establish protocols to study muscarinic receptors without hitches and falling to potential pitfalls. One of such pitfalls is the lack of subtype-selective ligands that makes studies targeted to specific subtype problematic. One of the methodological approaches for subtype identification in tissues and organs is immunohistochemical or Western blot analysis of muscarinic receptors. However, these methods have strong limitations as the selectivity of antibodies is usually poor and antibodies also target nonfunctional and degraded receptors. Thus these methods do not provide assessment of real number of receptors. Moreover, some artifacts can originate from tissue preparation. These can be avoided by studying receptors in their natural environment.

We start our book with methods for characterization of muscarinic receptor in crystallography studies that advanced our understanding of structural properties and activation mechanism of muscarinic receptors and are cornerstone in molecular modeling and computer-based approaches to study muscarinic receptors. Then we move to binding techniques that thanks to heterologous expression systems allow us to perform binding studies very accurately and easily and discuss overcoming difficulties arising from the lack of selective ligands. Next we provide protocols to investigate molecular properties of muscarinic receptors. Then we provide protocols to study muscarinic receptors in the central nervous system using autoradiography and PET studies. We end with protocols on animals with knock-out and knock-in muscarinic genes to study the role of muscarinic receptors in physiology and behavior.

Prague, Czech Republic

Jaromir Myslivecek
Jan Jakubik

Series Preface

Experimental life sciences have two basic foundations: concepts and tools. The *Neuromethods* series focuses on the tools and techniques unique to the investigation of the nervous system and excitable cells. It will not, however, shortchange the concept side of things as care has been taken to integrate these tools within the context of the concepts and questions under investigation. In this way, the series is unique in that it not only collects protocols but also includes theoretical background information and critiques which led to the methods and their development. Thus it gives the reader a better understanding of the origin of the techniques and their potential future development. The *Neuromethods* publishing program strikes a balance between recent and exciting developments like those concerning new animal models of disease, imaging, in vivo methods, and more established techniques, including, for example, immunocytochemistry and electrophysiological technologies. New trainees in neurosciences still need a sound footing in these older methods in order to apply a critical approach to their results.

Under the guidance of its founders, Alan Boulton and Glen Baker, the *Neuromethods* series has been a success since its first volume published through Humana Press in 1985. The series continues to flourish through many changes over the years. It is now published under the umbrella of Springer Protocols. While methods involving brain research have changed a lot since the series started, the publishing environment and technology have changed even more radically. Neuromethods has the distinct layout and style of the Springer Protocols program, designed specifically for readability and ease of reference in a laboratory setting.

The careful application of methods is potentially the most important step in the process of scientific inquiry. In the past, new methodologies led the way in developing new disciplines in the biological and medical sciences. For example, Physiology emerged out of Anatomy in the nineteenth century by harnessing new methods based on the newly discovered phenomenon of electricity. Nowadays, the relationships between disciplines and methods are more complex. Methods are now widely shared between disciplines and research areas. New developments in electronic publishing make it possible for scientists that encounter new methods to quickly find sources of information electronically. The design of individual volumes and chapters in this series takes this new access technology into account. Springer Protocols makes it possible to download single protocols separately. In addition, Springer makes its print-on-demand technology available globally. A print copy can therefore be acquired quickly and for a competitive price anywhere in the world.

Wolfgang Walz

Contents

Contributors

GLENDA ALQUICER • *Department of Neurochemistry, Institute of Physiology, Academy of Science of the Czech Republic, Prague, Czech Republic*

HIDETSUGU ASADA • *Department of Medical Chemistry and Cell Biology, Kyoto University Graduate School of Medicine, Kyoto, Japan*

KOICHI ASAKAWA • *School of Veterinary Medicine, Rakuno Gakuen University, Ebetsu, Hokkaido, Japan*

VÉRONIQUE BERNARD • *Université Pierre et Marie Curie UM 119 – CNRS UMR 8246 – INSERM U1130, Paris, France*

CHARLES D. BLAHA • *University of Memphis, Memphis, TN, USA*

MORITZ BÜNEMANN • *Institute of Pharmacology and Clinical Pharmacy, Philipps-University Marburg, Marburg, Germany*

DAVID K. CHALMERS • *Medicinal Chemistry, Monash Institute of Pharmaceutical Sciences, Monash University, Parkville, VIC, Australia*

VLADIMÍR DOLEŽAL • *Department of Neurochemistry, Institute of Physiology, Academy of Science of the Czech Republic, Prague, Czech Republic*

ESAM E. EL-FAKAHANY • *Department of Experimental and Clinical Pharmacology, University of Minnesota College of Pharmacy, Minneapolis, MN, USA*

VLADIMIR FARAR • *Institute of Medical Biochemistry and Laboratory Diagnostics, 1st Faculty of Medicine, Charles University, Prague, Czech Republic*

NAO HARADA • *School of Veterinary Medicine, Rakuno Gakuen University, Ebetsu, Hokkaido, Japan*

JAN JAKUBIK • *Department of Neurochemistry, Institute of Physiology, Academy of Sciences of the Czech Republic, Prague, Czech Republic*

SHINYA KANEGAE • *School of Veterinary Medicine, Rakuno Gakuen University, Ebetsu, Hokkaido, Japan*

CHARLES JAMES KIRKPATRICK • *Institute of Pathology, University Medical Center, Johannes-Gutenberg Universität, Mainz, Germany*

TAKIO KITAZAWA • *School of Veterinary Medicine, Rakuno Gakuen University, Ebetsu, Hokkaido, Japan*

TAKUYA KOBAYASHI • *Department of Medical Chemistry and Cell Biology, Kyoto University Graduate School of Medicine, Kyoto, Japan; Japan Science and Technology Agency (JST), Core Research for Evolutional Science and Technology (CREST), Kyoto, Japan; Platform for Drug Discovery, Informatics and Structural Life Science, Kyoto, Japan*

SEI-ICHI KOMORI • *Laboratory of Pharmacology, Faculty of Applied Biological Science, Gifu University, Gifu, Japan*

CORNELIUS KRASEL • *Institute of Pharmacology and Clinical Pharmacy, Philipps-University Marburg, Marburg, Germany*

TAKAYOSHI MASUOKA • *Department of Pharmacology, School of Medicine, Kanazawa Medical University, Uchinada, Ishikawa, Japan*

MARTIN C. MICHEL • *Department of Pharmacology, Johannes Gutenberg University, Mainz, Germany*

IKUNOBU MURAMATSU • *Department of Pharmacology, School of Medicine, Kanazawa Medical University, Uchinada, Ishikawa, Japan; Division of Genomic Science and Microbiology, School of Medicine, University of Fukui, Eiheiji, Fukui, Japan; Kimura Hospital, Awara, Fukui, Japan*
JAROMIR MYSLIVECEK • *Institute of Physiology, 1st Faculty of Medicine, Charles University, Prague, Czech Republic*
TATSURO NAKAMURA • *School of Veterinary Medicine, Rakuno Gakuen University, Ebetsu, Hokkaido, Japan*
MATOMO NISHIO • *Department of Pharmacology, School of Medicine, Kanazawa Medical University, Uchinada, Ishikawa, Japan*
SHINGO NISHIYAMA • *Central Research Laboratory, Hamamatsu Photonics K.K., Hamamatsu, Shizuoka, Japan*
KENTA OCHI • *School of Veterinary Medicine, Rakuno Gakuen University, Ebetsu, Hokkaido, Japan*
WISUIT PRADIDARCHEEP • *Department of Anatomy, Faculty of Medicine, Srinakharinwirot University, Bangkok, Thailand*
ANDREAS RINNE • *Department of Cardiovascular Physiology, Institute of Physiology, Ruhr-University Bochum, Bochum, Germany*
KIYONAO SADA • *Division of Genomic Science and Microbiology, School of Medicine, University of Fukui, Eiheiji, Fukui, Japan*
STEPHAN STEIDL • *Loyola University, Chicago, IL, USA*
RYOJI SUNO • *Department of Medical Chemistry and Cell Biology, Kyoto University Graduate School of Medicine, Kyoto, Japan; Platform for Drug Discovery, Informatics and Structural Life Science, Kyoto, Japan*
KAZUHIRO TAKAHASHI • *Department of Radiology and Nuclear Medicine, Research Institute of Brain and Blood Vessels-Akita, Akita, Japan*
TAKANOBU TANIGUCHI • *Department of Biochemistry, Asahikawa Medical University, Asahikawa, Hokkaido, Japan*
HIROKI TERAOKA • *School of Veterinary Medicine, Rakuno Gakuen University, Ebetsu, Hokkaido, Japan*
TRAYDER THOMAS • *Medicinal Chemistry, Monash Institute of Pharmaceutical Sciences, Monash University, Parkville, VIC, Australia*
HIDEO TSUKADA • *Central Research Laboratory, Hamamatsu Photonics K.K., Hamamatsu, Shizuoka, Japan*
TOSHIHIRO UNNO • *Laboratory of Pharmacology, Faculty of Applied Biological Science, Gifu University, Gifu, Japan*
JUNSUKE UWADA • *Department of Biochemistry, Asahikawa Medical University, Asahikawa, Hokkaido, Japan*
DAVID IAN WASSERMAN • *University of Guelph, Guelph, ON, Canada*
IGNAZ KARL WESSLER • *Institute of Pathology, University Medical Center, Johannes-Gutenberg Universität, Mainz, Germany*
MASAHISA YAMADA • *Common Resources Group, Okinawa Institute of Science and Technology, Okinawa, Japan*
NORIKO YAOSAKA • *School of Veterinary Medicine, Rakuno Gakuen University, Ebetsu, Hokkaido, Japan*
JOHN YEOMANS • *University of Toronto, Toronto, ON, USA*
HATSUMI YOSHIKI • *Division of Genomic Science and Microbiology, School of Medicine, University of Fukui, Eiheiji, Fukui, Japan*
ELIZABETH YURIEV • *Medicinal Chemistry, Monash Institute of Pharmaceutical Sciences, Monash University, Parkville, VIC, Australia*

Chapter 1

Towards the Crystal Structure Determination of Muscarinic Acetylcholine Receptors

Ryoji Suno, Hidetsugu Asada, and Takuya Kobayashi

Abstract

G protein-coupled receptors (GPCRs) constitute the largest family of receptors encoded by the human genome. Activation and inhibition of GPCRs under the physiological and pathophysiological conditions is largely mediated by chemical ligands (agonists and antagonists) that bind to the orthosteric binding pocket. Orthosteric ligands are, however, often nonspecific, binding to more than one GPCR subtype. In contrast to orthosteric agonists and antagonists, allosteric ligands do not directly compete with hormones and neurotransmitters for binding to the orthosteric binding pocket. Furthermore, allosteric ligands typically occupy structurally diverse regions of receptors and therefore are more selective for specific GPCRs, regulating receptor function in the more subtle ways by either enhancing or diminishing responses to natural ligands such as hormones or neurotransmitters. Recent X-ray crystallographic studies have provided detailed structural information regarding the nature of the orthosteric muscarinic binding site and an outer receptor cavity that can bind allosteric drugs. These new findings may guide the development of selective muscarinic receptor. The procedures involved in the production, purification, and crystallization of GPCRs are introduced here and facilitate a greater understanding of the structural basis of GPCR function.

Key words G-protein coupled receptor, X-ray crystal structure analysis, Muscarinic acetylcholine receptor, Antagonist, Agonist, Orthosteric binding site, Allosteric binding site

1 Background and Overview

G-protein coupled receptors (GPCRs) constitute the largest superfamily of cell surface receptors. GPCRs are seven-transmembrane domain receptors that mediate various cellular responses to specific ligands, including amines, eicosanoids, hormones, and peptides, as well as taste and light stimuli. Approximately 50 % of all currently available drugs act through GPCRs [1, 2], and GPCRs are the most important therapeutic targets for various disorders. Structure-guided drug development is important for the design of novel drugs devoid of side effects.

Electronic supplementary material The online version of this chapter (doi:10.1007/978-1-4939-2858-3_1) contains supplementary material, which is available to authorized users.

Jaromir Myslivecek and Jan Jakubik (eds.), *Muscarinic Receptor: From Structure to Animal Models*, Neuromethods, vol. 107,
DOI 10.1007/978-1-4939-2858-3_1, © Springer Science+Business Media New York 2016

Muscarinic acetylcholine receptors (mAChRs) belong to the GPCR family and five different subtypes of mAChRs (M_1, M_2, M_3, M_4, and M_5) have been cloned [3–7]. Muscarinic receptors are widely expressed in the cell membranes of brain and peripheral tissues [8, 9]. All mAChR subtypes contain a long third intracellular loop (ICL3) composed of about ~160–240 amino acids and mutagenesis studies have revealed that the N- and C-terminal portions of ICL3 play an important role in the specificity of the coupling of G-proteins to mAChRs [10]. The M_1, M_3, and M_5 subtypes couple with the G_q family, while the M_2 and M_4 subtypes couple with the G_i/G_o family.

Twenty-seven human GPCR structures including mAChRs have been determined to date and the crystal structures of the human M_2 and M_3 receptors were determined in 2012 [11, 12]. Structures of other classes of GPCRs such as glucagon receptor (class B), corticotropin releasing factor (CRF-1) receptor (class B), $GABA_A$ receptor (class C), metabotropic glutamate receptor (class C), and smoothened receptor (class F) were recently solved by X-ray crystallographic analysis. However, several technical problems still remain with regard to solve the crystal structure of membrane proteins such as GPCRs. Since milligram quantities of purified protein are required for crystallization and structural determination, one of the main obstacles for crystal structure determination is the preparation of sufficiently large amounts of functional GPCR protein [13].

Previous studies have attempted to express the human M_2 receptor with a deletion of the central portion of ICL3 from Ser234 to Arg381 using various expression systems, including *Escherichia coli* [14, 15], *Pichia pastoris* [16], and insect cells [17]. We previously identified 25 GPCRs expressed by *P. pastoris* [16] and Sf9 insect cells, which led us to suggest both systems as suitable hosts for GPCR crystal structure studies. To obtain large amounts of M_2 receptor, Sf9 insect cells and *P. pastoris* were utilized as mass production systems [18]. We expressed and purified the M_2 receptor lacking most of the central ICL3 region (Ser234–Arg381), and containing four asparagine residue mutations (Asn2, 3, 6, and 9 mutated to Asp) to prevent glycosylation. This ICL3 deletion M_2 receptor mutant was found to have the ability to bind agonists and activate G proteins [19], and ICL3 was shown to have a flexible structure [15]. The high-affinity inverse agonist (quinuclidinyl benzilate; QNB) was used and the QNB bound M_2 receptor was crystallized by hanging-drop vapor diffusion. This method yielded crystals that diffracted to around ~9 Å, and it was not possible to further improve the quality of the crystals using this method.

Exchanging a portion of the ICL3 of the M_2 receptor with T4 lysozyme (T4L) was initially used as a means of structure determination for the β_2 adrenergic receptor [20], and this method has also been used for structure determination of various GPCRs. The binding properties of M_2-T4L receptors with mAChR ligands were the same as those in the wild-type M_2 receptor, indicating that the introduction of T4L does not affect the overall architecture of M_2-T4L. The M_2-T4L receptor was subsequently crystallized in the lipidic cubic phase. A 3.0 Å structure was solved by molecular replacement from a data set obtained by merging diffraction data from 23 crystals [11]. The crystal structure of rat M_3 receptor with T4L was also determined by the lipidic cubic phase method, and diffraction data from more than 70 crystals were merged to create a data set to 3.4 Å resolution to allow for the structure to be solved by molecular replacement [12].

The structure determination experiments for the M_2 and M_3 receptors revealed that, like other biogenic amine receptors, the mAChR family exhibits the seven-transmembrane domain topology and overall fold of other GPCRs. The ligands bound to the M_2 and M_3 receptors in the two crystal structures (QNB and tiotropium, respectively) are both antagonists (inverse agonists). The orthosteric ligand-binding pocket is composed of various hydrophobic side chains of transmembrane domains (TMs) 3, 4, 5, 6, and 7 and the amino acid residues lining the orthosteric binding sites of all five (M_1–M_5) muscarinic receptor subtypes exhibit absolute sequence conservation. Some polar contacts exist between QNB and the receptors. One such contact is a pair of hydrogen bonds between $Asn^{6.52}$ and the hydroxyl and carbonyl groups in QNB, and the other is a charge–charge interaction between the cationic amine of the ligand and the conserved $Asp^{3.32}$ (superscripts denote Ballesteros-Weinstein [21] GPCR numbering, see Supplementary Fig. 1). $Asn^{6.52}$ seems to be a unique feature of the mAChR family and has been proposed to be an important factor in slow ligand dissociation from mAChRs. $Asn^{6.52}$ is important for the antagonist binding, and $Asp^{3.32}$ is important for both agonist and antagonist binding [11]. The QNB binding site forms a cavity, which is secluded from the extracellular solvent and a "lid" composed of three tyrosine residues located above the QNB-binding site to divide a large, solvent-accessible cavity into two distinct regions (Fig. 1) [11]. The lower region is the orthosteric ligand (QNB)-binding pocket, while the upper region is termed the extracellular vestibule and is implicated in the binding of allosteric modulators (allosteric site) [11].

Initial structures of the M_2 and M_3 receptors were obtained in complex with high-affinity antagonists (inverse agonists) representing inactive receptor conformations. However, obtaining crystals of active GPCRs has proved to be extremely challenging due to the conformational heterogeneity induced by agonist binding. This chapter will focus on the techniques used to solve the structure of inactive mAChR.

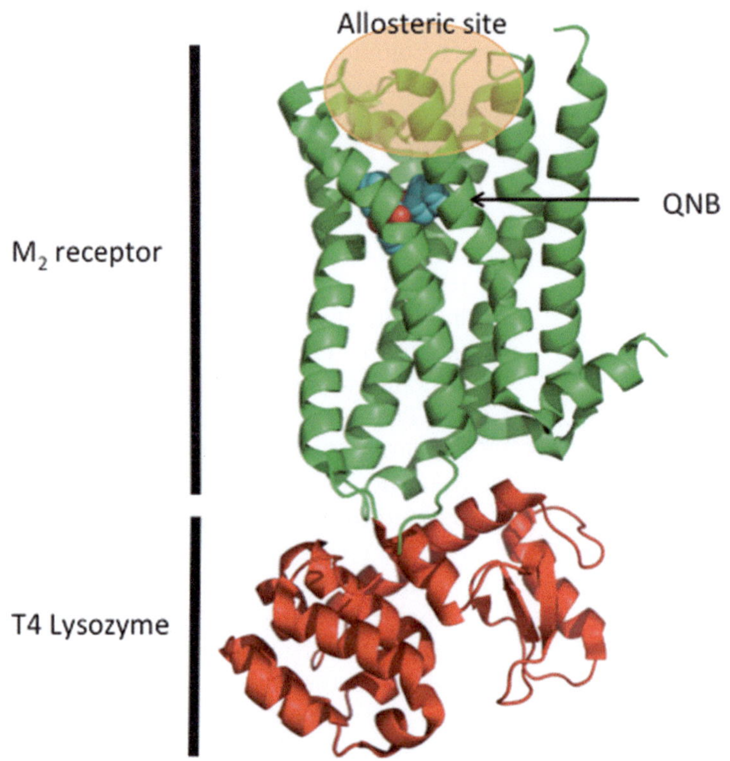

Fig. 1 The crystal structure of the inactive form of the M_2 receptor with bound antagonist, QNB, to the orthosteric binding site. M_2 receptor is colored in *green*, and the T4 lysozyme is in *red*. The allosteric site is shown above the orthosteric site

2 Materials

The Sf9 insect cell expression vector pFastBac1, the *P. pastoris* expression vector pPIC9K, and Sf9 insect cells were purchased from Life Technologies. IPL-41 insect medium was purchased from AppliChem GmbH and ESF-921 insect medium was purchased from Expression Systems. Tryptose phosphate, TC yeastolate, and Pluronic F-68 as supplements to IPL-41 medium were purchased from Life Technologies. Fetal bovine serum (FBS) was purchased from Biowest and penicillin–streptomycin was purchased from Wako. Protease inhibitor cocktail tablets (cOmplete) were purchased from Roche Diagnostics and the mAChR antagonist atropine was purchased from Sigma. Sf9 insect cells were maintained in ESF-921 medium containing 2.0 % heat-inactivated FBS. IPL-41 was supplemented with 0.1 % Pluronic F-68 and 2.0 % FBS. MD agar plates contained 1.34 % (w/v) yeast nitrogen base without amino acids, 2 % (w/v) dextrose, 0.00004 % (w/v) biotin, and 1.5 % (w/v) agar. G418-YPD agar plates contained 1 % (w/v) yeast extract, 2 % (w/v) peptone, 2 % (w/v) dextrose, 2 %

(w/v) agar, and 0.1 or 0.25 mg/ml G418. BMGY medium contained 1 % (w/v) yeast extract, 2 % (w/v) peptone, 1.34 % (w/v) yeast nitrogen base without amino acids, 0.00004 % (w/v) biotin, 1 % (w/v) glycerol, and 0.1 M phosphate buffer (pH = 6.0). BMMY medium contained 1 % (w/v) yeast extract, 2 % (w/v) peptone, 1.34 % (w/v) yeast nitrogen base without amino acids, 0.00004 % (w/v) biotin, 1 % (v/v) methanol, and 0.1 M phosphate buffer (pH = 6.0, 7.0, or 8.0). Tritium-labeled mAChR antagonist ($[^3H]$ QNB, quinuclindinyl benzilate, 1.0 mCi/ml) was obtained from PerkinElmer. GF/F glass fiber filters were purchased from Whatman. The M_2 receptor sequence was described as previously [11, 17]. For the flow cytometric analysis, anti-FLAG (M_2) antibody was purchased from Sigma and Alexa Fluor 488-conjugated goat anti-mouse IgG was purchased from Life Technologies. *N*-Dodecyl-β-d-maltopyranoside (DDM) was purchased from Affymetrix and cholesterol hemisuccinate (CHS) was purchased from Sigma. PEG 300, (±)-2-Methyl-2,4-pentanediol (MPD), and ammonium phosphate were purchased from Hampton research.

3 Methods

3.1 Construction of the Vector Encoding the M_2 Receptor for Expression in *P. pastoris*

The M_2 receptor cDNA was subcloned into the pPIC9K vector, digested by the restriction enzyme *Pme*I, and transfected into *P. pastoris* strain SMD1163 by electroporation (1500 V, 25 μF, and 600 Ω) using a Gene Pulser I (Bio Rad). Clone selection was performed as previously described [22, 23]. Briefly, histidine-positive clones were selected on MD agar plates. To select for multi-copy transformants, histidine-positive clones were grown on G418-YPD agar plates. Representative clones exhibiting resistance to G418 were tested for recombinant protein production by specific ligand-binding assays using $[^3H]$QNB. Highly functional clones were selected and stored as glycerol stocks at −80 °C.

3.2 Recombinant Expression in *P. pastoris*

For small-scale culture, a glycerol stock of a transformant of interest was inoculated onto a YPD agar plate containing 0.1 mg/ml G418. Cells were pre-cultured in 5 ml of BMGY medium at 30 °C with shaking at 250 rpm until an OD_{600} of 2.0–6.0 was reached. Induction of M_2 receptor expression was carried out in 5 ml of BMMY medium containing 0.04 % (w/v) histidine and 3 % (v/v) DMSO at 20 or 30 °C from an initial OD_{600} of 1.0. The procedure used for *P. pastoris* culture was described previously [16]. After induction of recombinant expression for an optimized amount of time (generally ~20–60 h), cells were harvested by centrifugation at 4000 ×*g* for 15 min, washed once with ice-cold water, and then either frozen or immediately used for membrane preparation.

For large-scale culture, a single *P. pastoris* colony exhibiting high expression levels of the M_2 receptor was picked from a YPD

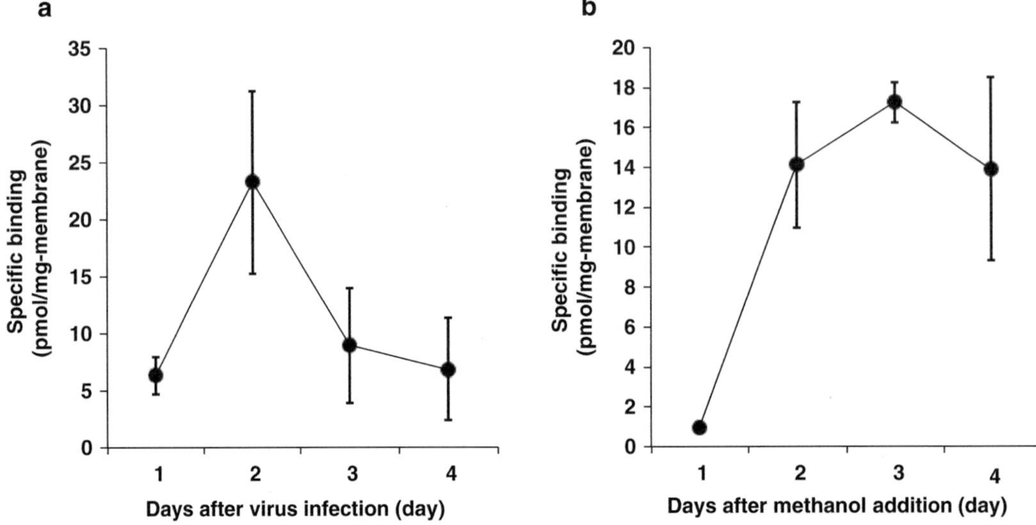

Fig. 2 Time course assessment of the M_2 expression levels in the insect cell and yeast expression systems based on the specific binding of [^{3}H]QNB to M_2 receptors in Sf9 insect cells (**a**) and *P. pastoris* (**b**)

plate containing 0.1 mg/ml G418 and was cultured in 5 ml BMGY medium overnight at 30 °C with shaking at 250 rpm. The cultured cells were inoculated into 200 ml BMGY medium and grown overnight at 30 °C with shaking at 250 rpm until an OD_{600} of 2.0–6.0 was reached. The overnight culture (200 ml) was inoculated into 1000 ml of BMGY medium for 4 h at 250 rpm and grown until an OD_{600} of 5.0–10.0 was reached. Cells were harvested by centrifugation at 4000 × *g* for 15 min, after which the cell pellet was washed with double-distilled water before being subjected to centrifugation again. The resulting cell pellet was resuspended in 1.0–8.0 l of BMMY medium to an OD_{600} of 1.0. The culture was incubated for 1–4 days at 20 °C shaking at 250 rpm, and 20 % methanol (50 ml/l culture) was added to maintain a final methanol concentration of 1 % until the end of the culture period. Culture media (1 ml) was sampled for the binding assay (Figs. 2 and 3, Table 1) and the remaining cells were harvested by centrifugation at 6000 × *g* for 10 min. The cell pellet was washed with 250 ml of double-distilled water and resuspended in 100 ml of double-distilled water containing a protease inhibitor cocktail tablet (Roche). Cells were quick-frozen in liquid nitrogen and stored at −80 °C.

3.3 Generation and Amplification of Baculovirus Encoding the M_2 Receptor for Expression in Sf9 Insect Cells

The Bac-to-Bac baculovirus expression system (Life Technologies) is a rapid and efficient system by which baculovirus can be generated for GPCR expression. An advantage of this system is the site-specific transposition property of the Tn7 transposon, which simplifies and enhances the process of generating recombinant bacmid DNA in *E. coli* [24, 25]. To solve the M_2 receptor structure, the M_2 receptor gene was constructed as described previously [11]. In brief, four N-linked glycosylation sites (Asn2, 3, 6, and 9)

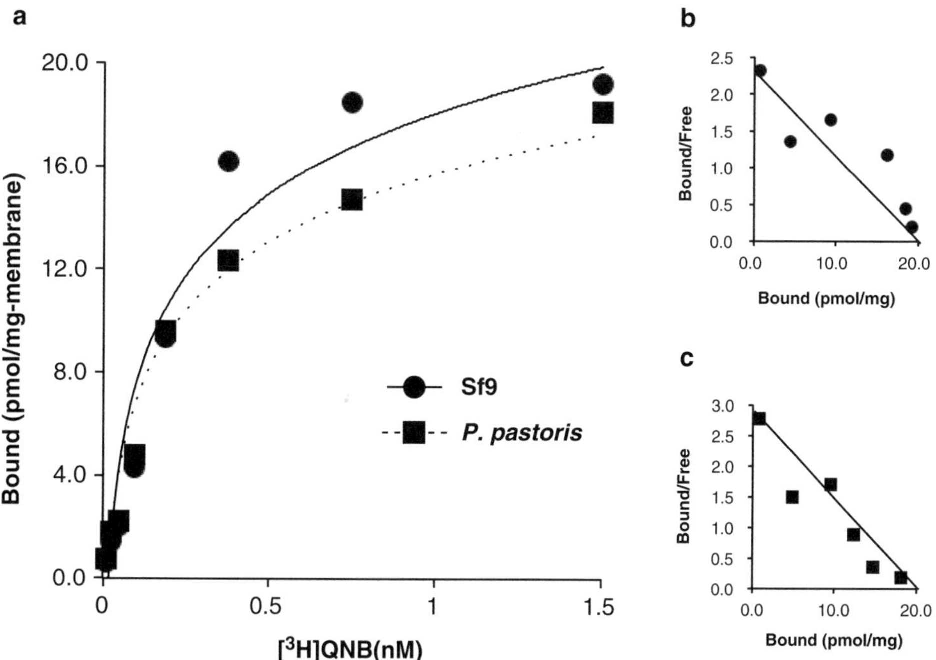

Fig. 3 Comparison of the binding properties of M_2 receptor expressed in Sf9 insect cells and *P. pastoris*. The saturation-binding curve of M_2 receptor from Sf9 insect cell and *P. pastoris* are shown in (**a**). Scatchard plots were calculated from the saturation-binding curves of M_2 receptor expressed in Sf9 insect cells (**b**) and in *P. pastoris* (**c**)

Table 1
The dissociation constant (K_D) and maximum specific binding (B_{MAX}) of M_2 receptor expressed in Sf9 insect cells and *P. pastoris*

	K_D (pM)	B_{MAX} (pmol/mg-membrane)
Sf9	86.2 ± 8.6	27.2 ± 5.6
P. pastoris	101.1 ± 15.1	33.5 ± 7.1

The M_2 receptor was expressed in Sf9 insect cells and *P. pastoris* on a small scale (Sf9: 10 ml, *P. pastoris*: 5 ml). Cell membranes were prepared to determine the specific binding of M_2 receptor using the radioactive M_2 receptor antagonist [^{3}H]QNB. K_D and B_{MAX} values were calculated from saturation-binding curves and Scatchard plots (Fig. 3). Data are presented as means ± standard deviation

at the N-terminus were substituted with aspartic acid and the third intracellular loop of the M_2 receptor was replaced with T4 lysozyme. The M_2 receptor gene was subcloned into the pFastBac1 vector and, to generate the recombinant bacmid encoding the M_2 receptor, the M_2 receptor gene in the pFastBac1 vector was transformed into the DH10Bac (*E. coli.*) cells. A 1.5 ml overnight culture inoculated from a single colony of this transformant yields sufficient amounts of recombinant bacmid DNA for several insect cell transfections.

Sf9 insect cells (2 ml, 1×10^6 cells/well) were seeded into 6-well plates and cultured for 24 h at 27 °C. Recombinant bacmid DNA (2 μl) was transfected into Sf9 insect cells using Cellfectin II (Life Technologies) and the baculovirus (P0 virus) was amplified in Sf9 insect cells. To amplify the P1 baculovirus, 50 ml of fresh Sf9 insect cells (1×10^6 cells/ml) were transfected with 0.5 ml P0 virus. The resulting baculovirus titer was determined by seeding Sf9 cells (1×10^6 cells/well) into 6-well plates and incubating them for 1 h at 27 °C. The cells were then transfected with serially diluted baculovirus (e.g., 1000–10,000 × dilutions), cultured for 24 h, harvested in test tubes, and washed twice with FACS buffer (PBS (−), 2 % FBS, 0.05 % NaN_3). Phycoerythrin (PE)-conjugated anti-gp64 antibody (0.1 μg; eBioscience) was then incubated with the cells for 30 min on ice in the dark. For flow-cytometric analysis, the labeled cells were washed with FACS buffer and resuspended in 100 μl of FACS buffer. For the expression of large amounts of M_2 receptor, 1 l virus supernatant was prepared by P1 virus infection at an MOI (multiplicity of infection) of 0.1. The virus supernatant (P2 virus) was collected and used for small- or large-scale expression of the M_2 receptor.

3.4 Expression Procedure for Sf9 Insect Cells

Small-scale culture was performed to transfer to large-scale culture using a wave bioreactor (GE Healthcare) or to determine optimal expression conditions such as MOI and culture period. Sf9 insect cells were cultured in 125 ml Erlenmeyer flasks shaken at 125 rpm at 27 °C and were passaged every 3–4 days. To optimize expression conditions, 10 ml Sf9 insect cells (1×10^6 cells/ml) were seeded into 125 ml Erlenmeyer flasks infected with several MOI of baculovirus (e.g., MOI = 0.5–5), and grown for 2–4 days at 27 °C with shaking at 125 rpm. The cells were stained with anti-FLAG antibody (primary antibody) and goat anti-mouse IgG conjugated with Alexa Fluor 488 (secondary antibody) for flow cytometric analysis. Ligand-binding activity was furthermore measured using [^{3}H]QNB (see Section 3.5 below).

For large-scale culture, Sf9 insect cells were suspended (1×10^6 cells/ml) in 5 l ESF921 insect media containing 2 % FBS. The cell suspension was transferred to a cell culture bag (CELLBAG 22 L/O, GE Healthcare) [26, 27] and cultured for 1 day under the following culture conditions: shaking at 20 rpm, rocking angle of 8.5°, 30 % O_2, air flow rate of 0.25 l/min, and 27 °C. After that, 100–300 ml baculovirus stock (in the case of the M_2 receptor, optimized MOI = 2) was transferred into the cell bag and infection was allowed to proceed under the following infection conditions: 22 rpm, rocking angle of 8.5°, 50 % O_2, air flow rate of 0.5 l/min, and 27 °C. After 2 days of infection, 1 ml culture media was sampled for the binding assay (Figs. 2 and 3, Table 1) and the remaining cells were harvested by centrifugation at $6000 \times g$ for 10 min. The cell pellet was washed with 250 ml of PBS (−) and resuspended in 100 ml of PBS (–) containing a protease inhibitor cocktail tablet (Roche). Cells were quick-frozen in liquid nitrogen and stored at –80 °C.

3.5 Radioligand Binding Assay

All experiments were performed in duplicate in a total volume of 200 μl. Membrane proteins were quantified using a bicinchoninic acid (BCA) assay (Thermo Fisher Scientific) with bovine serum albumin as a standard. Membrane (5 μg) was resuspended in assay buffer (20 mM potassium phosphate buffer) in a 1.5 ml of tube and was incubated for 30 min at room temperature in the presence of 1.5 nM [^{3}H]QNB for a single point binding assay carried out with a saturating radioligand concentration. Nonspecific binding was assessed by incubation in the presence of an excess of the non-radioactive ligand (1.5 μM QNB). GF/F glass filters were pre-soaked in 200 ml polyethyleneimine (0.3 %, v/v). Bound and free ligands were separated by rapid vacuum filtration through GF/F filters. Filtration was performed with a Brandel cell harvester at room temperature and filters were washed three times with 5 ml of deionized water. Residual radioactivity was measured using a LCS-5100 liquid scintillation counter (ALOKA).

3.6 Preparation of Cell Membranes

Cell membranes from *P. pastoris* were prepared at 4 °C. Harvested cells (1 g wet weight) were suspended in 4 ml lysis buffer (50 mM sodium phosphate buffer, pH = 7.4, 100 mM NaCl, 5 % (v/v) glycerol, and 2 mM EDTA) containing protease inhibitor cocktail (one tablet/100 ml lysis buffer). Suspended yeast cells were disrupted by vortex at 4 °C for 2 h with 0.5 mm glass beads. Lysis efficiency was assessed by light microscopy and was usually found to be >80 %. Intact cells and cell debris were separated from the membrane suspension by low-speed centrifugation (3000 × *g* for 5 min at 4 °C), after which membranes were snap-frozen in liquid nitrogen and stored at –80 °C.

For preparation of the cell membranes from the Sf9 insect cells, 1 l Sf9 biomass was centrifuged at 1500 × *g* for 10 min at 4 °C. The resulting cell pellet was washed with PBS (–) and resuspended in 100 ml of hypotonic buffer containing 10 mM HEPES at pH = 7.5, 20 mM potassium chloride, 10 mM $MgCl_2$, and protease inhibitor cocktail using a Dounce homogenizer. Insect cell membranes were centrifuged at 100,000 × *g* for 30 min and the resulting pellets were resuspended in 10 mM HEPES at pH = 7.5, 10 mM $MgCl_2$, 20 mM KCl, and 40 % glycerol. The suspensions were quick-frozen in liquid nitrogen and stored at –80 °C.

3.7 Solubilization and Purification of the M_2 Receptor

M_2-T4L membranes were solubilized with a digitonin/Na-cholate solution and purified using an affinity column with aminobenztropine (ABT) as a ligand [28]. The entire procedure was carried out at 4 °C. Sf9 cell membrane preparations with ~2 kg wet weight and ~1.5 μmol of M_2-T4L as measured by ligand binding assays using

[^{3}H]QNB were solubilized with 1 % digitonin, 0.35 % Na-cholate, 10 mM potassium phosphate-buffered saline (pH 7.0) (KPB), 50 mM NaCl, 1 mM EDTA, and protease inhibitor cocktail in a total volume of 300 ml. The suspension was stirred for 1 h at 4 °C and then centrifuged at 100,000 × *g* for 1 h. The resulting supernatant was stored at −80 °C. For M_2 purification, the supernatant was applied to two ABT columns run in parallel (500 ml each), which were then washed with 0.1 % digitonin, 0.1 % Na-cholate, 20 mM KPB, and 150 mM NaCl at a rate of ~90 ml/h. M_2-T4L was eluted from the ABT columns with 0.5 mM atropine, 0.1 % digitonin, 0.1 % Na-cholate, 20 mM KPB, and 150 mM NaCl. Eluate was applied to a column of hydroxyapatite (30 ml). The hydroxyapatite column was washed at a rate of 30–50 ml/h with a series of solutions as follows: (1) 0.1 % digitonin, 0.1 % Na-cholate, 20 mM KPB (100 ml); (2) 5 μM QNB, 0.1 % digitonin, 0.1 % Na-cholate, 20 mM KPB (600 ml); (3) 0.35 % Na cholate, 20 mM KPB (600 ml); (4) 0.2 % decylmaltoside, 20 mM KPB (500 ml); (5) 0.2 % decylmaltoside, 150 mM KPB (100 ml); [6] 0.2 % decylmaltoside, 500 mM KPB (60 ml). M_2-T4L–QNB was finally eluted with 0.2 % decylmaltoside, 1 M KPB (50 ml). The eluate was concentrated to ~1 ml (~30 mg protein per ml) using Amicon Ultra Centrifugal Filter Units (MILLIPORE), dialyzed against 0.2 % decylmaltoside, 20 mM Tris–HCl buffer (pH = 7.5) and then stored at −80 °C. The M_2 receptor yield was estimated at ~50 % based on the assumption that the recovered protein was pure M_2-T4L. Protein concentration was determined using the BCA method (PIERCE). Because M_2-T4L was purified in complex with QNB, the [^{3}H]QNB-binding activity could not be estimated because the dissociation rate of QNB is too slow. However, in preliminary experiments using [^{3}H]QNB or dissociable atropine as eluents, the receptor was confirmed to be purified to near homogeneity. The purity of M_2-T4L was confirmed by SDS-PAGE and gel permeation chromatography.

3.8 Crystallization of the M_2 Receptor

The solution including purified M_2-T4L with bound QNB was subjected to buffer exchange to 20 mM HEPES pH = 7.5, 100 mM NaCl, 0.1 % MNG and the sample was concentrated to 50 mg/ml. The protein sample was reconstituted in the lipidic cubic phase by combining it with a 1.5-fold weight excess of a 10:1 monoolein–cholesterol mixture by the twin-syringe method. The mixture was further mixed either by hand or using a Gryphon LCP robot (Art Robbins Instruments), and was dispensed using a ratchet device (Hamilton) or using the Gryphon LCP robot in drops to glass sandwich plates and overlaid with precipitant solution. Initial screening was carried out by in-house screening, after which single crystallization conditions were optimized. Crystals were grown in 100 mM HEPES pH 7.0–7.8, 25–35 % PEG 300, 100 mM ammonium phosphate, 2 % 2-methyl-2,4-pentanediol. The crystals reached full size and were harvested after 3–4 days at 20 °C, after which they were stored in liquid nitrogen. The crystals used for the data collection were grown in 100 mM HEPES pH 7.2–7.9, 20–80 mM EDTA pH 8.0, 10–20 % PEG 300, 1,2,3-heptanetriol.

4 Conclusion

The initial structures of the M_2 and M_3 receptors were obtained from receptors in an inactive state with high affinity inverse agonists bound to them. Determining the crystal structure of active GPCRs has thus far proved to be extremely difficult, because the large parts of agonist binding GPCRs are conformationally heterogeneous. The active-state structure of M_2 receptor with the agonist iperoxo has been obtained with the aid of a conformationally selective antibody fragment (nanobody) which mimicked G proteins and stabilized the active conformation of the receptor (Fig. 4) [29]. This approach may prove to be useful for obtaining active-state structures of other GPCRs in the future. Using this method, the

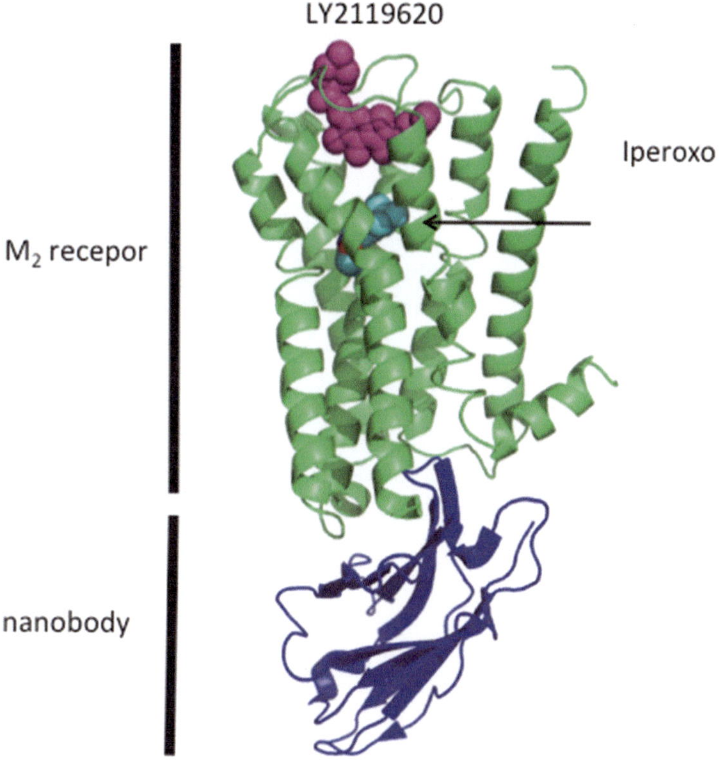

Fig. 4 The structure of the active form of M_2 receptor with bound agonist iperoxo and the positive allosteric modulator LY2119620. M_2 receptor is colored in *green* and nanobody is colored in *blue*. LY2119620 is bound in the allosteric site and is colored in *magenta*. Iperoxo is bound in the orthosteric site

positive allosteric modulator (PAM)-bound active-state M_2 receptor structure has been also determined. The allosteric ligand-binding site (extracellular vestibule) is also contracted along with the contraction of the orthosteric site and essentially, the conformation of the extracellular vestibule of the active M_2 receptor bound with LY2119620 is similar to that without LY2119620. This structural feature suggests the possibility that PAMs can shift the conformational equilibrium of the receptor towards active conformations not only of the allosteric site but also of the receptor as a whole [29].

To date, the crystal structures of two subtypes of mAChR have been solved. These structural data provide valuable information about ligand-binding and the differences, for example, between inactive and active conformations. The structures of all five subtypes of mAChR are likely to be solved in the future, both in inactive and active states bound to several kinds of ligands (orthosteric, allosteric, or bitopic ligands). Such structural information may be an important contribution to the development of new drugs that are highly selective for mAChR subtypes and thus have fewer side effects. Progress in the mAChR research field may be applicable to research on other GPCRs, offering new insights into improved design of therapeutic agents targeting the physiologically and pathophysiologically important GPCRs.

Acknowledgments

This work was supported by the Exploratory Research for Advanced Technology (ERATO) program of the Japan Science and Technology Agency (JST) (to T.K.), by the Toray Science Foundation (to T.K.), by Takeda Science Foundation (to T.K., R.S., and H.A.), by Ichiro Kanehara Foundation (to T.K.), by The Sumitomo Foundation (to T.K.), by the Core Research for Evolutional Science and Technology (CREST) program of the JST (to T.K.), and by the Platform for Drug Discovery, Informatics, and Structural Life Science from the Ministry of Education, Culture, Sports, Science and Technology, Japan (to T.K.).

References

1. Hopkins AL, Groom CR (2002) The druggable genome. Nat Rev Drug Discov 1(9): 727–730
2. Klabunde T, Hessler G (2002) Drug design strategies for targeting G-protein-coupled receptors. Chembiochem 3:455–459
3. Bonner TI, Buckley NJ, Young AC, Brann MR (1987) Identification of a family of muscarinic acetylcholine receptor genes. Science 237(4814):527–532
4. Peralta EG, Winslow JW, Peterson GL, Smith DH (1987) Primary structure and biochemical property of an M2 muscarinic receptor. Science 236(4801):600–605
5. Bonner TI, Young AC, Brann MR, Buckley NJ (1988) Cloning and expression of the human

and rat m5 muscarinic acetylcholine receptor genes. Neuron 1(5):403–410

6. Bonner TI (1989) The molecular basis of muscarinic receptor diversity. Trends Neurosci 12(4):148–151
7. Bonner TI (1989) New subtypes of muscarinic acetylcholine receptors. Trends Pharmacol Sci Suppl 11:5
8. Wess J, Bonner TI, Dorje F, Brann MR (1990) Delineation of muscarinic receptor domains conferring selectivity of coupling to guanine nucleotide-binding proteins and second messengers. Mol Pharmacol 38(4):517–523
9. Wess J, Bonner TI, Brann MR (1990) Chimeric m2/m3 muscarinic receptors: role of carboxyl terminal receptor domains in selectivity of ligand binding and coupling to phosphoinositide hydrolysis. Mol Pharmacol 38(6):872–877
10. Wess J, Liu J, Blin N et al (1997) Structural basis of receptor/G protein coupling selectivity studied with muscarinic receptors as model systems. Life Sci 60(13–14):1007–1014
11. Haga K, Kruse AC, Asada H et al (2012) Structure of the human M2 muscarinic acetylcholine receptor bound to an antagonist. Nature 482(7386):547–551
12. Kruse AC, Hu J, Pan AC et al (2012) Structure and dynamics of the M3 muscarinic acetylcholine receptor. Nature 482(7386):547–551
13. Alkhalfioui F, Magnin T, Wagner R (2009) From purified GPCRs to drug discovery: the promise of protein-based methodologies. Curr Opin Pharmacol 9(5):629–635
14. Furukawa H, Haga T (2000) Expression of functional M2 muscarinic acetylcholine receptor in *Escherichia coli.* J Biochem 127(1): 151–161
15. Ichiyama S, Oka Y, Haga K et al (2006) The structure of the third intracellular loop of the muscarinic acetylcholine receptor M2 subtype. FEBS Lett 580(1):23–26
16. Yurugi-Kobayashi T, Asada H, Shiroishi M et al (2009) Comparison of functional non-glycosylated GPCRs expression in *Pichia pastoris.* Biochem Biophys Res Commun 380(2):271–276
17. Hayashi MK, Haga T (1996) Purification and functional reconstitution with GTP-binding regulatory proteins of hexahistidine-tagged muscarinic acetylcholine receptors (m2 subtype). J Biochem 120(6):1232–1238
18. Asada H, Uemura T, Yurugi-Kobayashi T et al (2011) Evaluation of the *Pichia pastoris* expression system for the production of GPCRs for structural analysis. Microb Cell Fact 10:24
19. Kameyama K, Haga K, Haga T et al (1994) Activation of a GTP-binding protein and a GTP-binding-protein-coupled receptor kinase (β-adrenergic-receptor kinase-1) by a muscarinic receptor m2 mutant lacking phosphorylation sites. Eur J Biochem 226:267–276
20. Rosenbaum DM, Cherezov V, Hanson MA et al (2007) GPCR engineering yields high-resolution structural insights into β2-adrenergic receptor function. Science 318(5854):1266–1273
21. Ballesteros JA, Weinstein H (1995) Integrated methods for the construction of three dimensional models and computational probing of structure function relations in G protein-coupled receptors. Methods Neurosci 25:366–428
22. Scorer CA, Clare JJ, McCombie WR et al (1994) Rapid selection using G418 of high copy number transformants of *Pichia pastoris* for high-level foreign gene expression. Biotechnology (NY) 12(2):181–184
23. Weiss HM, Haase W, Michel H et al (1998) Comparative biochemical and pharmacological characterization of the mouse 5HT5A 5-hydroxytryptamine receptor and the human beta2-adrenergic receptor produced in the methylotrophic yeast Pichia pastoris. Biochem J 330(Pt 3):1137–1147
24. Ciccarone VC, Polayes DA, Luckow VA (1998) Generation of recombinant baculovirus DNA in *E. coli* using a baculovirus shuttle vector. Methods Mol Med 13:213–235
25. Luckow VA, Lee SC, Barry GF et al (1993) Efficient generation of infectious recombinant baculoviruses by site-specific transposon-mediated insertion of foreign genes into a baculovirus genome propagated in *Escherichia coli.* J Virol 67(8):4566–4579
26. Kadwell SH, Hardwicke PI (2007) Production of baculovirus-expressed recombinant proteins in wave bioreactors. Methods Mol Biol 388:247–266
27. Weber W, Weber E, Geisse S et al (2002) Optimisation of protein expression and establishment of the Wave Bioreactor for Baculovirus/insect cell culture. Cytotechnology 38(1–3):77–85
28. Haga K, Haga T (1983) Affinity chromatography of the muscarinic acetylcholine receptor. J Biol Chem 258(22):13575–13579
29. Kruse AC, Ring AM, Manglik A et al (2013) Activation and allosteric modulation of a muscarinic acetylcholine receptor. Nature 504(7478):101–106

Chapter 2

Homology Modeling and Docking Evaluation of Human Muscarinic Acetylcholine Receptors

Trayder Thomas, David K. Chalmers, and Elizabeth Yuriev

Abstract

The development of GPCR homology models for virtual screening is an active area of research. Here we describe methods for homology modeling of the acetylcholine muscarinic receptors M_1R–M_5R. The models are based on the β_2-adrenergic receptor crystal structure as the template and binding sites are optimized for ligand binding. An important aspect of homology modeling is the evaluation of the models for their ability to discriminate between active compounds and (presumed) inactive decoy compounds by virtual screening. The predictive ability is quantified using enrichment factors, area under the ROC curve (AUC), and an early enrichment measure, LogAUC. The models produce good enrichment capacity, which demonstrates their unbiased predictive ability. The optimized M_1R–M_5R homology models have been made freely available to the scientific community to allow researchers to use these structures, compare them to their results, and thus advance the development of better modeling approaches.

Key words Acetylcholine muscarinic receptor, Binding site optimization, Decoy, Docking, GPCR, Homology modeling, Virtual screening

1 Introduction

The use of structure-based design methods for G protein-coupled receptors (GPCRs) commenced in the early 2000s with the landmark report of the structure of bovine rhodopsin [1]. The first crystal structures of ligand-infusible GPCRs became available in 2007 [2–4], and, at the time of writing, the number of available structures has grown to a total of 119 crystal structures for 22 receptor subtypes [5]. Despite the considerable technical advances in the field, GPCR crystallization remains an area of highly specialized expertise, and the solved structures make up only a small fraction of the ~800 GPCRs present in the human genome (including 342 nonolfactory receptors) [6]. It is accepted that the prospect of solving the structures of all members of the GPCR superfamily is not realistic in the foreseeable future [7, 8]. Therefore, when receptor models are required for structure-based investigations and experimental data is

Jaromir Myslivecek and Jan Jakubik (eds.), *Muscarinic Receptor: From Structure to Animal Models*, Neuromethods, vol. 107,
DOI 10.1007/978-1-4939-2858-3_2, © Springer Science+Business Media New York 2016

lacking, researchers turn to homology models. A homology model of a protein (also known as a comparative model) is an atomic-resolution model of a protein (the "target") built based on its amino acid sequence and experimental three-dimensional structure of a related homologous protein (the "template").

GPCR homology models are important tools for understanding GPCR function and for structure-based drug design [7–11]. Virtual screening campaigns against GPCR homology models have identified novel active agents for a range of GPCR targets [12] in a prospective manner (i.e., where compounds initially identified through a virtual screen have been sourced and experimentally validated). A wider overview of GPCR modeling is provided by reference [13]; for brief summaries of GPCR docking studies (as well as other docking-related surveys), see [14, 15].

This chapter describes the procedural steps involved in building, optimizing, and evaluating models of muscarinic acetylcholine receptors (mAChRs). We start with the overview of muscarinic receptor modeling (Section 2) and then discuss the approaches we have used to develop refined GPCR homology models which are able to identify active compounds through virtual screening (Sections 3–4) [12, 16].

There are five subtypes of muscarinic acetylcholine receptors, denoted M_1R–M_5R [17]. Development of mAChR ligands (particularly, subtype-selective ligands) holds potential for the treatment of many diseases such as Alzheimer's, schizophrenia, drug addiction, type 2 diabetes, and cancer [18].

2 Overview of Muscarinic Receptor Modeling

Several muscarinic receptor models have been generated over the past few years. They have employed a variety of different templates as the basis for homology model construction. The template proteins used to generate the homology models in each case are listed in Table 1.

3 Homology Modeling of Muscarinic Acetylcholine Receptors

The modeling workflow is shown in Fig. 1 and described in detail in the following sections. We have built homology models of mAChRs M_1–M_5 [12], using the β_2-adrenergic receptor (β_2AR) crystal structure (PDB ID: 2RH1) [2] as the template and employing the induced fit docking (IFD) procedure [19] to optimize the models to improve their identification of compounds which bind to their orthosteric binding sites. The predictive quality of all five models was assessed through retrospective virtual screening investigations. The results obtained using property-matched decoy libraries demonstrated the unbiased predictive capacity of these models.

Table 1
Templates used for modeling muscarinic receptors

Template	Receptor modeled	Template PDB ID	References
Rhodopsin	M_1R	1U19	[22]
		1F88	[35, 36, 39]
	M_2R	1U19	[26]
	M_3R	1GZM	[23]
β_2AR	M_1R	2RH1	[16, 28–31]
	M_2R	3D4S	[26]
	M_2R	2RH1	[20, 21, 26]
	M_1R–M_5R	2RH1	[12]
M_2R	M_1R	3UON	[34]
M_3R	M_1R	4DAJ	[33, 37]
	M_2R	4DAJ	[26]
	M_5R	4DAJ	[25]
D_3R[a]	M_1R	3PBL	[32]
β_1AR	M_2R	2VT4	[26]
	M_3R	2VT4	[38]
	M_5R	2VT4	[24]

[a]The original D_3R-based model was edited to replace the extracellular loop 2 by fragments extracted from the structures of the human β_2AR (PDB ID: 2RH1) and $A_{2A}AR$ (PDB ID: 3EML) receptors

3.1 Software

Many software packages have been used for modeling of mAChRs including ICM [20, 21], MODELLER [22–30], MOE [31–35], Prime [12, 26, 35–38], YASARA [26], QUANTA [27–30], and VEGA [39]. Molecular modeling steps and options described in this protocol refer to the Schrödinger software suite [40], as used by us [12, 16]. Default settings were used, unless otherwise stated.

The following Schrödinger modules and programs were used for specific tasks (**Note 1**):

1. Homology modeling—Prime [41].
2. Multiple sequence alignment—ClustalW [42].
3. Ligand preparation—LigPrep [43].
4. Ligand docking—Glide [44, 45].
5. Binding site optimization—IFD [19].
6. Computation of physical descriptors for comparison of the decoy sets with the active compounds—ChemAxon Marvin Calculator (cxcalc) (http://www.chemaxon.com): The descriptors include molecular weight (MW), number of rotatable bonds, number of hydrogen bond donor and acceptor atoms, calculated logP (ClogP), polar surface area (PSA), and vdW volume.

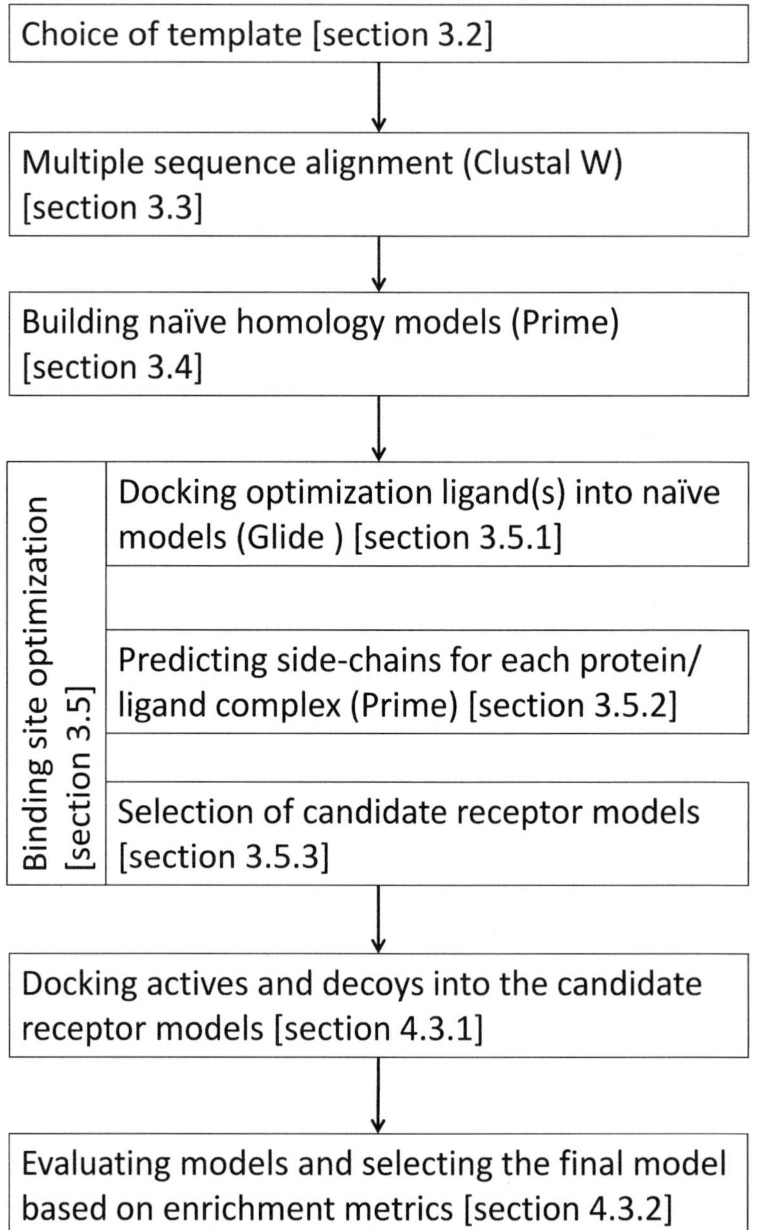

Fig. 1 Flow chart of homology modeling and model evaluation

7. Computation of the 2D Tanimoto score (using fragment sizes of 1–7 atoms, ignoring hydrogens) to demonstrate the diversity of the structures within the ligand sets—Silico [46].
8. Sorting of docked poses based on rank and calculating enrichment metrics—Silico [46].

3.2 Choice of Template

The choice of an appropriate template for GPCR homology modeling is an area of an ongoing debate [47–50]. While close sequence similarity is a very important factor in template selection, it has been shown that model refinement approaches such as binding site optimization and model enhancement based on experimental knowledge (discussed in more detail in Section 3.5 below) are capable of compensating for more distant sequence relationships [51]. Additionally, the ever-increasing number of available structures and improvements in methods for model refinement means that the most optimal choice of a template for a particular target cannot be determined once and for all and requires regular re-evaluation. As can be seen in Table 1, a variety of templates have been used as the basis for modeling of muscarinic acetylcholine receptors. Furthermore, it has also been shown that model quality can be improved by combining multiple templates [32].

A receptor model, built on a close sequence template, is usually considered to be preferable for structure-based drug design [26, 34]. However, when close sequence templates are not available (as still is the case for many GPCRs), knowledge-based optimization can be used to improve a model that is based on a more remote sequence template. We have previously demonstrated that an optimized model of the M_2R, based on a more remote template (β_2AR), outperformed a naïve (i.e., generated without binding site optimization) M_2R model, based on a close sequence template (M_3R), in virtual screening [12]. These results agree with similar observations for modeling the dopamine D_1 and D_2 receptors [48], the β_2AR [51], as well as a diverse panel of receptors (β_1A and β_2A, dopamine D_3, histamine H_1, muscarine M_2 and M_3, A_2A adenosine, S1P1, kappa-opioid, and C-X-C chemokine 4 receptors) [52].

3.3 Multiple Sequence Alignment

Due to a significant level of sequence conservation within the transmembrane regions (helices) of GPCRs, it is possible to align sequences by making use of highly conserved residues to identify the positions of gaps and inserts. GPCR sequences are available for download from the Universal Protein Resource (http://www.uniprot.org/). For effective receptor space coverage, representatives from different relevant GPCR subfamilies should be used for multiple sequence alignment. To establish the alignment between the mAChRs and the template (β_2AR) structure, we used ClustalW [12] (Fig. 1) to create a multiple sequence alignment of the muscarinic receptors with the human dopamine, serotonin, α- and β-adrenergic, adenosine, histamine, and bovine rhodopsin receptors. The ClustalW multiple sequence alignment required manual editing to remove gaps in helices and to ensure that highly conserved residues in each transmembrane helix were properly aligned.

3.4 Generation of a Starting (Naïve) Model

Using a well-aligned sequence of an appropriate template, a user can generate a starting (naïve) 3D receptor model. In our work, naïve homology models of the five human mAChRs were built in Prime (Fig. 1) from the multiple sequence alignment, using the β_2-adrenergic receptor (PDB ID: 2RH1) crystal structure [2] as the template.

3.5 Binding Site Optimization

A naïve model is not necessarily very good at identifying active ligands and can be improved by binding site optimization. Binding site optimization takes into account the structural plasticity of a binding site and its ability to adjust to the structural demands of the ligand. Binding site optimization via a variety of approaches has been widely used to improve model quality. These have been variously described as ligand-steered [53], ligand-guided [54, 55], ligand-adapted [49], or ligand-optimized [48] homology modeling. Binding site optimization via a variety of approaches—mainly those employing accessible experimental data relating to a target and its ligand knowledge (such as structure-activity relationships and/or site-directed mutagenesis)—has been commonly used and shown to improve model quality in GPCR Dock assessments [56–58].

The following steps describe binding site optimization via induced fit docking, as implemented by us for muscarinic receptors using the IFD module of the Schrödinger software suite [12] (Fig. 1).

3.5.1 Docking Optimization Ligands into Naïve Models

The mAChR homology models are first treated by the Maestro Protein Preparation Wizard workflow [43] to add and minimize hydrogen atoms using the OPLS_2005 force field. Following model preparation, the ligand-binding site is refined by docking an appropriate ligand (here referred to as the "optimization ligand") into each of the homology models using Glide within the IFD protocol. The optimization ligand should ideally be representative of the hits a user intends to find, e.g., by having a similar scaffold. In order to identify a larger range of hits, it is important to consider how adjusting the binding site around the optimization ligand will limit the ability of other ligands to dock. For example, optimization of the residues around a smaller ligand can reduce the volume of the binding site and preclude the ability of larger ligands to dock successfully.

Glide docks ligands within a predefined, cuboid region. This site should be roughly centered on the binding site, and be large enough, so as to allow the binding of large ligands. We chose to center the docking site upon residues Asp 3.32, Trp 6.48, Phe 6.52, and Tyr 7.43 (Ballesteros-Weinstein nomenclature [59]) (**Note 2**). Both the van der Waals (vdW) radii and the partial atomic charges were scaled by 0.5 in order to collect a more diverse range of poses. In the initial Glide docking step, up to 50 poses per ligand were collected.

Clozapine and atropine have been demonstrated as useful optimizing ligands for IFD [12] since they have high affinity for the

M_1–M_5 receptors; reported clozapine K_i values vary from 1.4 to 5.0 nM and atropine K_i values range between 0.2 and 1.5 nM [60]. Following the virtual screening evaluation procedure (described below), the atropine-optimized model for the M_1R gave the best enrichment, while the best M_2R–M_5R models were optimized using clozapine [12].

3.5.2 Predicting Conformations of Binding Site Side Chains

The user should select specific residues to include into the binding site refinement. We recommend selecting the side-chain conformations of the residues within 5 Å of ligand atoms, excluding Asp 3.32 and Trp 6.48, for optimization with Prime. Asp 3.32 and Trp 6.48 play a critical role in correctly orienting ligand molecules within receptor binding sites (**Note 3**). For M_1R–M_5R models, when Trp 6.48 and Asp 3.32 were omitted from binding site optimization, more credible ligand poses were obtained, which led to better enrichment in virtual screening [12, 49].

3.5.3 Selection of Candidate Receptor Models

Following binding site optimization with Prime, candidate receptor models are selected by evaluating how the optimization ligand fits within the binding site. The optimization ligand should be re-docked into the optimized receptors with Glide using default vdW and charge scaling parameters. Multiple ligand-receptor poses for each model should be generated and inspected. At this stage receptor models are chosen on the basis of the position and orientation of the ligand within the binding site, key hydrogen bonding and vdW interactions, and the relative energy of interaction, which is a composite of the protein and ligand energy scores (Eq. 1). The distance (ndist) between the ionizable or quaternary nitrogen of the ligand (for simplicity we will just refer to this atom as the "ionizable nitrogen") and the closest carboxylate oxygen of the conserved Asp 3.32 residue should also be taken into account (**Note 4** and **5**):

$$\text{IFDScore} = \text{GlideScore} + 0.05 \times \text{Prime\_Energy} \quad (1)$$

A maximum of 20 poses are required to be collected for further evaluation of receptor models by retrospective virtual screening.

4 Evaluation of Muscarinic Acetylcholine Receptor Models by Virtual Screening

The predictive quality of the candidate receptor models is established by measuring the ability of the candidate models to distinguish between known active compounds and chemically similar decoy compounds with physicochemical properties that closely match those of the actives. Methods for evaluating homology modeling and virtual screening protocols as applied to GPCRs are a focus of active research [56–58, 63, 64].

4.1 Actives

Known antagonists of muscarinic receptors (actives) can be obtained from the GLIDA database [65] (http://pharminfo.pharm.kyoto-u.ac.jp/services/glida/). We have used a set of 48 actives (Note 6) in our retrospective virtual screening studies [12]. Reference [16] contains the chemical structures of these actives and Table 2 lists their average physicochemical properties. We used the Maestro module LigPrep to generate compound 3D structures and to assign tautomeric states and formal charges at physiological pH (pH 7.4 ± 2.0) of active and decoy compounds with a single, likely structure per compound being retained for screening.

4.2 Decoys

Although there are many large decoy libraries available for retrospective virtual screening studies, it has been demonstrated that a set of approximately 1000 molecules is sufficient to detect enrichment trends. For example, it has been shown that little library size-dependent behavior is detected when screening with the entire Directory of Useful Decoys (DUD) set of approximately 100,000 molecules compared to a randomly selected subset of 1000 DUD molecules [66]. It is important that the decoys must not be readily distinguishable from the active compounds. Decoy sets where the physicochemical properties of the decoys differ substantially from those of the active ligands have been shown to lead to biased virtual screening results, and often artificially good enrichment [66].

In our studies we have evaluated three sets of decoys. Table 2 lists the properties of the decoy and active compounds: molecular weight MW (g/mol), number of rotatable bonds (NRB), polar surface area PSA (Å^2), calculated logP, number of hydrogen bond donors and acceptors (HBD and HBA, respectively), solvent accessible volume (Å^3), and 2D Tanimoto score. It can be seen that, generally, the properties of the active compounds are similar to those of the decoy libraries.

4.2.1 Set 1

The Schrödinger decoy set (http://www.schrodinger.com), previously used by us [12, 16], contains 1000 drug-like decoys, randomly selected from a library of one million compounds having

Table 2
Average ligand properties [12]

Ligand set	MW (g/mol)	NRB	PSA (Å^2)	ClogP	HBD	HBA	vdW volume (Å^3)	2D Tanimoto score
Actives	324	5.1	31	3.03	1.4	1.6	318	0.233
Decoy sets								
1: Schrödinger	360	5.0	84	2.90	2.0	4.2	316	0.125
2: ZINC	320	4.3	38	3.43	1.4	1.7	302	0.185
3: Refined Schrödinger	343	4.8	79	2.59	2.4	3.3	312	0.143

properties characteristic of drug molecules [44, 45]. The molecular weights of this set vary from 151 to 645 g/mol, with an average of 360 g/mol. These decoys were not specifically selected to mimic muscarinic antagonist compounds.

4.2.2 Set 2

We derived the ZINC decoy set from the ZINC database (http://zinc.docking.org/) [67] (7.2 million compounds, database version 7) by a process of successive eliminations, generating a subset of molecules closely matching the physicochemical properties of the actives (Table 2). 1000 molecules were randomly selected satisfying the physicochemical properties criteria (Note 7). This set is more challenging in terms of distinguishing between decoys and active compounds.

4.2.3 Set 3

The refined Schrödinger decoy set is a subset of the decoy Set 1, with molecular weight limited to be consistent with that of the active compounds (260–410 g/mol). To generate this set, all compounds from the Schrödinger decoy library with a molecular weight outside the range of the active compounds and without ionizable nitrogen were removed. This set contains 261 molecules and is more challenging than Set 1 from which it was derived.

4.3 Enrichment Studies

4.3.1 Docking

To establish the ability of the receptor to identify muscarinic antagonists, the decoys and actives are docked into the candidate receptor models (Note 8), generated and selected at Sections 3.5.1–3.5.3 above (Fig. 1). The top pose for each ligand (determined by GlideScore) is retained following post-docking minimization.

Following docking, models should always be visually inspected to ensure that the ligands bind within the defined binding pocket. Furthermore, the intermolecular interactions in these poses should be examined to ensure that important expected interactions, based on mutagenesis studies [68], are observed between ligand and receptor molecules.

4.3.2 Numeric and Graphic Assessment of Models

Both enrichment plots and receiver operating characteristic curve (ROC) plots have been used to establish the performance of homology models and crystal structures. It has been argued that ROC curves are superior to enrichment plots: not only do they reflect the selection of actives, but also the non-selection of decoys [69, 70]. One metric that can be derived from ROC plots is the area under the curve (AUC). The AUC, which has an ideal value of 1, gives an indication of the general ability of the model to distinguish active compounds (true positives) from decoy compounds (true negatives). It should be noted that the AUC does not specifically focus on the best docking scores being allocated to active compounds (early enrichment). In order to weight the AUC towards early enrichment, the ROC curve can instead be plotted with a logarithmic *x*-axis. The resulting LogAUC [71] is then calculated by computing the fraction of the ideal area under the semilog ROC curve. Another metric, NSQ_AUC [72], has also been developed to probe early, rather than overall, enrichment.

Calculating Curves

Due to the large scale of most virtual screening efforts there is significant advantage in automating the process of calculating ROC curves. There are many programs already designed to calculate ROC curves for popular docking packages. Below we detail the general procedure available in our scripts implemented in Silico [46].

The set of docked actives and decoys is ranked by GlideScore. The ROC curve is plotted by stepping through the list and plotting the cumulative fraction of actives encountered (true positives, *y*-value) against the cumulative fraction of decoys (true negatives, *x*-value).

An enrichment curve is generated similarly, differing in that the *x*-axis reflects the percentage of compounds encountered rather than decoys. The enrichment factors (EF) are the *y*-values corresponding to each % (*x*-value). Enrichment factors are usually calculated at 2, 5, and 10 % of the total number of compounds (N_{total}) screened, according to Eq. (2):

$$\text{EF} = \left(\frac{\text{Hits}_{\text{sampled}}}{N_{\text{sampled}}}\right) \div \left(\frac{\text{Hits}_{\text{total}}}{N_{\text{total}}}\right) \tag{2}$$

Area Under the Curve

The AUC is calculated by integrating the area under the ROC curve (Note 9) according to Eq. (3):

$$\text{AUC} = \sum_{1}^{d_{\text{T}}} \frac{1}{d_{\text{T}}} \times \frac{a_{\text{F}}}{a_{\text{T}}} \tag{3}$$

where d_{T} is the total number of decoys, a_{F} is the number of actives found, and a_{T} is the total number of actives.

LogROC curves follow a similar principle but weight the ROC curve to favor early enrichment. Because logarithmic curves have an asymptote at zero it is necessary to introduce a lower limit. This lower limit (λ) functions to restrict the length of the *x*-axis to a finite value. It is important to realize that LogAUC is only comparable for identical λ values (usually 0.001). There are also additional considerations: any coordinates smaller than λ should not be included in the calculations and unless one of the *x*-coordinates coincides with λ, an additional term needs to be added to calculate the first partial step.

In its reduced form, the formula for LogAUC is Eq. (4):

$$\text{for } \frac{d_{\text{F}}}{d_{\text{T}}} > \lambda, \text{LogAUC}_{\lambda} = \sum_{d=1}^{d_{\text{T}}} \frac{\log_{10}\left(\frac{d_{\text{F}}}{d_{\text{F}}-1}\right)}{-\log_{10}(\lambda)} \times \frac{a_{\text{F}}}{a_{\text{T}}} \tag{4}$$

However, this formula can be understood more intuitively when written as Eq. (5):

$$\text{for } \frac{d_{\mathrm{F}}}{d_{\mathrm{T}}} > \lambda, \text{LogAUC}_{\lambda} = \sum_{d=1}^{d_{\mathrm{T}}} \frac{\log_{10}\left(\frac{d_{\mathrm{F}}}{d_{\mathrm{T}}}\right) - \log_{10}\left(\frac{d_{\mathrm{F}}-1}{d_{\mathrm{T}}}\right)}{\log_{10}(1) - \log_{10}(\lambda)} \times \frac{a_{\mathrm{F}}}{a_{\mathrm{T}}} \quad (5)$$

To account for the first partial step, an additional term (Eq. 6) may need to be added to both of these previous formulas:

$$\text{if } \frac{d_{\mathrm{F}}}{d_{\mathrm{T}}} > \lambda \text{ and } \frac{d_{\mathrm{F}}-1}{d_{\mathrm{T}}} < \lambda, \frac{\log_{10}\left(\frac{d_{\mathrm{F}}}{d_{\mathrm{T}}}\right) - \log_{10}(\lambda)}{\log_{10}(1) - \log_{10}(\lambda)} \times \frac{a_{\mathrm{F}}}{a_{\mathrm{T}}} \quad (6)$$

where d_{F} is the number of decoys found, d_{T} is the total number of decoys, a_{F} is the number of actives found, and a_{T} is the total number of actives.

The intuitive formula above can be visualized (Fig. 2) as a ratio of two distances on the x-axis multiplied by the height of the curve (the fraction of actives found). The numerator is the distance

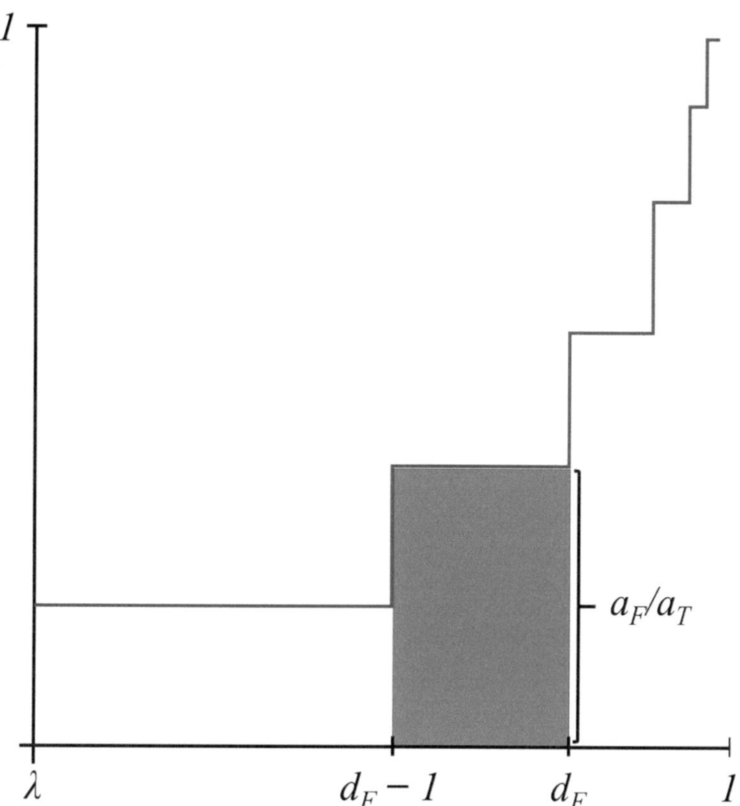

Fig. 2 Construction of a rectangle under a LogROC curve

between the decoy just found and the previous decoy (one horizontal step). The denominator is the length of the x-axis (λ to 1). The additional term is necessary because if none of the coordinates coincide with λ the first full horizontal step will precede λ.

Practically, this means that the LogAUC is being constructed in steps by a series of rectangles. Each step is calculated as a fraction of the Cartesian length of the logarithmic x-axis and multiplied by the height of the ROC curve at this point. In this way, the LogAUC of the perfect enrichment curve will be 1.

5 Representative Results

The receptor homology models are evaluated by testing their ability to rank active compounds above decoy molecules. In such evaluation, the decoy libraries (see Section 4.2 above) and active compounds (see Section 4.1 above) are docked into the receptor models. The optimization ligands, used for binding site optimization (see Section 3.5 above), should be excluded from virtual screening to remove any potential structural bias.

Representative ROC curves, enrichment plots, and semilogarithmic ROC curves for the M_5R model are shown in Fig. 3 and the corresponding metrics are listed in Table 3. The M_5R model shows excellent enrichment capacity and has particularly good early enrichment.

The main deficiency of the models is the failure to dock some of the actives, shown as a gap at the end of the ROC curves (Fig. 3). The properties of actives that did not dock or produced docked poses with a scoring energy greater than the set acceptable cutoff (a GlideScore of 0 kcal/mol) suggest that the most likely reason for this docking failure is the large size of these compounds. Therefore, a better model might be developed by using an alternative bulkier optimization ligand. Work is currently in progress in our laboratory which demonstrates that this is indeed the case and muscarinic receptor models generated using alternative optimization ligands are able to dock a wider range of compounds.

The metrics shown in Table 3 compare favorably with other similar docking studies (although such comparisons should not be over-interpreted given different actives, decoy sets, and receptor types used). For example, the MT_2 melatonin receptor models [49] which were based on the β_2AR and optimized for antagonists produced $EF_{2\%} = 3.1–18.7$; a range of antagonist-bound GPCR crystal structures gave $EF_{2\%} = 0.3–11.7$ and $EF_{10\%} = 1.5–3.9$ [64].

Both decoy sets 2 and 3 (ZINC and refined Schrödinger) were property-matched to actives. In addition, the decoy selection criteria for these sets included the requirement to contain only compounds with an ionizable nitrogen at physiological pH, based on the hypothesis that a ligand ionizable nitrogen should be able to

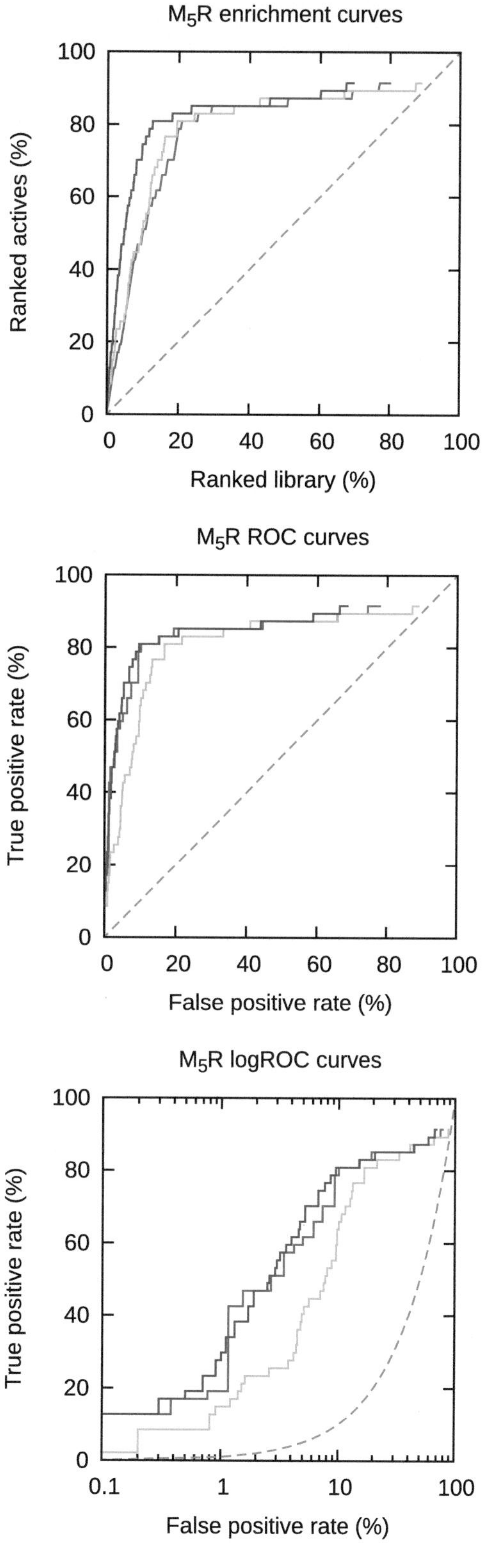

Fig. 3 Curves for the M_5R model. *Blue*: Set 1, Schrödinger; *green*: Set 2, ZINC; *red*: Set 3, refined Schrödinger. *Dotted line* indicates random choice (no enrichment) (Color figure online)

Table 3
Virtual screening metrics of the M_5R

Decoy set	AUC	$LogAUC_{0.001}$	EF (at x % of ranked database) 2	5	10
Set 1 (Schrödinger decoy set)	0.85	0.53	12.7	10.1	7.4
Set 2 (ZINC decoy set)	0.81	0.40	8.5	5.5	5.3
Set 3 (refined Schrödinger decoy set)	0.84	0.51	5.6	5.3	5.1

form the salt bridge with Asp 3.32. Thus, these decoys sets were designed to be challenging with respect to selecting for this interaction in actives ahead of decoys that also contained an ionizable nitrogen.

The enrichment metrics (Table 3) and ROC and enrichment curves (Fig. 3) demonstrate that indeed these sets of decoys are more challenging. However, they also show that the enrichment and early enrichment values are similar to values of non-property-matched decoys, indicating the model's capability to preferentially identify active compounds even amongst property-matched decoys.

The five subtypes M_1R–M_5R have pairwise sequence identities in the range of 50–70 %. For residues within 6 Å of the ligand (from the M_2R crystal structure) the sequence identity increases to 90–100 %. Due to this high similarity, compounds that act at one receptor subtype usually also have some affinity for the other subtypes [17]. A rigorous test of model quality would be to dock compounds with a high level of specificity for individual subtypes into all subtypes. However, a significant challenge encountered in homology modeling and evaluation of muscarinic acetylcholine receptors is to identify a sufficient number of compounds that are selective for one receptor over the other four subtypes.

6 Availability of Models

It is a significant problem for researchers interested in GPCR structure-based design that only a limited number of GPCR homology models (including models of the muscarinic acetylcholine receptors) are freely available for use and comparison. We aim to supplement the limited number of evaluated homology models that are available to the research community [12, 16]. The optimized M_1R–M_5R homology models, built and evaluated as described above, are freely available as part of the Supporting Information for

Refs. [12, 16]. ZINC-derived decoy sets (Sections 4.2.2 and 4.2.3) are available for users upon request. We consider such open access as crucial in our field since it allows researchers to use these structures, compare them to their results [34, 73], and thus advance the development of better modeling methods.

7 Conclusions

This chapter describes development of homology models of the muscarinic acetylcholine receptors M_1R–M_5R and their evaluation through retrospective virtual screening for the identification of antagonists. Model refinement, guided by experimental knowledge of active compounds and critical binding site residues, results in ligand-induced adaptation of the receptor binding sites and their optimization for antagonist recognition. Specifically, the generated homology models are capable of distinguishing known antagonists from matched decoy compounds. The confirmed predictive power of these models gives greater confidence in the use of these models for prospective virtual screening. The following aspects of the modeling procedure are particularly important: (1) binding site optimization is a critical step in model generation; (2) knowledge-based homology models of GPCRs are appropriate for prospective virtual screening, once confirmed in retrospective tests; and (3) property-matched decoys should be used in virtual screening evaluation of homology models. Future work is required to evaluate homology models in a flexible receptor scenario: by on-the-fly receptor flexibility [14, 15], molecular dynamics [74], or using receptor ensembles [75–78].

8 Notes

1. For specific version numbers of programs and modules, readers should refer to original publications. For example, for homology modeling of the five muscarinic M_1–M_5 acetylcholine receptors we employed Prime (versions 3.0 and 3.1) using the Maestro interface (versions 9.2 and 9.3). Ligand molecules were prepared with LigPrep (version 2.5) and docked into the homology models using Glide (versions 5.7 and 5.8).
2. In the Ballesteros-Weinstein residue numbering system, the first number corresponds to the helix number and the second number represents the position relative to the most conserved residue in that helix (assigned the arbitrary number "50"). This nomenclature is not used for the variable loop regions, where receptor sequence numbering is used, usually following the crystal structure (PDB) numbering.

3. Trp 6.48 is a key residue of the aromatic cluster of transmembrane helices 5 and 6. It has been suggested that it acts as a "micro-switch" for receptor activation and inactivation [61]. The IFD protocol consistently caused Trp 6.48 to undergo a conformational "flip" during the Prime step [12], which forced the bulky indole side chain down and away from the binding pocket.
4. During the IFD optimization of the binding sites, monitoring the distance (ndist) between the ionizable or quaternary nitrogen of the ligand and the closest carboxylate oxygen of the conserved Asp 3.32 residue is advisable. The Asp 3.32 residue crucial for ligand binding of all aminergic GPCRs has been determined by site-directed mutagenesis [62]. The distance ndist is a quantitative measure of this important ionic interaction and receptors with ndist >3.0 Å should be excluded from further analysis.
5. In the new IFD protocol (2013-1 release: label "IFD"; 2013-3 release: label "Extended Sampling"), the re-optimized IFDScore is

   $$\text{IFDScore} = 1.0 \times \text{Prime\_Energy} + 9.057 \times \text{GlideScore} + 1.428 \times \text{Glide\_Ecoul}$$

   Source: http://www.schrodinger.com/kb/307.
6. Actives used in model evaluation via retrospective virtual screening: atropine, benzquinamide, benztropine, biperiden, buclizine, carbinoxamine, chlorpromazine, chlorprothixene, clidinium, clozapine, cyclizine, cyclopentolate, cycrimine, desipramine, dicyclomine, diphenidol, dosulepin, doxepin, doxylamine, ethopropazine, flavoxate, glycopyrrolate, homatropine methyl bromide, hyoscyamine, methantheline, methotrimeprazine, metixene, metoclopramide, olanzapine, orphenadrine, oxybutynin, oxyphencyclimine, oxyphenonium, pirenzepine, procyclidine, promazine, promethazine, propantheline, propiomazine, quinacrine, scopolamine, solifenacin, thiethylperazine, tolterodine, tridihexethyl, triflupromazine, trihexyphenidyl, and trospium. Where atropine was used as the induced fit ligand, it was excluded from the virtual screen, and likewise for clozapine.
7. To generate the ZINC-based decoy library, molecules were required to fall within a similar normal distribution as the active compounds (MW = 265–434 g/mol; mean 322 g/mol; standard deviation 40 g/mol). They were also required to contain an ionizable nitrogen and not to contain more than three hydrogen bond donors or four hydrogen bond acceptors. Finally, to ensure topological diversity, each decoy was required to have a Tanimoto score <0.8 with respect to all other molecules within the set.

8. The docking site is centered as per Section 3.5.1. If using Glide, both the Standard Precision (SP) and the Extra Precision (XP) scoring functions could be used. XP gives marginally better results [12].
9. Calculating the AUC can be done during, or separately from, the curve plotting process. In general the *x*, *y*-coordinates are used to construct a series of geometric shapes that fit the curve and sum their areas. For an ROC curve the geometric shapes will always be rectangles but if one wants to calculate the area under an enrichment curve (note that this is generally less meaningful or comparable as the ideal area can change with the size of the library) then trapezoids will be necessary.

Acknowledgments

T.T. is a recipient of an Australian Postgraduate Award (APA) scholarship. This work was supported by the Victorian Life Sciences Computation Initiative (VLSCI, grant number VR0004), and by the National Computational Infrastructure (grant number: y96), which is supported by the Australian Commonwealth Government.

References

1. Palczewski K, Kumasaka T, Hori T, Behnke CA, Motoshima H, Fox BA, Le Trong I, Teller DC, Okada T, Stenkamp RE, Yamamoto M, Miyano M (2000) Crystal structure of rhodopsin: a G protein-coupled receptor. Science 289(5480):739–745
2. Cherezov V, Rosenbaum DM, Hanson MA, Rasmussen SG, Thian FS, Kobilka TS, Choi HJ, Kuhn P, Weis WI, Kobilka BK, Stevens RC (2007) High-resolution crystal structure of an engineered human beta2-adrenergic G protein-coupled receptor. Science 318(5854):1258–1265
3. Rasmussen SG, Choi HJ, Rosenbaum DM, Kobilka TS, Thian FS, Edwards PC, Burghammer M, Ratnala VR, Sanishvili R, Fischetti RF, Schertler GF, Weis WI, Kobilka BK (2007) Crystal structure of the human beta2 adrenergic G-protein-coupled receptor. Nature 450(7168):383–387
4. Rosenbaum DM, Cherezov V, Hanson MA, Rasmussen SG, Thian FS, Kobilka TS, Choi HJ, Yao XJ, Weis WI, Stevens RC, Kobilka BK (2007) GPCR engineering yields high-resolution structural insights into beta2-adrenergic receptor function. Science 318(5854):1266–1273
5. Yang J, Zhang Y (2014) GPCRS-EXP: a database for experimentally solved GPCR structures. http://zhanglab.ccmb.med.umich.edu/GPCR-EXP/. Accessed 3 Dec 2014
6. Fredriksson R, Lagerstrom MC, Lundin LG, Schioth HB (2003) The G-protein-coupled receptors in the human genome form five main families. Phylogenetic analysis, paralogon groups, and fingerprints. Mol Pharmacol 63(6):1256–1272. doi:10.1124/mol.63.6.1256
7. Shoichet BK, Kobilka BK (2012) Structure-based drug screening for G-protein-coupled receptors. Trends Pharmacol Sci 33(5):268–272. doi:10.1016/j.tips.2012.03.007
8. Stevens RC, Cherezov V, Katritch V, Abagyan R, Kuhn P, Rosen H, Wuthrich K (2013) The GPCR Network: a large-scale collaboration to determine human GPCR structure and function. Nat Rev Drug Discov 12(1):25–34. doi:10.1038/nrd3859
9. Mason JS, Bortolato A, Congreve M, Marshall FH (2012) New insights from structural biology into the druggability of G protein-coupled receptors. Trends Pharmacol Sci 33(5):249–260. doi:10.1016/j.tips.2012.02.005
10. Granier S, Kobilka B (2012) A new era of GPCR structural and chemical biology. Nat Chem Biol 8(8):670–673. doi:10.1038/nchembio.1025
11. Kooistra AJ, Roumen L, Leurs R, de Esch IJ, de Graaf C (2013) From heptahelical bundle to hits from the haystack: structure-based virtual screening for GPCR ligands. Methods Enzymol 522:279–336. doi:10.1016/B978-0-12-407865-9.00015-7

12. Thomas T, McLean KC, McRobb FM, Manallack DT, Chalmers DK, Yuriev E (2014) Homology modeling of human muscarinic acetylcholine receptors. J Chem Inf Model 54(1):243–253. doi:10.1021/ci400502u

13. Costanzi S (2013) Modeling G protein-coupled receptors and their interactions with ligands. Curr Opin Struct Biol 23(2):185–190. doi:10.1016/j.sbi.2013.01.008

14. Yuriev E, Agostino M, Ramsland PA (2011) Challenges and advances in computational docking: 2009 in review. J Mol Recognit 24:149–164

15. Yuriev E, Ramsland PA (2013) Latest developments in molecular docking: 2010–2011 in review. J Mol Recognit 26(5):215–239. doi:10.1002/jmr.2266

16. McRobb FM, Capuano B, Crosby IT, Chalmers D, Yuriev E (2010) Homology modeling and docking evaluation of aminergic G protein-coupled receptors. J Chem Inf Model 50:626–637

17. Wess J, Eglen RM, Gautam D (2007) Muscarinic acetylcholine receptors: mutant mice provide new insights for drug development. Nat Rev Drug Discov 6(9):721–733. doi:10.1038/nrd2379

18. Kruse AC, Weiss DR, Rossi M, Hu J, Hu K, Eitel K, Gmeiner P, Wess J, Kobilka BK, Shoichet BK (2013) Muscarinic receptors as model targets and antitargets for structure-based ligand discovery. Mol Pharmacol 84(4):528–540. doi:10.1124/mol.113.087551

19. Sherman W, Day T, Jacobson MP, Friesner RA, Farid R (2006) Novel procedure for modeling ligand/receptor induced fit effects. J Med Chem 49(2):534–553

20. Gregory KJ, Hall NE, Tobin AB, Sexton PM, Christopoulos A (2010) Identification of orthosteric and allosteric site mutations in M2 muscarinic acetylcholine receptors that contribute to ligand-selective signaling bias. J Biol Chem 285(10):7459–7474

21. Valant C, Gregory KJ, Hall NE, Scammells PJ, Lew MJ, Sexton PM, Christopoulos A (2008) A novel mechanism of G protein-coupled receptor functional selectivity. Muscarinic partial agonist McN-A-343 as a bitopic orthosteric/allosteric ligand. J Biol Chem 283(43):29312–29321. doi:10.1074/jbc.M803801200

22. Marquer C, Fruchart-Gaillard C, Letellier G, Marcon E, Mourier G, Zinn-Justin S, Menez A, Servent D, Gilquin B (2011) Structural model of ligand-G protein-coupled receptor (GPCR) complex based on experimental double mutant cycle data: MT7 snake toxin bound to dimeric hM1 muscarinic receptor. J Biol Chem 286(36):31661–31675. doi:10.1074/jbc.M111.261404

23. Martinez-Archundia M, Cordomi A, Garriga P, Perez JJ (2012) Molecular modeling of the M3 acetylcholine muscarinic receptor and its binding site. J Biomed Biotechnol 2012:789741. doi:10.1155/2012/789741

24. Huang X, Zheng G, Zhan CG (2012) Microscopic binding of M5 muscarinic acetylcholine receptor with antagonists by homology modeling, molecular docking, and molecular dynamics simulation. J Phys Chem B 116(1):532–541. doi:10.1021/jp210579b

25. Zheng G, Smith AM, Huang X, Subramanian KL, Siripurapu KB, Deaciuc A, Zhan CG, Dwoskin LP (2013) Structural modifications to tetrahydropyridine-3-carboxylate esters en route to the discovery of M5-preferring muscarinic receptor orthosteric antagonists. J Med Chem 56(4):1693–1703. doi:10.1021/jm301774u

26. Jakubik J, Randakova A, Dolezal V (2013) On homology modeling of the M2 muscarinic acetylcholine receptor subtype. J Comput Aided Mol Des 27(6):525–538. doi:10.1007/s10822-013-9660-8

27. Blaney FE, Raveglia LF, Artico M, Cavagnera S, Dartois C, Farina C, Grugni M, Gagliardi S, Luttmann MA, Martinelli M, Nadler GM, Parini C, Petrillo P, Sarau HM, Scheideler MA, Hay DW, Giardina GA (2001) Stepwise modulation of neurokinin-3 and neurokinin-2 receptor affinity and selectivity in quinoline tachykinin receptor antagonists. J Med Chem 44(11):1675–1689

28. Lebon G, Langmead CJ, Tehan BG, Hulme EC (2009) Mutagenic mapping suggests a novel binding mode for selective agonists of M1 muscarinic acetylcholine receptors. Mol Pharmacol 75(2):331–341. doi:10.1124/mol.108.050963

29. Avlani VA, Langmead CJ, Guida E, Wood MD, Tehan BG, Herdon HJ, Watson JM, Sexton PM, Christopoulos A (2010) Orthosteric and allosteric modes of interaction of novel selective agonists of the M1 muscarinic acetylcholine receptor. Mol Pharmacol 78(1):94–104. doi:10.1124/mol.110.064345

30. Kaye RG, Saldanha JW, Lu ZL, Hulme EC (2011) Helix 8 of the M1 muscarinic acetylcholine receptor: scanning mutagenesis delineates a G protein recognition site. Mol Pharmacol 79(4):701–709. doi:10.1124/mol.110.070177

31. Xu J, Chen H (2012) Interpreting the structural mechanism of action for MT7 and human muscarinic acetylcholine receptor 1 complex by modeling protein-protein interaction. J Biomol Struct Dyn 30(1):30–44. doi:10.1080/07391102.2012.674188

32. Daval SB, Valant C, Bonnet D, Kellenberger E, Hibert M, Galzi JL, Ilien B (2012) Fluorescent derivatives of AC-42 to probe

bitopic orthosteric/allosteric binding mechanisms on muscarinic M1 receptors. J Med Chem 55(5):2125–2143. doi:10.1021/jm201348t

33. Daval SB, Kellenberger E, Bonnet D, Utard V, Galzi JL, Ilien B (2013) Exploration of the orthosteric/allosteric interface in human M1 muscarinic receptors by bitopic fluorescent ligands. Mol Pharmacol 84:71–85. doi:10.1124/mol.113.085670
34. Jójárt B, Balint AM, Balint S, Viskolcz B (2012) Homology modeling and validation of the human M1 muscarinic acetylcholine receptor. Mol Inf 31(9):635–638. doi:10.1002/minf.201200062
35. Jacobson MA, Kreatsoulas C, Pascarella DM, O'Brien JA, Sur C (2010) The M1 muscarinic receptor allosteric agonists AC-42 and 1-[1′-(2-methylbenzyl)-1,4′-bipiperidin-4-yl]-1,3-dihydro-2H-benzimidazol-2-one bind to a unique site distinct from the acetylcholine orthosteric site. Mol Pharmacol 78(4):648–657. doi:10.1124/mol.110.065771
36. Ma L, Seager MA, Wittmann M, Jacobson M, Bickel D, Burno M, Jones K, Graufelds VK, Xu G, Pearson M, McCampbell A, Gaspar R, Shughrue P, Danziger A, Regan C, Flick R, Pascarella D, Garson S, Doran S, Kreatsoulas C, Veng L, Lindsley CW, Shipe W, Kuduk S, Sur C, Kinney G, Seabrook GR, Ray WJ (2009) Selective activation of the M1 muscarinic acetylcholine receptor achieved by allosteric potentiation. Proc Natl Acad Sci U S A 106(37):15950–15955. doi:10.1073/pnas.0900903106
37. Chin SP, Buckle MJC, Chalmers DK, Yuriev E, Doughty SW (2014) Towards activated homology models of the human M_1 muscarinic acetylcholine receptor. J Mol Graph Model 49:91–98
38. McMillin SM, Heusel M, Liu T, Costanzi S, Wess J (2011) Structural basis of M3 muscarinic receptor dimer/oligomer formation. J Biol Chem 286(32):28584–28598. doi:10.1074/jbc.M111.259788
39. Espinoza-Fonseca LM, Pedretti A, Vistoli G (2008) Structure and dynamics of the full-length M1 muscarinic acetylcholine receptor studied by molecular dynamics simulations. Arch Biochem Biophys 469(1):142–150. doi:10.1016/j.abb.2007.09.002
40. Suite 2012: Maestro, version 9.3; LigPrep, version 2.5; Schrödinger Suite 2012 Protein Preparation Wizard; Schrödinger Suite 2012 Induced Fit Docking protocol; Glide version 5.8; Prime version 3.1, Schrödinger, LLC (2012). New York, NY
41. Jacobson MP, Pincus DL, Rapp CS, Day TJ, Honig B, Shaw DE, Friesner RA (2004) A hierarchical approach to all-atom protein loop prediction. Proteins 55(2):351–367
42. Thompson JD, Higgins DG, Gibson TJ (1994) CLUSTAL W: improving the sensitivity of progressive multiple sequence alignment through sequence weighting, position-specific gap penalties and weight matrix choice. Nucleic Acids Res 22(22):4673–4680
43. Sastry GM, Adzhigirey M, Day T, Annabhimoju R, Sherman W (2013) Protein and ligand preparation: parameters, protocols, and influence on virtual screening enrichments. J Comput Aided Mol Des 27(3):221–234. doi:10.1007/s10822-013-9644-8
44. Friesner RA, Banks JL, Murphy RB, Halgren TA, Klicic JJ, Mainz DT, Repasky MP, Knoll EH, Shelley M, Perry JK, Shaw DE, Francis P, Shenkin PS (2004) Glide: a new approach for rapid, accurate docking and scoring. 1. Method and assessment of docking accuracy. J Med Chem 47(7):1739–1749
45. Halgren TA, Murphy RB, Friesner RA, Beard HS, Frye LL, Pollard WT, Banks JL (2004) Glide: a new approach for rapid, accurate docking and scoring. 2. Enrichment factors in database screening. J Med Chem 47(7):1750–1759
46. Chalmers DK, Roberts BP (2011) Silico—a Perl Molecular Modelling Toolkit, Monash University: Melbourne
47. Mobarec JC, Sanchez R, Filizola M (2009) Modern homology modeling of G-protein coupled receptors: which structural template to use? J Med Chem 52(16):5207–5216. doi:10.1021/jm9005252
48. Kolaczkowski M, Bucki A, Feder M, Pawlowski M (2013) Ligand-optimized homology models of D_1 and D_2 dopamine receptors: application for virtual screening. J Chem Inf Model 53:638–648. doi:10.1021/ci300413h
49. Pala D, Beuming T, Sherman W, Lodola A, Rivara S, Mor M (2013) Structure-based virtual screening of MT_2 melatonin receptor: influence of template choice and structural refinement. J Chem Inf Model 53(4):821–835. doi:10.1021/ci4000147
50. Rataj K, Witek J, Mordalski S, Kosciolek T, Bojarski AJ (2014) Impact of template choice on homology model efficiency in virtual screening. J Chem Inf Model 54(6):1661–1668. doi:10.1021/ci500001f
51. Tang H, Wang XS, Hsieh JH, Tropsha A (2012) Do crystal structures obviate the need for theoretical models of GPCRs for structure-based virtual screening? Proteins 80(6):1503–1521. doi:10.1002/prot.24035
52. Beuming T, Sherman W (2012) Current assessment of docking into GPCR crystal structures and homology models: successes,

challenges, and guidelines. J Chem Inf Model 52(12):3263–3277. doi:10.1021/ci300411b

53. Phatak SS, Gatica EA, Cavasotto CN (2010) Ligand-steered modeling and docking: a benchmarking study in class a g-protein-coupled receptors. J Chem Inf Model 50(12): 2119–2128
54. Neves MA, Simoes S, Sáe Melo ML (2010) Ligand-guided optimization of CXCR4 homology models for virtual screening using a multiple chemotype approach. J Comput Aided Mol Des 24(12):1023–1033
55. Katritch V, Kufareva I, Abagyan R (2011) Structure based prediction of subtype-selectivity for adenosine receptor antagonists. Neuropharmacology 60(1):108–115. doi:10.1016/j.neuropharm.2010.07.009
56. Kufareva I, Rueda M, Katritch V, Stevens RC, Abagyan R (2011) Status of GPCR modeling and docking as reflected by community-wide GPCR Dock 2010 assessment. Structure 19(8): 1108–1126. doi:10.1016/j.str.2011.05.012
57. Michino M, Abola E, Brooks CL III, Dixon JS, Moult J, Stevens RC (2009) Community-wide assessment of GPCR structure modelling and ligand docking: GPCR Dock 2008. Nat Rev Drug Discov 8(6):455–463
58. Kufareva I, Katritch V, Participants of GPCR Dock 2013, Stevens RC, Abagyan R (2014) Advances in GPCR modeling evaluated by the GPCR Dock 2013 assessment: meeting new challenges. Structure 22(8):1120–1139. doi:10.1016/j.str.2014.06.012
59. Ballesteros JA, Weinstein H, Stuart CS (1995) Integrated methods for the construction of three-dimensional models and computational probing of structure-function relations in G protein-coupled receptors. Methods Neurosci 25:366–428
60. Bymaster FP, Felder CC, Tzavara E, Nomikos GG, Calligaro DO, McKinzie DL (2003) Muscarinic mechanisms of antipsychotic atypicality. Prog Neuro Psychopharmacol Biol Psychiatry 27(7):1125–1143. doi:10.1016/j.pnpbp.2003.09.008
61. Holst B, Nygaard R, Valentin-Hansen L, Bach A, Engelstoft MS, Petersen PS, Frimurer TM, Schwartz TW (2010) A conserved aromatic lock for the tryptophan rotameric switch in TM-VI of seven-transmembrane receptors. J Biol Chem 285(6):3973–3985. doi:10.1074/jbc.M109.064725
62. Spalding TA, Birdsall NJ, Curtis CA, Hulme EC (1994) Acetylcholine mustard labels the binding site aspartate in muscarinic acetylcholine receptors. J Biol Chem 269(6):4092–4097
63. Anighoro A, Rastelli G (2013) Enrichment factor analyses on G-protein coupled receptors with known crystal structure. J Chem Inf Model 53(4):739–743. doi:10.1021/ci4000745
64. Gatica EA, Cavasotto CN (2012) Ligand and decoy sets for docking to G protein-coupled receptors. J Chem Inf Model 52(1):1–6. doi:10.1021/Ci200412p
65. Okuno Y, Tamon A, Yabuuchi H, Niijima S, Minowa Y, Tonomura K, Kunimoto R, Feng C (2008) GLIDA: GPCR—ligand database for chemical genomics drug discovery—database and tools update. Nucleic Acids Res 36(Suppl 1):D907–D912. doi:10.1093/nar/gkm948
66. Huang N, Shoichet BK, Irwin JJ (2006) Benchmarking sets for molecular docking. J Med Chem 49(23):6789–6801
67. Irwin JJ, Sterling T, Mysinger MM, Bolstad ES, Coleman RG (2012) ZINC: a free tool to discover chemistry for biology. J Chem Inf Model 52:1757–1768. doi:10.1021/ci3001277
68. Shi L, Javitch JA (2002) The binding site of aminergic G protein-coupled receptors: the transmembrane segments and second extracellular loop. Annu Rev Pharmacol Toxicol 42:437–467. doi:10.1146/annurev.pharmtox.42.091101.144224
69. Hawkins PCD, Warren GL, Skillman AG, Nicholls A (2008) How to do an evaluation: pitfalls and traps. J Comput Aided Mol Des 22(3–4):179–190
70. Nicholls A (2008) What do we know and when do we know it? J Comput Aided Mol Des 22(3–4):239–255
71. Mysinger MM, Shoichet BK (2010) Rapid context-dependent ligand desolvation in molecular docking. J Chem Inf Model 50(9): 1561–1573
72. Katritch V, Rueda M, Lam PC, Yeager M, Abagyan R (2010) GPCR 3D homology models for ligand screening: lessons learned from blind predictions of adenosine A2a receptor complex. Proteins 78(1):197–211
73. Lin X, Huang XP, Chen G, Whaley R, Peng S, Wang Y, Zhang G, Wang SX, Wang S, Roth BL, Huang N (2012) Life beyond kinases: structure-based discovery of sorafenib as nanomolar antagonist of 5-HT receptors. J Med Chem 55(12):5749–5759. doi:10.1021/jm300338m
74. Miao Y, Nichols SE, Gasper PM, Metzger VT, McCammon JA (2013) Activation and dynamic network of the M2 muscarinic receptor. Proc Natl Acad Sci U S A 110(27):10982–10987. doi:10.1073/pnas.1309755110
75. Bottegoni G, Rocchia W, Rueda M, Abagyan R, Cavalli A (2011) Systematic exploitation of multiple receptor conformations for virtual ligand screening. PLoS One 6(5), e18845. doi:10.1371/journal.pone.0018845

76. Korb O, Olsson TS, Bowden SJ, Hall RJ, Verdonk ML, Liebeschuetz JW, Cole JC (2012) Potential and limitations of ensemble docking. J Chem Inf Model 52(5):1262–1274. doi:10.1021/ci2005934
77. Rueda M, Totrov M, Abagyan R (2012) ALiBERO: evolving a team of complementary pocket conformations rather than a single leader. J Chem Inf Model 52(10):2705–2714. doi:10.1021/ci3001088
78. Xu M, Lill MA (2012) Utilizing experimental data for reducing ensemble size in flexible-protein docking. J Chem Inf Model 52(1):187–198. doi:10.1021/ci200428t

Chapter 3

Radioligand Binding at Muscarinic Receptors

Esam E. El-Fakahany and Jan Jakubik

Abstract

Five subtypes of muscarinic acetylcholine receptors denoted M_1 through M_5 have been cloned. Muscarinic receptors mediate a wide array of physiological functions and impairment of muscarinic signaling is involved in numerous pathological conditions including Alzheimer's disease and schizophrenia. Reliable radioligand binding techniques allow study of involvement of individual muscarinic receptor subtypes in the physiology and pathology of muscarinic signaling, and study of the structure of muscarinic receptors and structure-activation relationship of muscarinic ligands. Here we discuss the current state of knowledge of radioligand binding experiments at muscarinic receptors from the perspective of available radioligands and selective unlabeled muscarinic ligands. We relate binding properties of muscarinic ligands to experimental design (e.g., nonspecific binding determination, incubation conditions, buffers, temperature). We also list tissue/cell sources of muscarinic receptors suitable for radioligand binding studies and describe procedures of cell and tissue preparation for radioligand binding experiments. We also describe several techniques of receptor-bound ligand separation applicable at muscarinic receptors and provide basic information for binding data analysis.

Key words Muscarinic acetylcholine receptors, Radioligand binding

1 Historical Background

Radioligand binding methods are a cornerstone of receptor pharmacology, taking muscarinic acetylcholine receptors as an example. The main principle of the method is to allow a radiolabeled compound specific to a given receptor to incubate with a biological sample enriched with that receptor, and then separate the bound and free radioligand. Many radiolabeled muscarinic ligands with high affinity and specific activity are currently available. Development of reliable radioligand binding technique at muscarinic receptors cleared the way for identification, purification, and subsequent sequencing of the first muscarinic receptor [1] that enabled ensuing cloning of five subtypes of muscarinic receptors (M_1 through M_5) [2] that revolutionized the field of research of muscarinic receptor pharmacology. Furthermore, radioligand binding played a key role in identifying the orthosteric ligand-binding site that is located in a

Jaromir Myslivecek and Jan Jakubik (eds.), *Muscarinic Receptor: From Structure to Animal Models*, Neuromethods, vol. 107, DOI 10.1007/978-1-4939-2858-3_3, © Springer Science+Business Media New York 2016

pocket formed by transmembrane helices [3, 4], receptor subtype-specific ligands, allosteric modulators, and structure-activation relationship. Radioligand binding may also be employed in sophisticated experiments to study kinetics of drug-receptor interactions or in a combination with site-directed mutagenesis to determine the role of specific residues and domains in ligand binding. Similar to other receptor targets, radioligand binding studies at muscarinic receptors are simple and if performed correctly are very sensitive and highly accurate.

2 Principles of Ligand Binding

2.1 Ligand Binding Definition

Interaction of a small molecule (ligand) with a protein (receptor) is mediated by four chemical forces: electrostatic force, hydrogen bonding, van der Waals interactions, and hydrophobic bonds. Electrostatic force mediates attraction between opposed charged groups or repulsion between similarly charged groups that is proportional to the net sum of charges and inversely proportional to the square of distance between charges (as described by Coulomb's law). van der Waals interactions are attractive *and* repulsive forces between dipoles approximated by Lennard-Jones function that has its minimum (strongest attraction) at certain distance of the components. A hydrogen bond is a special case of the electrostatic attractive interaction between polar molecules, in which hydrogen is bound to a highly electronegative atom like nitrogen or oxygen. A hydrogen bond is weaker than electrostatic force but stronger than a van der Waals interaction. Hydrophobic bonds are entropy-driven interactions between nonpolar groups to avoid interaction with polar groups, mainly water. Hydrophobic bonds are stronger than van der Waals interaction. Because these forces vary in their strength and dependence on the distance between components, the combination of all these forces directs the positioning of a ligand on the receptor-binding site with minimal free energy. A measure of attraction of the ligand to the binding site is termed affinity. Thermodynamic movement does not allow ligands to sit still in the binding site and makes them associate and dissociate from the receptor, even at equilibrium. Thus, the probability with which a ligand is bound to (stays at) the binding site of a receptor is given by ligand concentration, temperature, and strengths of interactions. Dependence of the ligand binding on its concentration (Fig. 1 black curve) is defined by Langmuir isotherm:

$$\text{binding} = \frac{[\text{L}]}{[\text{L}] + K_\text{D}} \tag{1}$$

where square brackets designate concentration and K_D is the equilibrium dissociation constant that is equal to concentration at which

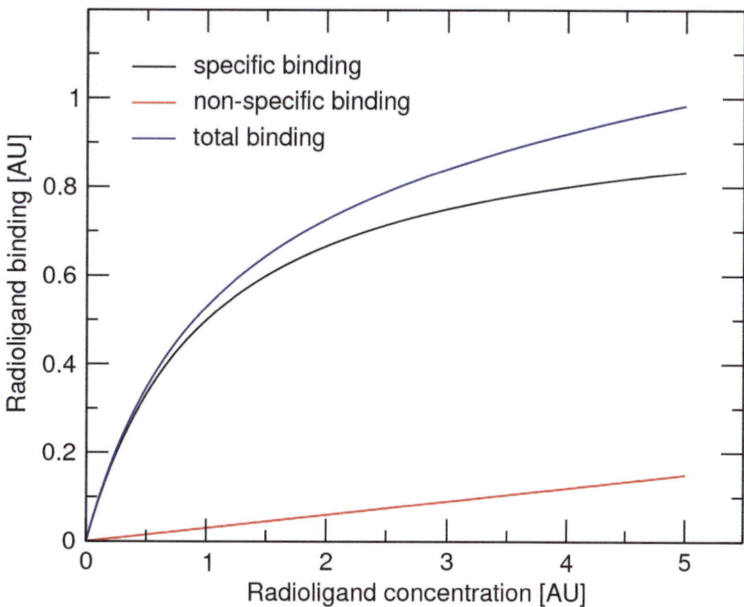

Fig. 1 Radioligand binding. *Lines* represent hypothetical radioligand binding (*blue curve*) that consists of saturable specific binding (*black curve*) defined by the binding isotherm and nonspecific binding (*red line*) that is linearly proportional to radioligand concentration and is non-saturable

the binding site is occupied with 50 % probability (for $[L] = K_D$ expression is equal to 1/2 while for $[L] \gg K_D$ it limits to 1).

2.2 Radioligands

The aim of binding experiments is to quantify ligand binding to the receptor under given conditions. Labeling the ligand with a radioactive isotope allows easy and sensitive quantification of binding and (unlike fluorescent labeling) does not interfere with ligand binding. High specific radioactivity is required, so ligand binding translates to high signal. Low nonspecific binding is required for high signal-to-noise ratio. Finally, a good radioligand should have high affinity for the receptor to prevent ligand dissociation from the receptor during the separation of free and bound radioligand. High affinity also affords the use of low concentrations of expensive radioligands.

2.3 Ligand-Specific Binding

Ligand binding to a receptor is a dynamic process of attraction mediated by chemical forces and disruption of binding by thermal movement of molecules. As a result a ligand incessantly associates with and dissociates from the receptor with time. Ligand (L) binding to the receptor (R) and formation of ligand-receptor complexes (LR) can be described as a reversible bimolecular reaction:

$$L + R \underset{k_{On}}{\overset{k_{Off}}{\rightleftarrows}} LR \qquad \text{(Schema 1)}$$

where k_{On} is the association rate constant and k_{Off} is the dissociation rate constant. The observed rate of association is directly proportional to the concentration of receptor, ligand, and k_{On}. While the *rate* of dissociation of a ligand from the receptor (k_{Off}) is only a property of the affinity of binding, the *amount* of remaining ligand-receptor complex at any moment is dependent on the starting amount of the complexes prior to the onset of dissociation. Thus, for any given moment the magnitude of change of the concentration of the ligand-receptor complex is given by the difference between formation and decay of ligand-receptor complexes:

$$\frac{d[LR]}{dt} = k_{On} \times [L] \times [R] - k_{Off} \times [LR] \tag{2}$$

Under binding equilibrium the change in LR is nil and formation of LR happens at the same rate as its decay:

$$k_{On} \times [L] \times [R] = k_{Off} \times [LR] \tag{3}$$

The ratio of k_{On} and k_{Off} defines ligand affinity. Affinity (also known as equilibrium association constant) is a measure of attraction between ligand and receptor and thus is directly proportional to k_{On} and inversely proportional to k_{Off}. Equilibrium dissociation constant K_D is the reciprocal value of equilibrium association constant and defines the ligand concentration necessary to occupy 50 % of receptors (see Eq. (1)). Restating Eq. (3) gives K_D as

$$K_D = \frac{k_{Off}}{k_{On}} = \frac{[L] \times [R]}{[LR]} \tag{4}$$

At any time the total number of receptors $[R_T]$ is the sum of free receptors and receptors in complex with the ligand:

$$[R_T] = [R] + [LR] \tag{5a}$$

or

$$[R] = [R_T] - [LR] \tag{5b}$$

Substitution of R in Eq. (4) according Eq. (5b) gives Eq. (6a):

$$K_D = \frac{[L] * ([R_T] - [LR])}{[LR]} \tag{6a}$$

or

$$K_D = \frac{[L] \times [R_T]}{[LR]} - \frac{[L] \times [LR]}{[LR]} \tag{6b}$$

or

$$K_D = \frac{[L] \times [R_T]}{[LR]} - [L] \tag{6c}$$

or

$$K_D + [L] = \frac{[L] \times [R_T]}{[LR]} \tag{6d}$$

and finally

$$[LR] = \frac{[L] \times [R_T]}{K_D + [L]} \tag{6e}$$

that is actually Eq. (1) of specific binding related to concentration (or number) of binding sites R_T.

2.4 Ligand Nonspecific Binding

Apart from the specific binding site on the receptor a ligand may also bind to other sites on the biological sample by nonspecific interaction that is by orders of magnitude weaker than specific binding. Nonspecific binding is linearly proportional to ligand concentration and in principle has infinite binding capacity (non-saturable) (Fig. 1, red curve). Nonspecific binding is not only given by chemical properties of the radioligand but also by arrangement of the experiment (e.g., sample washing, removal of tissue components that do not express the receptor). The magnitude of nonspecific binding is determined in the presence of excess of a highly specific non-labeled ligand sufficient to fully occupy the receptor. For example 1 μM atropine is commonly used for determination of nonspecific binding of muscarinic ligands. Higher concentrations should be avoided as it may slow down tracer dissociation [5] and so lead to overestimation of nonspecific binding. A ligand from the same pharmacological class but different from the radiolabeled ligand is preferred. A non-labeled ligand of the same chemical structure may bind to the same nonspecific sites and protect them from tracer binding so that these nonspecific binding sites are erroneously counted as specific binding sites. Ideally, several different unlabeled competitors should all yield statistically indistinguishable estimates of nonspecific binding. The ratio of nonspecific to specific binding should be as low as possible (less then 1 % of total added radioactivity to the sample and less than 10 % of total binding in case of muscarinic receptors). Washing is the least accurate step in the radioligand binding assay. Since nonspecific binding is affected by washing, variations in washing lead to variations in nonspecific binding. Thus, nonspecific binding should be determined in each filtration.

3 Radioligand Binding Experiments at a Glance

A radioligand binding experiment consists of these steps: (1) preparation of samples from tissues, cell cultures, or cell lines; (2) incubation of samples with radioligand; (3) separation of free radioligand from the bound one; (4) scintillation counting to determine radioactivity of individual samples; and (5) data analysis. Therefore, two pieces of specialized devices (besides common laboratory equipment) are needed to conceive radioligand binding experiments: first, an apparatus for separation of free and bound radioligand for which purpose cell harvesters are most commonly used; second, a scintillation counter compatible with the format of apparatus used for separation of free and bound radioligand. For example, a scintillation counter that can read filtration plates is needed when filtration plates are used in the cell harvester.

4 Available Radioligands

Nowadays available muscarinic radioligands cover almost all experimenter needs. Available radioligands include both reversible antagonists and agonists as well as covalent ligands, antagonists with fast as well as slow kinetics, and antagonists selective to M_1, M_2, and M_3 subtypes. A list of common muscarinic radioligands is shown in Table 1. Their structures and chemical names are depicted in Fig. 2.

4.1 Antagonist Radioligands

Muscarinic receptor antagonists are preferred over agonists as tracers because of their higher affinity (Table 1). The most commonly used tracers are *N*-methylscopolamine (NMS) and quinuclidinyl benzilate (QNB). NMS exhibits slightly lower affinity at M_2 and M_5 receptor subtypes (Table 1). Tritiated NMS is commercially available at high specific radioactivity (80 Ci/mmol). The half-life of ligand-receptor complex is around 15 min at M_1, M_3, and M_4 receptors, 3 min at the M_2 receptor, and 53 min at the M_5 receptor [6]. Combined with a rather fast rate of association, NMS is suitable for most common radioligand binding studies. Half-life of the complexes is short enough to reach equilibrium and slow enough for ligand separation by simple filtration. The advantage of QNB over NMS is its higher affinity and lower nonspecific binding (mainly to glass-fiber filters thanks to the absence of the positive charge). However, commercially available tritiated QNB has lower specific radioactivity (50 Ci/mmol). Moreover, QNB has extremely slow kinetics. Half-life of QNB in complex with muscarinic receptors is around 100 min, at M_5 receptors even 180 min [6]. Slow kinetics of QNB may be problematic when attaining the equilibrium quickly is needed but may be of advantage when samples are

Table 1
List of common muscarinic radioligands

Ligand	Type	K_D (nM)				
		M_1	M_2	M_3	M_4	M_5
4-DAMP[a]	Antagonist	0.58	3.8	0.52	1.2	1.0
Acetylcholine[b]	Agonist	23–30	21–26	19–24	18–23	19–23
ACM	Agonist	Covalent binding				
Atropine[c]	Antagonist		1.35	1.48		
NMPB[d]	Antagonist		2.29			
NMQNB[e]	Antagonist	0.13	0.45	0.65		
NMS[a]	Antagonist	0.08–0.15	0.2–0.4	0.15–0.2	0.05–0.1	0.5–0.7
Oxotremorine[f]	Agonist	900	70	390	220	510
PBCM	Antagonist	Covalent binding				
Pirenzepine[a]	Antagonist	0.003–0.015	400–10,000	200–2500	25–1200	125–630
QNB[a]	Antagonist	0.015–0.060	0.02–0.05	0.03–0.09	0.02–0.08	0.02–0.06

4-DAMP 4-diphenylacetoxy-*N*-methylpiperidine, *ACM* acetylcholine mustard (*N*-2-chloroethyl-*N*-methyl-2-acetoxyethylamine), *NMPB* *N*-methylpiperidyl benzilate, *NMQNB* *N*-methylquinuclidinebenzilate, *NMS* *N*-methylscopolamine, *PBCM* propylbenzilylcholine mustard (2-[2-chloroethyl(propyl)amino]ethyl 2-hydroxy-2,2-diphenylacetate), *QNB* quinuclidinylbenzilate

Source:

[a]Alexander et al. [32]

[b]Jakubik et al. [8]

[c]Melchiorre et al. [33]

[d]Hejnová et al. [34]

[e]Visser et al. [35]

[f]Dallanoce et al. [36]

washed from free radioligand (e.g., washing tissues for radioimaging) or ligand separation is slow (e.g., gel filtration, see Protocol F). Finally, the lipophilic nature of QNB results in its uptake into cells, which causes significantly high levels of nonspecific binding in intact cell studies. Pirenzepine is M_1-selective antagonist (Table 1) that is commonly used for selective labeling of M_1 receptors in samples with mixture of receptor subtypes. The use of other selective as well as nonselective radiolabeled antagonists has been reported; however, they have in general lower affinity than NMS (Table 1) and thus are less suitable for routine radioligand binding experiments.

4.2 Agonist Radioligands

There are two commercially available tritiated muscarinic agonists, the nonselective acetylcholine and the M_2 preferring oxotremorine. In general, muscarinic agonists are not good tracers because

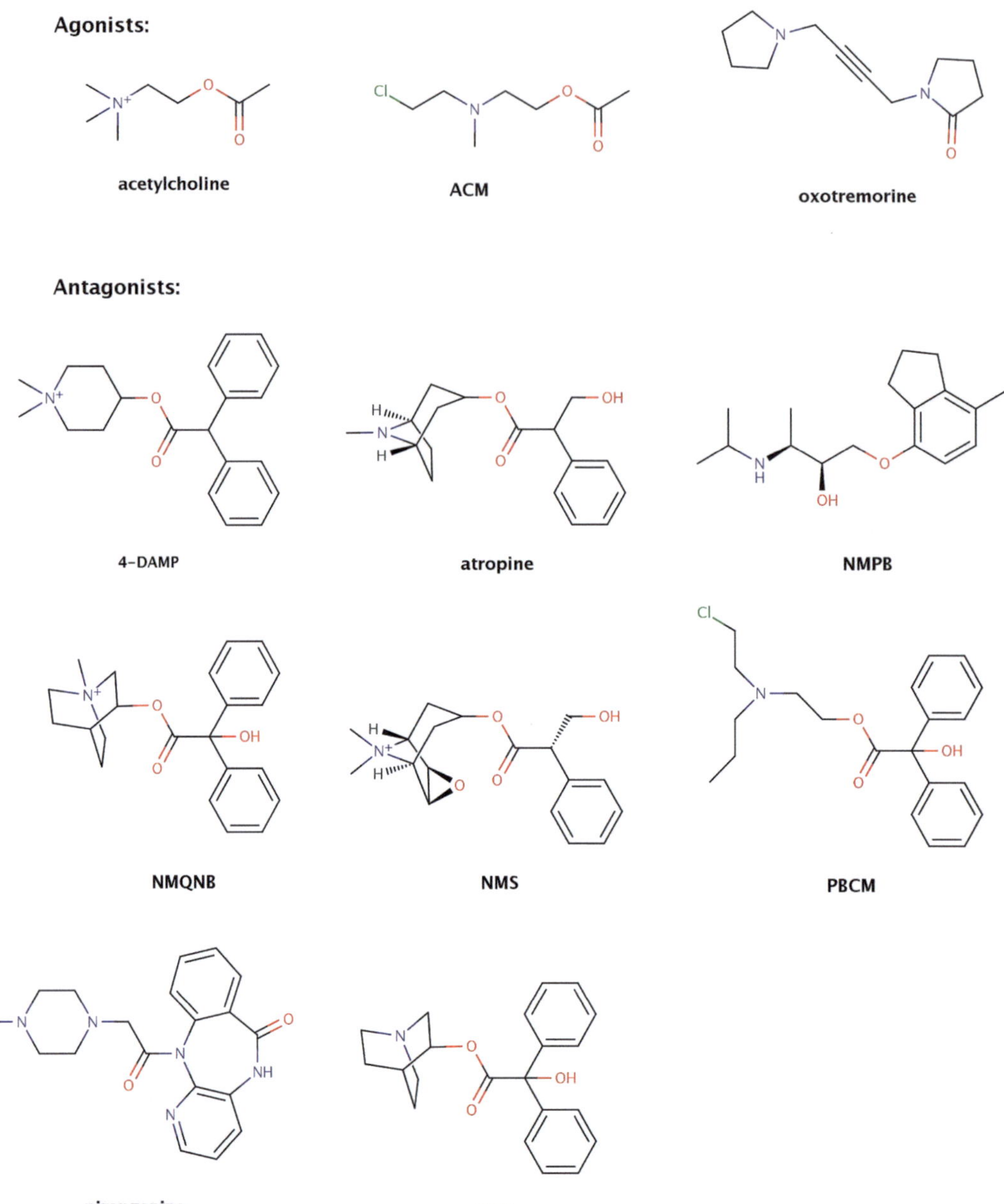

Fig. 2 Structures of muscarinic radioligands. Structures of radiolabeled muscarinic agonists and antagonists

of their low affinity (Table 1). Moreover, only receptors in high-affinity states (receptors in complexes with GDP-free G-protein) [7] can be detected by radiolabeled agonists. Thus agonists as tracers are employed only when specific effects on agonist binding have to be determined [8]. The use of acetylcholine as a tracer at muscarinic receptors is limited by several factors. First, acetylcholine binds both to muscarinic and nicotinic types of acetylcholine receptors. Second, acetylcholine is readily cleaved by cholinesterases (acetylcholinesterase and butylcholinesterase) that are omnipresent in animal tissues either in a free form in body fluids or anchored to plasma membranes [9]. To use acetylcholine as a tracer samples has to be devoid of cholinesterases (e.g., membrane preparations of non-neural cell lines) or activity of cholinesterases has to be blocked by specific inhibitors. However, many of acetylcholinesterase inhibitors are allosteric modulators of muscarinic receptors [10] and affect binding of muscarinic ligands. Moreover, the ester bond of acetylcholine is not chemically stable and acetylcholine decays spontaneously in aqueous solution to acetate and choline. Commercially available tritiated acetylcholine is radiolabeled either at acetate or choline group. It is easier (and cheaper) to label acetylcholine to high specific radioactivity at the choline group. However, acetylcholine labeled in this manner has high nonspecific binding due to traces of labeled choline from acetylcholine chemical decay that is hard to wash from nonspecific sites due to its positive charge. In contrast, radiolabeling of acetylcholine at the acetate group gives low nonspecific binding. Carbachol may be considered as an alternative to acetylcholine. It is similar to acetylcholine in structure, has similar affinity, and is resistant to cholinesterases. Unfortunately, radiolabeled carbachol is not commercially available. Unlike acetylcholine oxotremorine is chemically stable, is not a substrate for cholinesterases, and is specific for the muscarinic type of acetylcholine receptors. However, oxotremorine has much lower affinity than acetylcholine (Table 1).

4.3 Irreversible Radioligands

Two radiolabeled ligands, the agonist acetylcholine mustard (ACM) [4] and the antagonist propylbenzilyl choline mustard (PBCM) [3], form covalent bonds with muscarinic receptors. The advantage of covalently bound, practically irreversible tracers is that binding withstands long-lasting sample manipulation (like receptor isolation, electrophoresis, or immunoprecipitation) without ligand dissociation. The disadvantage of these tracers is that their nonspecific reactivity (binding) cannot be prevented by reversible antagonists (irreversible ligand always wins over reversible one in competition for binding) nor removed by washing (because tracer is bound irreversibly). Thus these tracers are suitable only for at least partly purified receptors.

5 Source of Muscarinic Receptors

5.1 Tissues

Muscarinic receptors are expressed throughout the body at the central nervous system, peripheral neurons, as well as target tissues innervated by cholinergic neurons (primarily parasympathetic neurons) [11] (Table 2). Almost all tissues express a mixture of subtypes of muscarinic receptors; for example smooth muscles express

Table 2
Mammalian tissues with significant expression of muscarinic receptors

Subtype	Location	Reference
M_1	Cortex, hippocampus, striatum, salivary glands	Levey [37]
	Lymphocytes	Kawashima and Fujii [14]
M_2	Brainstem, cerebellum, thalamus, heart, ileum, lung	Levey [37]
	Smooth muscles	Caulfield [38]
M_3	Salivary glands	Levey [37]
	Smooth muscles	Caulfield [38]
	Hypothalamus	Gautam et al. [39]
	Hippocampus	Poulin et al. [40]
M_4	Striatum, lung	Levey [37]
M_5	Ventral tegmental area, substantia nigra	Eglen and Nahorski [12]

M_2 and M_3 receptors, and lungs express M_2 and M_4 receptors. Natural expression of M_5 receptor is limited only to certain parts of brain like the ventral tegmental area and substantia nigra [12, 13]. Muscarinic receptors were also reported in non-neuronal non-innervated cells like lymphocytes [14]. Tissues may serve as a source of membranes or purified receptors (Sections 5.3 and 5.4) or used in binding experiments in whole, as slices, dissipated cells, or tissue culture.

5.2 Cell Lines

Besides expressing a mixture of muscarinic receptors another drawback of animal tissues as a source of muscarinic receptors is the relatively low density of receptor expression (less than 1 pmol of binding sites per mg of membrane proteins). Since cloning of all five subtypes of muscarinic receptors CHO cell lines stably expressing individual subtypes of muscarinic receptors in high density have become available and widely used [15]. Although muscarinic receptors were detected in naive CHO-K1 cells by highly sensitive second messenger assays [16] they cannot be detected in binding studies. Thus CHO cell lines are a widely used source of individual subtypes of muscarinic receptors in radioligand binding studies. Currently CHO cells stably expressing individual subtypes are available also from commercial sources (e.g., Perkin Elmer; Missouri S&T cDNA Resource Center). Muscarinic receptors are also routinely transiently expressed in COS-7 or HEK-293 cell lines [17, 18]. Since these cells have to be prepared anew for each binding experiment, stable transfection is usually used for studies of native receptors. Transiently transfected cells become quite useful for studying the binding characteristics of a large number of engineered receptor constructs, e.g., in receptor mutagenesis studies.

5.3 Membranes

The use of membranes instead of whole tissue or cells in binding experiments facilitates separation of bound and free radioligand and eliminates nonspecific binding to tissues and cellular components that do not express the receptor. The use of membranes rather than intact cells removes GTP that uncouples the receptor from G proteins. This increases agonist affinity and enables the use of radiolabeled agonist. Protocols A and B yield a mixture of plasma membranes, light membrane vesicles, and mitochondria, and are a practical compromise between purity of sample and receptor yield. Membrane preparation according to Protocol A results in 10–30 % of high-affinity sites for agonists (receptors in complex with GDP-free G-protein). Protocol B is modification of Protocol A that facilitates GDP dissociation [19] and subsequent formation of high-affinity complexes of receptor. This makes membranes suitable for experiments with radiolabeled agonists.

5.4 Purified Receptors

Isolation of purified muscarinic receptors may be desired for specific purposes (like determination of effects of membrane or membrane composition on receptor binding properties). Preparation of purified muscarinic receptors is described in Protocol C. It should be noted that a source with high expression density of muscarinic receptors like transfected Sf9 cells [20] has to be used. Purified receptors can be reconstituted in artificial lipid vesicles [21]. Protocol D describes reconstitution of purified receptors into artificial vesicles with a simplified composition of common membranes (cholesteryl hemisuccinate:phosphatidyl choline:phosphatidyl inositol, 4:48:48) that can be varied as desired [22] and optionally purified G-proteins may be added [23].

Protocol A: Preparation of membranes for general use

1. *Put tissue or cells of your choice in ice-cold homogenization medium (e.g., 100 mM NaCl, 10 mM EDTA, 20 mM HEPES buffer pH = 7.4). Keep on ice during steps 1 and 2. Homogenization medium should contain EDTA to stop calcium-dependent proteases. Protease inhibitors should be used with caution as they may modify muscarinic receptors and affect their binding properties [24, 25]. Homogenization medium should have more or less normal ionic strength for steps 3 and 4 to work. If for any reason low ionic strength medium has to be used centrifugal force and duration of centrifugation in steps 3 and 4 need to be increased and extended because low ionic strength improves membrane dispersion that results in increased membrane flotation and thus greater centrifugal forces and longer times are needed to sediment the membranes.*

2. *Homogenize the sample by the method of your choice (e.g., in Ultra-Turrax homogenizer by two 30-s strokes with 30-s pause between strokes) while cooling them on ice.*
3. *To remove unbroken cells, cell nuclei, cytoskeleton, and extracellular matrix proteins spin down the samples at 1000×g for 5 min and take the supernatant for the next step. During preparation of certain tissues rich in lipids (like brain cortex) the lipid foam that forms on top of the water phase needs to be discarded.*
4. *To remove the cytosolic fraction spin down the supernatant from step 3 at 30,000×g for 30 min. Remove the supernatant and dissolve pellet in incubation medium (e.g., 100 mM NaCl, 10 mM* $MgCl_2$*, 20 mM HEPES buffer pH = 7.4).*
5. *Leave samples for 30 min at 4 °C.*
6. *Spin down samples at 30,000×g for 30 min and discard supernatant.*
7. *Pellets may be stored for limited time (couple of months) frozen at −20 °C or below.*

Protocol B: Preparation of membranes for experiments with radiolabeled agonists

Steps 1–3 are the same as in Protocol A.

4. *To remove the cytosolic fraction spin down supernatant from step 3 at 30,000×g for 30 min. Remove supernatant and dissolve pellet in 1 M ammonium sulfate.*
5. *Allow samples to denature for 3 h at 4 °C.*
6. *Spin down samples at 60,000×g for 60 min, discard supernatant, and dissolve pellet in incubation medium containing 20 % glycerol.*
7. *Leave samples for 1 h at 4 °C to renaturate.*
8. *Spin down samples at 60,000×g for 60 min, discard supernatant, and dissolve pellet in incubation medium.*
   Continue with steps 5–7 from Protocol A.

Protocol C: Preparation of purified muscarinic receptors

1. *Harvest Sf9 cells by centrifugation at 1500×g for 10 min.*
2. *Prepare crude membranes according to steps 1–4 of Protocol A.*
3. *Dilute crude membranes in 20 mM HEPES buffer pH = 7.4, 5 mM imidazole, 1 mM EDTA, 1 % digitonin, and 0.1 % sodium cholate to a protein concentration of 1 mg/ml.*
4. *Incubate stirred for 1 h at 4 °C.*
5. *Centrifuge at 100,000×g for 90 min at 4 °C and take the supernatant fraction.*
6. *Apply 1 l of supernatant fraction from step 5 to ABT-agarose [26] column (300 ml) at 4 °C at a flow rate of 70 ml/h.*

7. *Wash the column with 15 l of washing medium (0.2 M NaCl, 20 mM potassium phosphate buffer (pH 7.0), 0.1 % digitonin).*
8. *Connect hydroxyapatite column (1 ml) to the outflow of ABT-agarose column.*
9. *Apply 700 ml of the washing medium supplemented with 0.1 mM carbachol (to dissociate receptors from affinity column) to the ABT-agarose column at the same rate as above (carbachol is preferred over atropine in this step for easier removal in step 11).*
10. *Disconnect hydroxyapatite column from ABT-agarose column and elute purified receptors with 0.5 M potassium phosphate buffer (pH = 7.0) containing 0.1 % digitonin.*
11. *Prior to performing the radioligand binding experiment remove bound carbachol by 50-fold dilution in 0.5 M potassium phosphate buffer (pH = 7.0) containing 0.1 % digitonin. Allow carbachol dissociate for 1 h at 4 °C and then concentrate receptors by centrifugation through Centricon-30 membranes (Amicon Co., Ltd.).*

Protocol D: Reconstitution of receptors into artificial vesicles

Steps 10 and 11 are optional.

1. *Prepare 4 mg of lipids by combining 16 μl of cholesteryl hemisuccinate (10 mg/ml in methanol), 48 μl of phosphatidyl choline (20 mg/ml in chloroform), and 48 μl of phosphatidyl inositol (20 mg/ml in chloroform) in glass tube.*
2. *Form lipid film on the wall of tube by evaporation with N_2.*
3. *Add 1 ml of solution A (100 mM NaCl, 1 mM ETDA, 20 mM HEPES pH = 7.4) supplemented with 1 % sodium cholate.*
4. *Sonicate on ice for 20 min.*
5. *Combine 66 μl of receptors (300 pmol/ml), 32 μl of solution A, and 100 μl of lipid vesicles from step 4. Vortex vigorously and leave on ice for 30 min.*
6. *Wash 2 ml Sephadex G-50 fine-grade columns with 5 ml of solution A.*
7. *Apply mixture from step 5 to Sephadex G-50 column.*
8. *Wash column three times with 0.2 ml of solution A.*
9. *Add 0.4 ml of solution A and collect.*
10. *Mix 0.2 ml of receptor vesicles (~10 pmol of receptors) from step 9 with 28 μl of G-proteins (~50 pmol), 1.25 μl of 2 M $MgCl_2$, 1.25 μl of 1 M dithiothreitol, and 19.5 μl of solution A and vortex vigorously.*
11. *Leave on ice for 60 min.*
12. *Dilute with 5 volumes of solution A and use 50 μl per sample in binding assay.*

6 Incubation Conditions

6.1 Buffers

Many types of buffers may be used for radioligand binding experiments. However, it should be noted that buffer composition influences ligand affinity. Low concentrations of sodium (low ionic strength in general) lead to higher affinity and slower dissociation of the tracer [27]. This is desired in experiments with tracers with low affinity and fast dissociation like agonists. Agonists bind to complexes of receptor and GDP-free G-protein. Formation of these complexes is conditioned by the presence of magnesium ions [23]. For routine measurements on membranes thus simple buffer consisting of 100 mM NaCl, 5 mM $MgCl_2$, and 20 mM HEPES buffer pH = 7.4 is suitable for a wide range of tracers. For binding experiments on whole cells iso-osmotic buffer (like Krebs-HEPES buffer: 138 mM NaCl, 4 mM KCl, 1.3 mM $CaCl_2$, 1 mM $MgCl_2$, 1.2 mM NaH_2PO_4, 10 mM glucose, 20 mM HEPES pH = 7.4; 340 mOsm/l) has to be used. An advantage of Krebs-HEPES buffer is that many functional assays like accumulation of inositol phosphates, inhibition of cAMP synthesis, or microfluorometric determination of intracellular calcium can be conducted in it. This allows the comparison of ligand affinity in binding and functional studies under similar conditions [28].

Another consideration is the addition of chelating agents for their beneficial effects. Chelating agents may inhibit possible contamination with proteases. Chelating agents in combination with low ionic strength promote membrane dispersion and thus improve handling properties of membrane preparations. On the other hand chelating agents significantly perturb ligand interactions by removing multivalent ions. As stated above magnesium ions are essential for agonist high-affinity binding.

6.2 Sample Size

For radioactivity of the sample to be counted accurately it should be about 1000 cpm (about 2000 dpm for tritiated ligands on the assumption of 50 % efficiency of counting). Typical specific radioactivity of tritiated commercial grade muscarinic radioligand is 160 dpm/fmol that translates to 12.5 fmol radioligand occupied sites per sample to achieve 2000 dpm of specific binding. The best ratio of specific to nonspecific binding is observed around ligand K_D. When radioligand in concentration equal to K_D is used 50 % of binding sites are occupied by radioligand. Thus 25 fmol of receptors per sample is needed to get 2000 dpm of specific binding. Membranes prepared according to Protocol A from CHO cells stably expressing muscarinic receptors usually have 1–10 fmol of receptors per microgram of proteins; thus 10–20 μg of protein per sample is generally used. However, agonists bind with high affinity only to receptors in complex with GDP-free G-protein that represent only a fraction of total receptors. Thus, up to five times more membranes are needed in comparison to antagonist binding. The capacity limit of filtration on 96-well plates (filter diameter 5 mm) is 100 μg of protein per sample. This means that only membranes from high-expression systems like cell lines or brain cortex (0.5 fmol/μg of protein) can be used in this compact and economic assay. Tissue homogenates or preparations from low-expressing systems (0.1 fmol/μg of protein or lower) like lung [29] have to be filtered through filters with higher capacity like 24-tube cell harvester (filter diameter 15 mm) which maximum capacity is about 1 mg of proteins.

6.3 Incubation Volume

Incubation volume is determined by the method of separation of bound and free radioligand. In separation on gel filter the incubation volume is equal to the loading volume, which is dependent on the volume of the column (e.g., for 2 ml G-50 Sephadex column 50 μl of loading volume is required). In scintillation proximity assay the incubation volume is only limited by the size of well or tube. In separation on filters the incubation volume is related to the size of the filter. Larger filters require larger washing volumes and thus larger incubation volumes. The larger the area of the filter, the higher the capacity and more membranes, cells, or tissue can be and should be used. Large filters (like in 24-tube Brandel cell harvester) are thus suitable for preparations with low receptor density, and therefore require a large amount of biological sample, e.g., non-neural tissues. For 24-tube filtration (filter diameter 15 mm) the optimal incubation volume is about 3 ml; for 96-well filtration (filter diameter 5 mm) the optimal incubation volume is about 0.4 ml.

Another aspect that influences the size of incubation volume is the combination of tracer amount, tracer affinity, and number of binding sites in the sample. If the amount of the tracer in relation to the tracer affinity and the number of binding sites is low a substantial part (>10 %) of the tracer is bound to the receptors. Such conditions are termed tracer depletion and may happen usually in saturation binding experiments. Tracer depletion complicates data analysis because the free tracer concentration is significantly lower than that inferred from the amount of added tracer (radioactivity) and final incubation volume. The incubation volume should be increased if it is estimated that there is a risk of tracer depletion. For such cases 1.2 ml 96-well plates are available. When the incubation volume of samples intended for gel filtration is increased the volume of the gel column has to be proportionally increased. On the other hand when tracers with low affinity (like agonists) are used the incubation volume can be reduced to save expensive radioligand. However, in a 96-well plate Brandel cell harvester samples with volume smaller than 0.2 ml are difficult to apply and wash reliably. In case of gel filtration a smaller incubation volume can be diluted with ice-cold buffer to the desired volume and applied to the gel column of the standard size.

6.4 Temperature

As explained above in Section 2 temperature affects ligand binding. An increase in temperature potentiates thermal movement of molecules including the receptor and the ligand but the strength of intermolecular forces remains the same. Temperature dependence of the affinity constant K_A and equilibrium dissociation constant K_D is described by the Van't Hoff equation:

$$K_A = \frac{1}{K_D} = e^{-(\Delta G/RT)} = e^{-(\Delta H/RT)} \times e^{(\Delta S/R)} \quad (7)$$

where ΔG is free binding energy, ΔH is enthalpy of binding, ΔS is entropy of binding, and R and T are gas constant and absolute temperature, respectively. Enthalpy contribution to equilibrium ligand binding is small. The major contribution to the thermodynamics of ligand binding is change in entropy (mainly ligand desolvation on association with receptor). Temperature dependence of equilibrium constants is thus relatively small; however it is still about twofold change over 10 °C (the difference between room temperature and body temperature). An increase in temperature usually leads to a decrease in affinity unless hydrophobic effects (which strengthen with temperature) contribute to ligand binding substantially [30]. Thanks to relatively low temperature dependence of equilibrium binding incubation temperature can be often chosen to be the same as in other types of assays in the conducted study (e.g., 37 °C as in study of functional response). For study of agonist binding 30 °C appears to be optimal as the fraction of high-affinity sites for agonists is at its maximum [31].

Temperature dependence of the association rate constant k_{On} is described by the Arrhenius equation:

$$k_{On} = \frac{k_B \times T}{\hbar} \times e^{-(\Delta G/k_B T)} = \frac{k_B \times T}{\hbar} \times e^{-(\Delta H/k_B T)} \times e^{-(\Delta S/k_B)} \quad (8)$$

where k_B and $\hbar$ are Boltzmann's and Planck's constants, respectively. Usually there is large enthalpic contribution to kinetics of ligand association with receptor that gives it large temperature sensitivity. Thus thermodynamics of ligand binding may be inferred by assessing temperature dependence of the rate of association. For muscarinic ligands a 10 °C increase in temperature may lead to a tenfold increase in the rate of association. Thus, time of incubation of a radioligand with the receptor source in equilibrium binding studies must be increased if a lower temperature is employed. For tracers with fast kinetics like muscarinic agonists the very short time steps required to accurately determine the rate of association may be unattainable by available technique and lowering incubation temperature to slow down the rate of association may be considered.

7 Radioligand Separation

As stated above separation of free radioligand from the bound one is the most crucial step of the procedure. Separation, on the one hand, must not disturb the formed ligand-receptor complex (has to be quick). On the other hand it has to be complete (since any remains of free radioligand counts as nonspecific binding). Thus optimization of separation step is about finding a balance between the duration of separation and intensity of washing.

7.1 Radioligand Binding in Cell Membranes

The simplest method to separate free radioligand from the bound one is filtration through glass-fiber filters (Protocol E) where due to difference in the size membranes with bound radioligand are retained on the filter and free radioligand passes through it. Tracer nonspecific binding is reduced in the filtration assay by washing the filters. Nonspecific binding should be determined for each filtration as it may vary among filtrations due to variations in the process of washing (that is the step with lowest precision and main source of variation). Solubilized receptors and purified receptors reconstituted in artificial lipid vesicles are small enough to pass through glass-fiber filters. For radioligand separation either gel filtration (Protocol F) or scintillation proximity assay (Protocol G) can be used. Filtration times on gel filter are long; thus radioligands with slow dissociation (e.g., QNB) have to be used. Although filtration on gel filters can be expedited by centrifugation of gel columns it is not suitable for radioligands with fast kinetics like agonists and certain antagonists. In gel filtration free radioligand is trapped in the gel pores and receptors while bound radioligand is eluted in void volume. In scintillation proximity assays receptors with bound radioligand are coprecipitated with scintillation beads by antibodies, so only bound radioligand is close enough to scintillation beads and scintillates.

Protocol E: Filtration through glass-fiber filters in a 96-well plate setup

1. *If the used radioligand has positive charge (e.g., NMS, NMQNB, acetylcholine) soak filters in 0.5 % solution of polyethylenimine to lower radioligand adsorption to filters.*
2. *Place GF/C filter or filtration plate into Brandel filtration apparatus and wash it with ice-cold deionized water to remove bubbles from tubing.*
3. *Place incubation 96-well plate on Brandel filtration apparatus, harvest the samples, and immediately wash the samples with the ice-cold deionized water (QNB and NMQNB for 9 s, other antagonists for 6 s, agonists for 3 s).*
4. *Let harvest vent open for at least 30 s to remove excess moisture from filter.*

5. *Take filter or filter plate out of Brandel filtration apparatus and dry it in microwave oven at maximum power for 2 min.*
6. *For glass-fiber filters, melt on Metltilex A solid scintillator for 90 s on 105 °C hot plate. For filtration plates, seal the bottom of the plate with transparent tape, add 50 μl of liquid scintillator to each well, and seal top of the plate with transparent tape.*

Protocol F: Gel filtration

1. *Prepare 2 ml Sephadex G-50 fine-grade columns and wash them with 5 ml of washing solution (100 mM NaCl, 1 mM EDTA, 20 mM HEPES buffer pH = 7.4, 0.05 % Lubrol PX).*
2. *Add 0.2 ml of washing solution to samples (incubation volume 50 μl) and apply to column immediately.*
3. *Add 0.1 ml of washing solution to columns.*
4. *Place 6 ml scintillation vials under the columns. Elute with 1 ml of washing solution.*
5. *Add 4 ml of water-compatible scintillation cocktail (e.g., EcoLite, Rotiszint, OptiPhase) to scintillation vials.*

Protocol G: Scintillation proximity assay

1. *If membranes were incubated solubilize them by the addition of 20 μl of 10 % Nonidet P-40 and shake the samples for 20 min.*
2. *Add 10 μl of rabbit polyclonal IgG antibody against muscarinic receptor in a final dilution of 1:5000 and incubate for 1 h.*
3. *Dilute one batch of anti-rabbit IgG-coated scintillation beads in 40 ml of incubation medium. Add 50 μl of the scintillation bead suspension to each sample and incubate for 3 h.*
4. *Centrifuge samples for 15 min at 1000 × g and count samples using the scintillation proximity assay protocol. If the background radioactivity due to scintillation of free ligand is too high filter samples through GF/C filter plate according to Protocol E.*

7.2 Radioligand Binding in Intact Cells

For separation of free and bound radioligand in case of the cells in suspension (e.g., Sf9 cells, detached CHO cells, dissociated tissues, or tissue cultures) filtration through glass-fiber filters with large pores (Whatman GF/A) according to Protocol D is the most straightforward approach. Alternatively cells may be centrifuged for 3 min at 250 × *g*. This method does not allow complete removal of the free radioligand and is therefore associated with high non-specific binding. It is preferred for radioligands with very fast dissociation that does not allow washing that is necessary in case of filtration.

Protocol H: Processing of attached cells grown on 24-well plate

1. *Remove cell culture media and wash the cells with 0.5 ml of Krebs-HEPES buffer (KHB; final concentrations in mM: NaCl*

138; KCl 4; $CaCl_2$ 1.3; $MgCl_2$ 1; NaH_2PO_4 1.2; Hepes 20; glucose 10; pH adjusted to 7.4).

2. *Incubate cells with radioligand in 0.5 ml (final volume) of KHB for 20 min (NMS) or 12 h (QNB).*
3. *Remove incubation medium and quickly wash cells twice with 0.5 ml of KHB removing it immediately.*
4. *Dissolve cells in 0.4 ml of 1 M NaOH and shake the plate for 15 min at room temperature.*
5. *Pipet 0.2 ml aliquot to 4 ml scintillation vials and add 3 ml of water-compatible scintillation cocktail (e.g., EcoLite, Rotiszint, OptiPhase).*

8 Experimental Arrangement

Arrangement of the binding experiment must conform to the specific parameters intended to be determined. In so-called kinetic experiments time of incubation varies while other parameters remain constant and the association rate constant k_{On} or dissociation rate constant k_{Off} is determined. In so-called equilibrium experiments time of incubation is constant and long enough to achieve binding equilibrium and the concentration of ligand is varied to determine the equilibrium dissociation constant K_D and number of receptors R_T.

8.1 Measurement of the Rate of Association

Usually association experiments are performed first to determine the time needed to achieve equilibrium that is a prerequisite for dissociation and equilibrium binding experiments. In association experiments samples are incubated with a constant concentration of tracer for different periods of time and dependence of ligand binding on time is evaluated. Association is usually started by the addition of tracer to free receptor and terminated by tracer removal. When tracer is added to free receptors the concentration of ligand-receptor complexes rises according to Eq. (9) that is the integral of Eq. (2) over time:

$$[\mathrm{LR}] = \left[\mathrm{LR}_{\mathrm{Eq}}\right] \times \left(1 - e^{-(k_{\mathrm{On}} \times [\mathrm{L}] + k_{\mathrm{Off}}) \times t}\right) \tag{9}$$

$[LR_{Eq}]$ is the concentration of LR under equilibrium according to Eq. (6e). In association experiments the tracer both associates with and dissociates from the receptors. Thus neither k_{On} nor k_{Off} can be determined. Instead the *observed* rate association constant k_{Obs} is calculated according Eq. (10):

$$[\mathrm{LR}] = \left[\mathrm{LR}_{\mathrm{Eq}}\right] \times \left(1 - e^{-k_{\mathrm{Obs}} \times t}\right) \tag{10}$$

Theoretical association curves are shown in Fig. 3. The association rate constant k_{On} is calculated after subtraction of k_{Off} determined

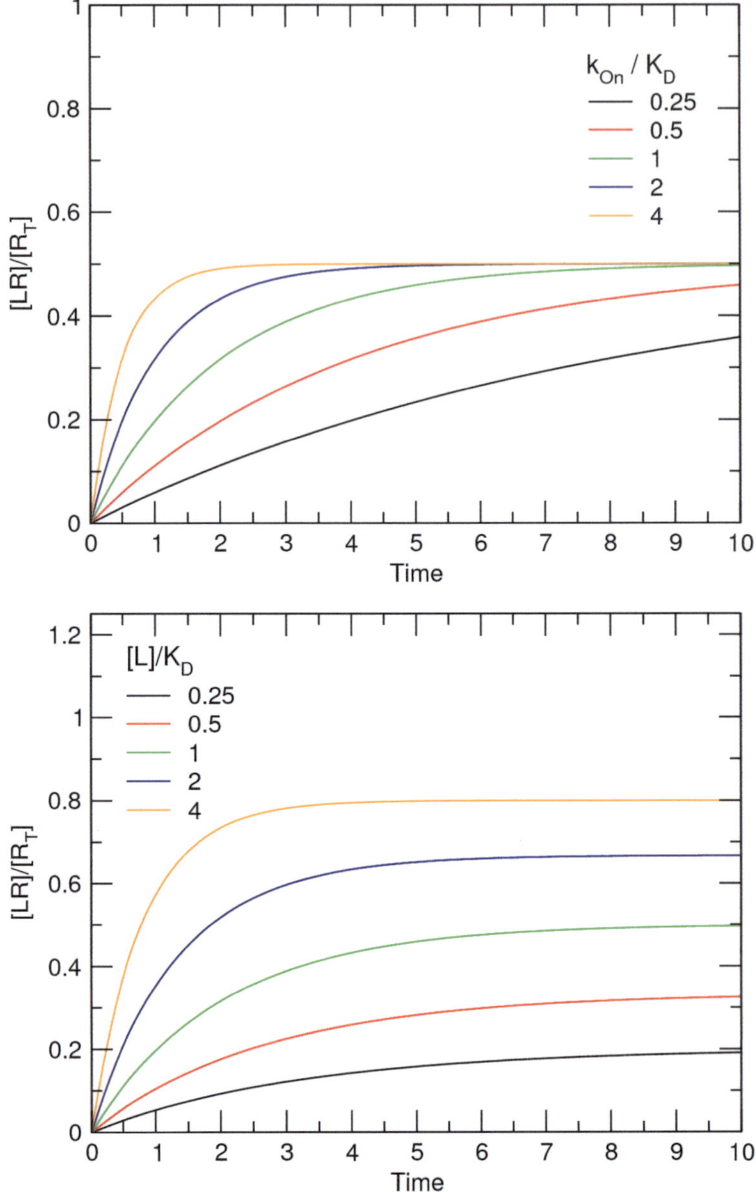

Fig. 3 Theoretical association curves. Abscissa, time. Ordinate, tracer binding expressed as a fraction of total receptor R_T. *Upper graph*: Relationship between the association rate constants k_{On} (indicated in legend as the ratio to equilibrium dissociation constant K_D) on the association of L in a concentration equal to K_D. When L is equal to K_D equilibrium binding represents 50 % of total receptors regardless of the association rate. Increasing k_{On} accelerates association and equilibrium is achieved earlier. *Lower graph*: Effects of changing the concentration of L (indicated in legend as ratio to equilibrium dissociation constant K_D) for ligand with k_{On} equal to $0.25 \times K_D$. Increasing concentration of L both accelerates association and increases equilibrium binding

in dissociation experiments and division by ligand concentration. Association of muscarinic antagonists is easy to measure as NMS reaches equilibrium within minutes and QNB in hours [5]. However, the logarithmic shape of the association curve dictates that the initial time intervals of measurement should be in the fractions of minute for NMS. Kinetics of muscarinic agonists are much faster than kinetics of antagonists. Acetylcholine reaches equilibrium from within 1 (M_5 receptors) to 3 min (M_3 receptors) [8]. Thus time steps should be as short as possible. With the aid of pipetting robots the initial steps can be as short as 2 s. The association rate is strongly dependent on the temperature, so lowering incubation temperature to slow down association may be considered. Due to the rushed nature of association experiments samples of nonspecific binding should be preincubated with unlabeled ligand to allow association with all specific sites. When atropine is used for determination of nonspecific binding 30-min preincubation is sufficient. Formation of nonspecific binding is instant and is not time dependent. Thus determination of nonspecific binding at a single time point is sufficient. In determination of nonspecific binding a non-labeled, chemically distinct, highly specific ligand is used at a receptor saturating concentration (e.g., 1 μM atropine). Much higher concentrations should not be used to avoid blockade of nonspecific binding sites. The latter appears as a fraction of binding sites with extremely fast rate of association.

8.2 Measurement of Dissociation

In dissociation experiments samples are first preincubated with tracer. Equilibrium binding should be reached prior to initiation of dissociation. To safely reach equilibrium preincubation should last at least 5 min for acetylcholine, 20 min for NMS or atropine, and 3 h for QNB or NMQNB. Dissociation of tracer can be achieved by one of the two ways: (1) by removal of the tracer and (2) by addition of the excess of unlabeled ligand that prevents tracer association. The unlabeled ligand used for detection of nonspecific binding should be of a different chemical nature not to protect nonspecific binding sites that would appear as extremely fast dissociating sites. Tracer can be removed either by replacement of incubation medium with tracer-free medium (that is easily achievable by centrifugation or suction in case of whole tissues, cells in suspension, attached cell lines, and the like) or by dilution of incubation medium to lower tracer concentration substantially (at least 100 times). Medium for replacement or dilution should have the same temperature as preincubation medium to prevent temperature effects on dissociation. When dissociation is initiated binding starts to decline according to Eq. (11) that is a modification of Eq. (9) for zero concentration of L:

$$[\mathrm{LR}] = [\mathrm{LR}_0] \times e^{-k_{\mathrm{Off}} \times t} \tag{11}$$

where $[LR_0]$ is tracer binding in the start of dissociation and is equal to $[LR_{Eq}]$ in Eq. (10) and [LR] in Eq. (6e) if equilibrium was reached. As is apparent from Eq. (11) the rate of dissociation is independent from used concentration of the tracer.

8.3 Measurement of Binding Saturation

Equilibrium dissociation constant K_D and the number of binding sites in sample R_T can be determined in saturation binding experiment where samples are incubated with various concentrations of the tracer. Under equilibrium tracer-specific binding depends on tracer concentration according to Eq. (6e). It should be stressed that Eqs. (6a), (6b), (6c), (6d), and (6e) were derived on the assumption that the concentration of tracer is much higher than the concentration of receptors and thus there is no ligand depletion. Therefore the concentration of free tracer is constant during the experiment. In practice this is not always true and free tracer concentration has to be calculated by subtraction of ligand binding from initial ligand concentration that is calculated as total radioactivity added to the sample divided by specific radioactivity of the tracer. It should be noted that this simple correction does not work for large tracer depletion that takes place at concentrations of the radioligand significantly below its K_D. In such case the incubation volume has to be increased. Theoretical curves of saturation binding are shown in Fig. 4.

Because tracer nonspecific binding also depends on the concentration of the tracer it must be determined for each tracer concentration used and subtracted from total binding (Fig. 1). For K_D and R_T to be defined and reliably used the concentrations of tracer should be evenly distributed around K_D. Using only high concentrations of the tracer leads to erroneous estimates of K_D (usually underestimation) and using only low concentrations of the tracer leads to erroneous estimates of R_T (usually overestimation). Extremely low and high concentrations of the tracer (far from K_D) should be avoided as the ratio of specific to nonspecific binding is unfavorable (Fig. 1). If only R_T is of interest a single saturating (several times K_D) concentration of the tracer can be used (e.g., 1 nM NMS) as at concentrations saturating for tracer variation in K_D has small effect on tracer binding. However, equilibrium time and K_D should be determined in preliminary experiment. In typical saturation experiment eight concentrations of NMS ranging from 58 pM to 1 nM (starting with 1 nM and diluting it 3:2 in each step) are used.

8.4 Displacement (Competition) Binding

As shown in Table 1 the number of available radiolabeled muscarinic ligands is limited. However, there are means to determine the binding affinity of non-labeled ligands. For this purpose the ability of a non-labeled ligand to compete for specific binding of a radioligand and decrease tracer binding is utilized. In practice binding of the tracer at a fixed concentration is measured in the presence of various concentrations of the non-labeled ligand. If the binding of

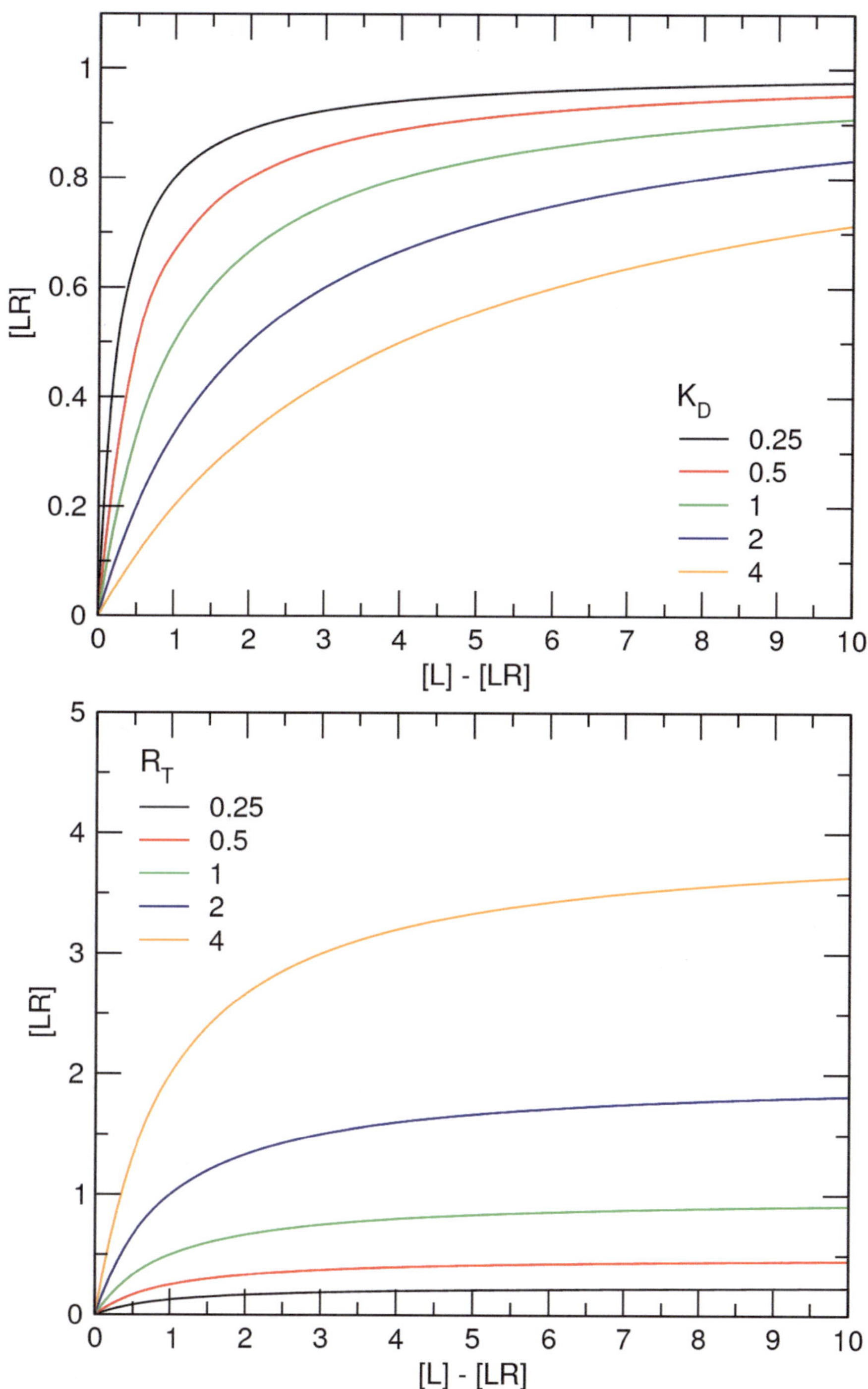

Fig. 4 Theoretical saturation binding curves. Abscissa, concentration of free tracer. Ordinate, concentration of ligand-receptor complexes. *Upper graph*: Effects of tracer equilibrium dissociation constant K_D (indicated in legend) on tracer binding for R_T equal to 1. *Lower graph*: Effects of total receptor number R_T (indicated in legend) on binding of tracer with equilibrium dissociation constant K_D equal to 1

the non-labeled ligand X and tracer L is mutually exclusive then the amount of tracer-receptor complexes is given by Eq. (12) that is a combination of two equations (Eq. 6e) (one for tracer L and one for competitor X):

$$\frac{[\mathrm{LR}]}{[\mathrm{R_T}]} = \frac{[\mathrm{L}]/K_\mathrm{D}}{[\mathrm{L}]/K_\mathrm{D} + [\mathrm{X}]/K_\mathrm{X} + 1} \tag{12}$$

where K_X is the equilibrium dissociation constant of non-labeled ligand X. For practical purposes the tracer binding can be expressed as its fraction in the absence of X:

$$\frac{[\mathrm{LR}]}{[\mathrm{LR_0}]} = 1 - \frac{[\mathrm{X}]}{[\mathrm{X}] + K_\mathrm{X} \times ([\mathrm{L}]/K_\mathrm{D} + 1)} \tag{13}$$

where $[LR_0]$ is the tracer binding in the absence of X. Equation (13) can be further simplified for practical purposes by the introduction of IC_{50} value that represents the concentration of X that decreases the tracer binding to 50 %:

$$\mathrm{IC_{50}} = K_\mathrm{X} \times \left(\frac{[\mathrm{L}]}{K_\mathrm{D}} + 1 \right) \tag{14}$$

Equation (13) then becomes Eq. (15):

$$\frac{[\mathrm{LR}]}{[\mathrm{LR_0}]} = 1 - \frac{[\mathrm{X}]}{[\mathrm{X}] + \mathrm{IC_{50}}} \tag{15}$$

Theoretical curves of competitive binding are shown in Fig. 5, upper graph. As can be inferred from Eq. (14) IC_{50} is dependent on the ratio of L to K_D. As the K_D is constant IC_{50} depends on the concentration of L. The higher the concentration of L the higher the IC_{50}; in other words the bigger is the ratio of the IC_{50} and K_X. Equation (14) dictates that the ratio of IC_{50} to K_X is equal to the ratio of L to K_D plus 1 (Fig. 5, lower graph). Nonlinearity of the dependence of IC_{50} on K_X on the ratio of L to K_D implies that the interaction between the tracer and non-labeled ligand is not competitive, for example allosteric (see Chap. 6) or irreversible.

The application of competition binding study to determine equilibrium dissociation constant of non-labeled compound is obvious. Measurement of competition binding of tracer and non-labeled ligand with preferential affinity at individual receptor subtypes may also be applied to determine receptor subtypes and their proportion in analyzed sample [29]. A list of selective muscarinic ligands is shown in Table 3. Antagonists with varied degrees of selectivity are available for all receptor subtypes. However, no true binding selectivity was found in the case of muscarinic agonists.

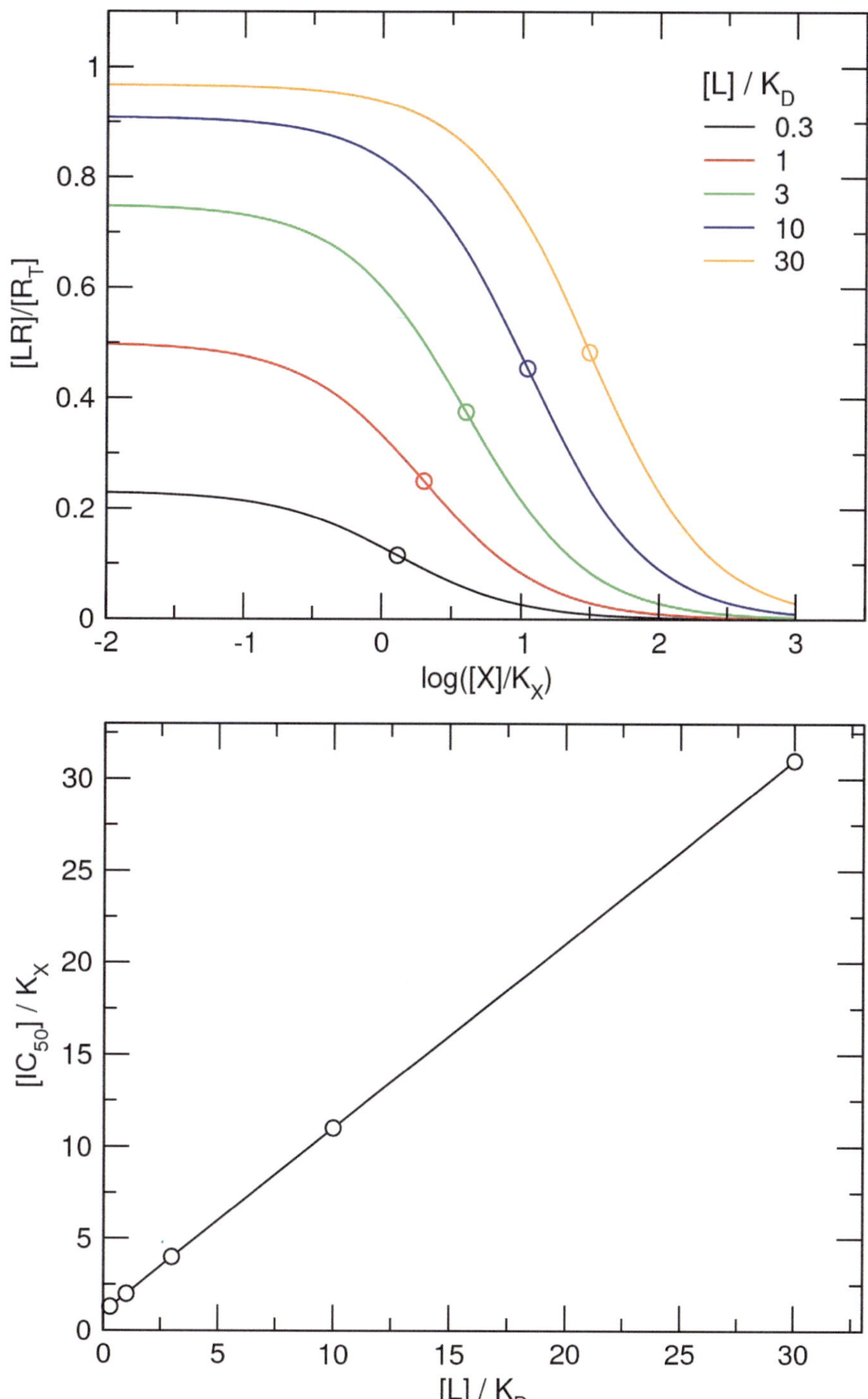

Fig. 5 Competition binding. *Upper graph*: Competition binding of tracer L and competitor X. Abscissa, logarithm of ratio of competitor concentration [X] to its equilibrium constant [K_X]. Ordinate, tracer binding [LR] expressed as fraction of total receptors [R_T]. Ratio of the tracer concentration [L] to its equilibrium dissociation constant K_D is indicated in the legend. *Circles* are IC_{50} at individual binding curves. Increasing tracer concentration leads to higher binding in the absence of competitor and to increase in IC_{50}. *Lower graph*: Dependence of IC_{50} on tracer concentration. IC_{50} values (*circles*) from *upper graph* are plotted against tracer concentration. Abscissa, ratio of tracer concentration [L] to its equilibrium dissociation constant K_D. Ordinate, ratio of IC_{50} concentration to competitor dissociation constant K_X. Dependence is linear with slope equal to 1 and constant equal to 1

Table 3
Selective and partially selective muscarinic antagonists

	Selectivity	M_1	M_2	M_3	M_4	M_5
4-DAMP	M_3/M_2	9.2	8.3	9.3	8.9	9.0
AFDX-116	M_2, M_4	6.2	6.7–7.3	6.1	7.0–8.7	5.3–5.6
AFDX-384	M_2, M_4	7.3–7.5	8.0–9.0	7.2–7.8	8.0–8.7	6.3
Darifenacin	M_3	8.3	7.3–7.6	9.1	8.1	8.6
Guanylpirenzepine	M_1	7.7	5.6	6.5	6.5	6.8
Himbacine	M_2, M_4	7.1	7.9–8.4	6.9–7.2	7.9–8.2	5.4–6.5
MTX3	M_4	7.1	<6	<6	8.7	<6
MTX7	M_1	10.9	<5	<5	<5	<5
Pirenzepine	M_1	8.3	4.9–6.4	5.6–6.7	5.9–7.6	6.2–6.9
Tripitramine	M_2	8.8	9.6	7.1–7.4	7.8–8.2	7.3–7.5
VU0255035	M_1	7.8	6.2	6.1	5.9	5.6
VU0488130[a]	M_5	<5	<5	<5	<5	6.5

Data adapted from Alexander et al. [32] unless otherwise indicated

Inhibition constants K_I of muscarinic antagonists are expressed as negative logarithms

4-DAMP, 4-diphenylacetoxy-*N*-methylpiperidine; AFDX116 (otenzepad), 1-[2-[2-(diethylaminomethyl)piperidin-1-yl]acetyl]-5*H*-pyrido[2,3-b][1,4]benozodiazepin-6-one; AFDX384, (±)-5,11-dihydro-11-([(2-[2-[dipropylamino) methyl]-1-piperidinyl)ethyl)amino)carbonyl)-6*H*-pyrido[2,3-b](1,4)benzodiazepine-6-one; darifenacin, 2-[(3*S*)-1-[2-(2,3-dihydro-1-benzofuran-5-yl)ethyl]pyrrolidin-3-yl]-2,2-diphenylacetamide; guanylpirenzepine, 4-[2-oxo-2-(6-oxo-5H-pyrido[2,3-b][1,4]benzodiazepin-11-yl)ethyl]piperazine-1-carboximidamide;himbacine,(3*S*,3a*R*,4*R*,4a*S*,8a*R*,9a*S*)-4-[(E)-2-[(2*S*,6*R*)-1,6-dimethylpiperidin-2-yl]ethenyl]-3-methyl-3a,4,4a,5,6,7,8,8a,9,9a-decahydro-3H-benzo[f][2] benzofuran-1-one; MTX3 and MTX7, the Eastern green mamba (*Dendroaspis angusticeps*) venom toxins (Liang et al. [41]; Fruchart-Gaillard et al. [42]; VU0255035, *N*-(3-oxo-3-(4-(pyridine-4-yl)piperazin-1-yl)propyl)-benzo[c][1,2,5] thiadiazole-4sulfonamide;VU0488130,5-(3-acetylphenoxymethyl)-*N*-methyl-*N*-[(1*S*)-1-(pyridin-2-yl)ethyl]-1,2-oxazole-3-carboxamide

[a]Gentry et al. [43]

8.5 Determination of Radioligand-Specific Radioactivity

The specific radioactivity is the amount of radiolabeled mass in a sample expressed as Ci/mol or Bq/mol. The specific radioactivity is required to compute mass amounts (e.g., total receptor number R_T in saturation binding) from radioactivity measures of the sample. Specific radioactivity of commercial radioligands is provided by the manufacturer. For radioligands synthesized in-house the specific radioactivity can be estimated by comparing the K_D value obtained by a tracer saturation binding with the value from a homologous competition experiment in which the same nonlabeled ligand is used to displace the binding of the tracer. In saturation binding the concentration of radioactivity [dpm/l] at K_D is assessed according to Eq. (6e). Then in a homologous competition experiment the concentration of radioligand [mol/l] at K_D is assessed according to Eq. (13). Specific radioactivity [in dpm/

mol] is then calculated by division of radioactive concentration [dpm/l] by radioligand concentration [mol/l]. Figure 6 shows a typical saturation experiment and homologous competition for the determination of specific radioactivity. It should be noted that the requirement of equal affinity necessitates that the non-labeled and labeled ligands are chemically identical. Thus, if the ligand is labeled with [^{125}I], the non-labeled ligand must also be in the iodinated form.

9 Data Analysis

9.1 Regression Analysis

Parameters of ligand binding are determined by fitting the appropriate equation to the data by nonlinear regression. The most common assumption is that data points are randomly scattered on both sides of a curve. The goal of regression is to adjust parameters of the equation to find the curve that minimizes the sum of the squares of the differences in *y*-values of points and curve. Simple least square method weights each point equally. However there are methodological reasons to weight points differently. If the replicates show that standard deviation is dependent on *y*-value (the most common situation) then data points should be weighted according to *y*-value. When the standard deviation is proportional to the *y*-value (relative error to *y*-value is constant) like in Fig. 6 then it is appropriate to perform *relative* weighting (weighting by $1/Y^2$). When the standard deviation follows Poisson distribution (e.g., error from radioactive counting) then *Poisson* weighting (weighting by $1/Y$) should be performed. Error coming only from radioactive counting is rarely the case. In practice the source of variation is a mix of sources and *general* weighting (weighting by $1/Yk$), where k is the slope of regression between standard deviation and *y*-value and ranges from 0 to 2. When k is zero or close to it then there is no correlation of standard deviation and *y*-value and no weighting is needed. It may be tempting at first glance to weight the data by standard deviation (weighting by $1/SD^2$) but a very large number of replicates (dozens of samples) are needed for weighting to be correct. However, the use of such large number of replicates is usually not the case in radioligand binding studies.

Distribution of binding parameter estimates from nonlinear regression follows data distribution along axes. Thus parameters determined from semilogarithmic plots (e.g., competition binding) are log normally distributed. This implies that the logarithms of these parameters (e.g., EC_{50}) should be compared and statistically analyzed. Also in case of linear plots (e.g., saturation binding) uneven data distribution along the abscissa may skew distribution of binding parameter (e.g., K_D) estimates. Thus it should be checked for normality.

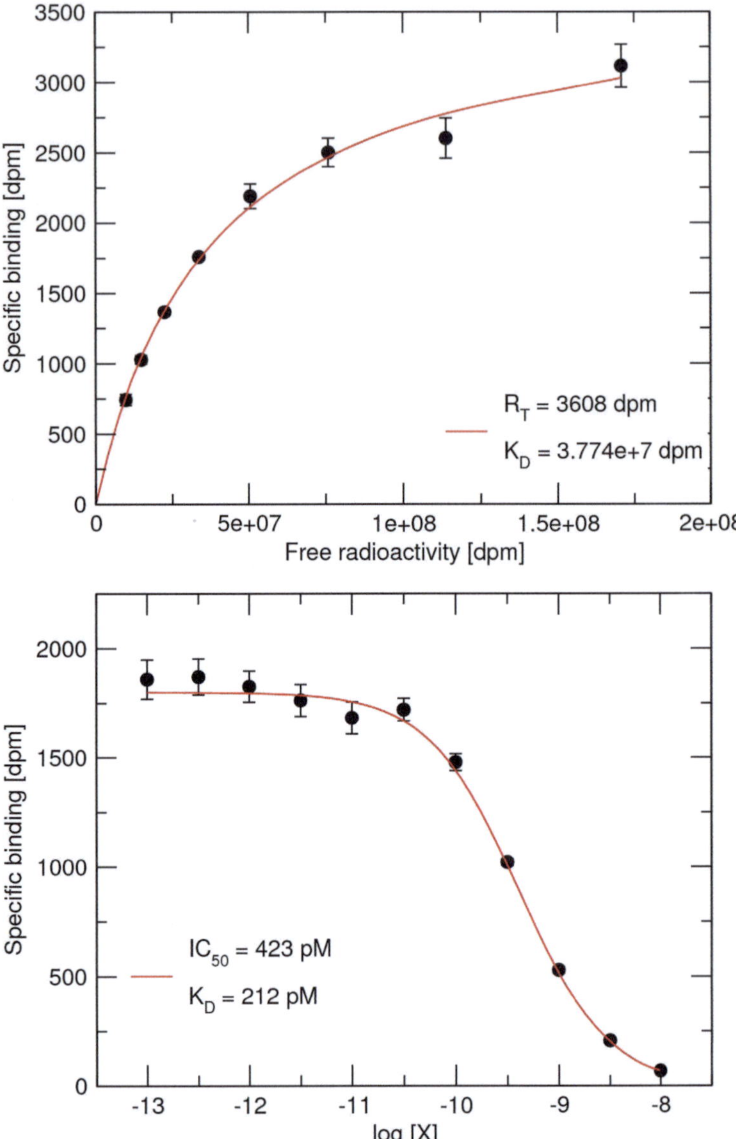

Fig. 6 Determination of specific radioactivity. *Upper graph*: Saturation binding of radiolabeled ligand L with unknown specific radioactivity. Abscissa, free radioactivity of the sample expressed in dpm. Ordinate, specific binding of the tracer expressed in dpm. Fitting Eq. (6e) to data gives maximum binding capacity R_T at about 3600 dpm and equilibrium dissociation constant K_D of tracer at about 38 million dpm. *Lower graph*: Homologous competition of the labeled ligand L with non-labeled chemically identical ligand X. Abscissa, logarithm of concentration of non-labeled ligand X. Ordinate, specific binding of labeled ligand L in dpm. Labeled ligand was used in a concentration close to its equilibrium dissociation constant K_D as indicated by specific binding around 1900 dpm that is half of total receptor number R_T in *upper graph*. Fitting Eq. (15) to data gives IC_{50} of 423 pM and equilibrium dissociation constant K_D 212 pM. Dividing K_D from *upper graph* by K_D from *lower graph* gives specific radioactivity 178 dpm/fmol

9.2 Software

Data have to be preprocessed before fitting. Namely, replicates are averaged and standard deviations calculated. Specific binding is calculated by subtraction of nonspecific binding from total binding. Specific binding is converted to amounts of substance [mol] by division of radioactivity of the sample by specific radioactivity of the ligand. Then specific binding may be related to protein content of the sample, mass of tissue, etc. Used concentration of radioligand is determined by division of the amount of total radioactivity added to the sample by specific radioactivity of the radioligand and sample volume. The easiest way of data preprocessing is to use a spreadsheet software one is familiar with.

It is good practice to perform basic data analysis (like sample variation analysis or outlier identification) prior to nonlinear regression analysis. Nonlinear regression analysis to extract binding parameters and subsequent statistical analysis can be performed using various software ranging from software specialized to analysis of binding data, pharmacological or biochemical experiments, many plotting and curve fitting programs, as well as any general-purpose mathematical package. Table 4 lists several software packages suitable

Table 4
List of software suitable for fitting binding equations to data

Name	Type	Operating systems	License	Reference
COPASI	Biochemical network simulator with fitting functionality	Linux MacOS X Windows	Free and Commercial	www.copasi.org
CurveExpert	Plotting, curve fitting, and statistical analysis	Linux MacOS X Windows	Shareware	www.curveexpert.net
DataFit	Plotting and curve fitting	Windows	Commercial	www.oakdaleengr.com
GraphPad Prism	Pharmacological experiments	MacOS X Windows	Commercial	www.graphpad.com
GTK/Grace	Plotting and curve fitting	Linux Windows	GPL	sourceforge.net/projects/gracegtk
Lab Fit	Plotting, curve fitting, and statistical analysis	Windows	Shareware	zeus.df.ufcg.edu.br/labfit/
SciDAVis	Plotting and curve fitting	Linux MacOS X Windows		scidavis.sourceforge.net
Scilab	General-purpose mathematical	Linux MacOS X Windows	Free CeCILL	www.scilab.org

Disclaimer. Table lists currently available software known to authors. It is not intended to be full list neither recommendation to use for analysis of binding experiments. A major advantage of specialized pharmacological programs is that they are easy to use and have built-in data preprocessing routines, pre-regression checks, predefined equations for all types of binding experiments, and implement various post-regression tests. The main disadvantage of these programs is their relatively high price and inability of scripting and incorporation into workflows with other software. On the other hand general-purpose mathematical and plotting programs are more knowledge demanding to the user but allow scripting and creation of various workflows. Some of them are open source and free of charge. For example for teaching purposes or when one does not have any specialized program at hand a simple least-sum-of-squares regression can be done even using common spreadsheet software (Protocol I).

Protocol I: Nonlinear regression analysis in spreadsheet

1. *Input variable (x values) in column A, and dependent variable (y values) in column B.*
2. *Enter initial values for binding function parameters in column E.*
3. *In column C enter binding function calling binding parameters from column E.*
4. *In column D calculate square of deviations between cells in columns C and B in the current row.*
5. *In the cell F1 calculate sum of values in column D.*
6. *Open solver function of the spreadsheet and instruct it to minimize value in the cell F1 by changing values in column E. Choose Levenberg-Marquardt method if available.*

10 Conclusions

Overall, the current status of radioligand binding experiments allows very accurate and detailed study of equilibrium binding, binding kinetics, and structure-activation relationship at muscarinic receptors. Binding experiments may be performed in various forms ranging from tissue cultures via whole cells to purified receptors when criteria discussed in this chapter are met. Their main limitation remains to be the lack of selective radiolabeled agonists and antagonists for some receptor subtypes.

Acknowledgments

This research was supported by Academy of Sciences of the Czech Republic support RVO: 67985823, and Grant Agency of the Czech Republic grant P304/12/G069.

References

1. Kubo T, Fukuda K, Mikami A, Maeda A, Takahashi H, Mishina M, Haga T, Haga K, Ichiyama A, Kangawa K, Masayasu K, Hisayuki M, Tadaaki H, Shosaku N (1986) Cloning, sequencing and expression of complementary DNA encoding the muscarinic acetylcholine receptor. Nature 323:411–416
2. Bonner TI, Buckley NJ, Young AC, Brann MR (1987) Identification of a family of muscarinic acetylcholine receptor genes. Science 237:527–532
3. Birdsall NJ, Burgen AS, Hulme EC (1979) A study of the muscarinic receptor by gel electrophoresis. Br J Pharmacol 66:337–342
4. Spalding TA, Birdsall NJ, Curtis CA, Hulme EC (1994) Acetylcholine mustard labels the binding site aspartate in muscarinic acetylcholine receptors. J Biol Chem 269:4092–4097
5. Jakubík J, El-Fakahany EE, Tuček S (2000) Evidence for a tandem two-site model of ligand binding to muscarinic acetylcholine receptors. J Biol Chem 275:18836–18844
6. Jakubík J, Bačáková L, El-Fakahany EE, Tuček S (1995) Subtype selectivity of the positive allosteric action of alcuronium at cloned m1–m5 muscarinic acetylcholine receptors. J Pharmacol Exp Ther 274:1077–1083
7. Jakubík J, Janíčková H, El-Fakahany EE, Doležal V (2011) Negative cooperativity in binding of muscarinic receptor agonists and GDP as a measure of agonist efficacy. Br J Pharmacol 162:1029–1044
8. Jakubík J, Randaková A, El-Fakahany EE, Doležal V (2009) Divergence of allosteric effects of rapacuronium on binding and function of muscarinic receptors. BMC Pharmacol 9:15
9. Massoulié J, Pezzementi L, Bon S, Krejci E, Vallette FM (1993) Molecular and cellular biology of cholinesterases. Prog Neurobiol 41: 31–91
10. Jakubík J, El-Fakahany EE (2010) Allosteric modulation of muscarinic acetylcholine receptors. Pharmaceuticals 9:2838–2860
11. Eglen RM (2012) Overview of muscarinic receptor subtypes. Handb Exp Pharmacol 208: 3–28
12. Eglen RM, Nahorski SR (2000) The muscarinic M_5 receptor: a silent or emerging subtype? Br J Pharmacol 130:13–21
13. Felder CC, Bymaster FP, Ward J, DeLapp N (2000) Therapeutic opportunities for muscarinic receptors in the central nervous system. J Med Chem 43:4333–4353
14. Kawashima K, Fujii T (2008) Basic and clinical aspects of non-neuronal acetylcholine: overview of non-neuronal cholinergic systems and their biological significance. J Pharmacol Sci 106:167–173
15. Buckley NJ, Bonner TI, Buckley CM, Brann MR (1989) Antagonist binding properties of five cloned muscarinic receptors expressed in CHO-K1 cells. Mol Pharmacol 35:469–476
16. Wang SZ, Zhu SZ, El-Fakahany EE (1995) Expression of endogenous muscarinic acetylcholine receptors in Chinese hamster ovary cells. Eur J Pharmacol 291:R1–R2
17. Krejčí A, Tuček S (2001) Changes of cooperativity between N-methylscopolamine and allosteric modulators alcuronium and gallamine induced by mutations of external loops of muscarinic M(3) receptors. Mol Pharmacol 60:761–767
18. Arden JR, Nagata O, Shockley MS, Philip M, Lameh J, Sadée W (1992) Mutational analysis of third cytoplasmic loop domains in G-protein coupling of the HM1 muscarinic receptor. Biochem Biophys Res Commun 188:1111–1115
19. Ferguson KM, Higashijima T, Smigel MD, Gilman AG (1986) The influence of bound GDP on the kinetics of guanine nucleotide binding to G proteins. J Biol Chem 261:7393–7399
20. Rinken A, Kameyama K, Haga T, Engström L (1994) Solubilization of muscarinic receptor subtypes from baculovirus infected sf9 insect cells. Biochem Pharmacol 48:1245–1251
21. Jakubík J, Haga T, Tuček S (1998) Effects of an agonist, allosteric modulator, and antagonist on guanosine-gamma-[35s]thiotriphosphate binding to liposomes with varying muscarinic receptor/go protein stoichiometry. Mol Pharmacol 54:899–906
22. Berstein G, Haga T, Ichiyama A (1989) Effect of the lipid environment on the differential affinity of purified cerebral and atrial muscarinic acetylcholine receptors for pirenzepine. Mol Pharmacol 36:601–607
23. Shiozaki K, Haga T (1992) Effects of magnesium ion on the interaction of atrial muscarinic acetylcholine receptors and GTP-binding regulatory proteins. Biochemistry 31:10634–10642
24. Jakubík J, Tuček S (1994) Protection by alcuronium of muscarinic receptors against chemical inactivation and location of the allosteric binding site for alcuronium. J Neurochem 63:1932–1940
25. Jakubík J, Tuček S (1995) Positive allosteric interactions on cardiac muscarinic receptors:

effects of chemical modifications of disulphide and carboxyl groups. Eur J Pharmacol 289: 311–319

26. Haga K, Haga T (1985) Purification of the muscarinic acetylcholine receptor from porcine brain. J Biol Chem 260:7927–7935
27. Lysíková M, Fuksová K, Elbert T, Jakubík J, Tuček S (1999) Subtype-selective inhibition of [methyl-3H]-N-methylscopolamine binding to muscarinic receptors by alpha-truxillic acid esters. Br J Pharmacol 127:1240–1246
28. Santrůčková E, Doležal V, El-Fakahany EE, Jakubík J (2014) Long-term activation upon brief exposure to xanomleline is unique to m1 and m4 subtypes of muscarinic acetylcholine receptors. PLoS One 9:e88910
29. Jakubík J, Tuček S (1994) Two populations of muscarinic binding sites in the chick heart distinguished by affinities for ligands and selective inactivation. Br J Pharmacol 113:1529–1537
30. Jakubík J, Tucek S, El-Fakahany EE (2004) Role of receptor protein and membrane lipids in xanomeline wash-resistant binding to muscarinic m1 receptors. J Pharmacol Exp Ther 308:105–110
31. Aronstam RS, Narayanan TK (1988) Temperature effect on the detection of muscarinic receptor-g protein interactions in ligand binding assays. Biochem Pharmacol 37:1045–1049
32. Alexander SPH, Mathie A, Peters JA (2011) Guide to receptors and channels (GRAC), 5th edition. Br J Pharmacol 164(Suppl 1):S1–S324
33. Melchiorre C, Quaglia W, Picchio MT, Giardinà D, Brasili L, Angeli P (1989) Structure-activity relationships among methoctramine-related polymethylene tetraamines. Chain-length and substituent effects on m-2 muscarinic receptor blocking activity. J Med Chem 32:79–84
34. Hejnová L, Tucek S, El-Fakahany EE (1995) Positive and negative allosteric interactions on muscarinic receptors. Eur J Pharmacol 291: 427–430
35. Visser TJ, van Waarde A, Jansen TJ, Visser GM, van der Mark TW, Kraan J, Ensing K, Vaalburg W (1997) Stereoselective synthesis and biodistribution of potent [11c]-labeled antagonists for positron emission tomography imaging of muscarinic receptors in the airways. J Med Chem 40:117–124
36. Dallanoce C, De Amici M, Barocelli E, Bertoni S, Roth BL, Ernsberger P, De Micheli C (2007) Novel oxotremorine-related heterocyclic derivatives: synthesis and in vitro pharmacology at the muscarinic receptor subtypes. Bioorg Med Chem 15:7626–7637
37. Levey AI (1993) Immunological localization of m1–m5 muscarinic acetylcholine receptors in peripheral tissues and brain. Life Sci 52:441–448
38. Caulfield MP (1993) Muscarinic receptors—characterization, coupling and function. Pharmacol Ther 58:319–379
39. Gautam D, Jeon J, Li JH, Han S, Hamdan FF, Cui Y, Lu H, Deng C, Gavrilova O, Wess J (2008) Metabolic roles of the m3 muscarinic acetylcholine receptor studied with m3 receptor mutant mice: a review. J Recept Signal Transduct Res 28:93–108
40. Poulin B, Butcher A, McWilliams P, Bourgognon J, Pawlak R, Kong KC, Bottrill A, Mistry S, Wess J, Rosethorne EM, Charlton SJ, Tobin AB (2010) The m3-muscarinic receptor regulates learning and memory in a receptor phosphorylation/arrestin-dependent manner. Proc Natl Acad Sci U S A 107:9440–9445
41. Liang JS, Carsi-Gabrenas J, Krajewski JL, McCafferty JM, Purkerson SL, Santiago MP, Strauss WL, Valentine HH, Potter LT (1996) Anti-muscarinic toxins from dendroaspis angusticeps. Toxicon 34: 1257–1267
42. Fruchart-Gaillard C, Mourier G, Marquer C, Ménez A, Servent D (2006) Identification of various allosteric interaction sites on m1 muscarinic receptor using 125i-met35-oxidized muscarinic toxin 7. Mol Pharmacol 69: 1641–1651
43. Gentry PR, Kokubo M, Bridges TM, Cho HP, Smith E, Chase P, Hodder PS, Utley TJ, Rajapakse A, Byers F, Niswender CM, Morrison RD, Daniels JS, Wood MR, Conn PJ, Lindsley CW (2014) Discovery, synthesis and characterization of a highly muscarinic acetylcholine receptor (mAChR)-selective M5-orthosteric antagonist, VU0488130 (ML381): a novel molecular probe. ChemMedChem 9:1677–1682

Chapter 4

Binding Method for Detection of Muscarinic Acetylcholine Receptors in Receptor's Natural Environment

Ikunobu Muramatsu, Hatsumi Yoshiki, Kiyonao Sada, Junsuke Uwada, Takanobu Taniguchi, Takayoshi Masuoka, and Matomo Nishio

Abstract

The pharmacological and biochemical properties of G protein-coupled receptors have been recently revealed to be more complex than originally supposed. Especially, in natural environment or in vivo, some receptors including muscarinic acetylcholine receptor (mAChR) are frequently modified by many factors, so that the receptors may exhibit multiple pharmacological profiles and biochemical functions, which are different from relatively constant and uniform properties originally reported in cell-free preparations and recombinant system. In order to detect the native properties of receptors occurring in tissues and cells without altering their natural environment and also to solve discrepancy between the functional affinity obtained by a bioassay approach and the binding affinity estimated from the conventional binding method with membrane preparations, the tissue segment binding method without homogenization has been recently developed as a new approach. In this chapter, the detailed protocol of tissue segment binding method and some unique properties of mAChRs observed in tissue segments are described.

Key words Radioligand binding, Tissue segments, Tissue homogenates, Muscarinic acetylcholine receptor (mAChR), Natural tissue environment, Affinity and density, Subcellular distribution

1 Introduction

The radioligand-binding method has been one of the most important techniques in studying the pharmacological characterization and biochemical identification of many types of receptors [1, 2]. This method was pioneered by Paton and Rang in 1965 [3], who incubated intact strips of intestinal smooth muscle with [^{3}H]atropine, in order to study the ligand binding properties of mAChRs. However, the binding method was thereafter applied to homogenates or membrane fractions prepared from tissue, because receptor density is high in the membrane-rich preparations and any binding-interfering substances such as endogenous neurotransmitters could be removed in the fractionated preparations ([2], also see other chapters in this volume). Since the pharmacological

Jaromir Myslivecek and Jan Jakubik (eds.), *Muscarinic Receptor: From Structure to Animal Models*, Neuromethods, vol. 107, DOI 10.1007/978-1-4939-2858-3_4, © Springer Science+Business Media New York 2016

profiles of receptors obtained by the conventional membrane binding method are generally uniform among many tissues and relatively well consistent with those of recombinant receptors [2, 4], the conventional binding method with membrane preparations and the recombinant receptors have been widely employed for many purposes including the identification of receptors and the screening of drug candidates. However, there is emerging evidence that a receptor can show multiple pharmacological or biochemical properties under different conditions or states, in contrast to the uniform property supposed originally [5–10]. In particular, pharmacological specificity for some receptors may differ markedly between intact cells and cell-free preparations [11–13]. For example, M_3-mAChR subtype in rat cerebral cortex has relatively low affinity to M_3-selective antagonists (darifenacin and solifenacin) in the natural environment, while the antagonists can recognize the M_3-mAChRs with high affinity after homogenization [14]. Furthermore, there are often discrepancies in the native mAChRs between the functional affinities for antagonists obtained by a bioassay approach and the binding affinities estimated from the conventional membrane binding assay [15, 16]. Moreover, M_1-mAChR subtype has been recently demonstrated to exist and operate not only on the cell surface but also in intracellular sites in the central nervous system [17, 18]. The existence of functional intracellular mAChRs is inconsistent with a classical concept that mAChRs are representative cell-surface receptor. As tissue homogenization may cause disintegration of cell structure and/or dissociation with receptor and other membrane proteins and may result in significant changes of the pharmacological or biochemical properties, it seems very important to keep receptor's natural environment as possible. Recently, we have developed "intact tissue segment binding method" without homogenization [7, 19]. In this chapter, the detailed protocol of intact segment binding method and some unique profiles for native mAChRs are described.

2 Materials

2.1 Solution for Tissue Isolation

In order to maintain the native tissue environment, isotonic nutrient solutions have been used. We employ a modified Krebs–Henseleit solution that is commonly used in the functional bioassay. The composition is as follows: 121 mM NaCl, 5.9 mM KCl, 1.2 mM $MgCl_2$, 2.0 mM $CaCl_2$, 1.2 mM NaH_2PO_4, 25.5 mM $NaHCO_3$, and 11.5 mM glucose. The pH of the solution is maintained at 7.4 by gassing with 95 % O_2 and 5 % CO_2. It is better to partially freeze the solution (0 °C) before tissue isolation in order to stop rapidly tissue/cell metabolism upon tissue isolation. Especially, brain must be placed in 0 °C solution immediately after isolation.

2.2 Incubation Buffer

The goal of using intact tissue in the tissue segment binding method is to keep the receptor environment closely as possible to that in tissues in vivo. Therefore, isotonic nutrient solutions are used during binding experiments. We use an incubation buffer (136 mM NaCl, 5.9 mM KCl, 1.2 mM $MgCl_2$, 2.0 mM $CaCl_2$, 1.2 mM NaH_2PO_4, 10.5 mM $NaHCO_3$, and 11.5 mM glucose, pH 7.4 *in air*), whose composition is essentially the same as the modified Krebs–Henseleit solution described above. However, the incubation buffer cannot be aerated during the incubation period, in contrast to the case of a functional bioassay where the nutrient solution is bubbled with 95 % O_2 and 5 % CO_2. Therefore, the $NaHCO_3$ concentration in the modified Krebs–Henseleit solution is reduced from 25.5 to 10.5 mM to adjust the pH to 7.4 under equilibration in room air, and the osmolality is compensated by adding NaCl. This bicarbonate buffer solution is sufficient to keep the pH constant under the incubation conditions at low temperature. Other isotonic solutions buffered with HEPES or Tris might be used instead of the bicarbonate buffer, but it should be tested before use whether the buffer composition affects the binding properties of the target receptors.

2.3 Radioligands

Two distinct radioligands have been used for the identification of mAChRs: 1-quinuclidinyl-[phenyl-4-^{3}H]-benzilate ([^{3}H]QNB) and 1-[N-methyl-^{3}H]-scopolamine methyl chloride ([^{3}H]NMS). [^{3}H]QNB is hydrophobic and permeable through plasma membrane. Thus, as the proportion of nonspecific binding is significantly higher than that of hydrophilic [^{3}H]NMS, [^{3}H]NMS is useful to detect cell surface mAChRs. On the contrast, hydrophobic [^{3}H]QNB can bind not only cell surface but also intracellular mAChRs [20–22]. In addition to radioligands, surface and intracellular mAChRs may be differently recognized by membrane-permeable atropine, or impermeable N-methylatropine or non-radioactive NMS, resulting in distinct proportions of nonspecific binding sites. Therefore, different combinations with radioligand and nonspecific ligand must be selected depending on research purposes (see Sect. 3) Radioligands and all tested drugs are diluted with incubation buffer before use. H_2O must not be used for drug dilution. Glass tubes, but not plastic tubes, must be used for drug dilution and incubation (Fig. 1, step A), because *plastic tubes* may rapidly absorb radioligands and other tested drugs, resulting in a rapid reduction in their concentrations during dilution and incubation in contrast to glass tubes.

3 Methods

3.1 Preparation of Tissue Segments

The tissue segment binding method can be applied to all tissues isolated from animals including humans. Under a stereoscopic microscope and at 4 °C, surrounding unnecessary parts such as fat

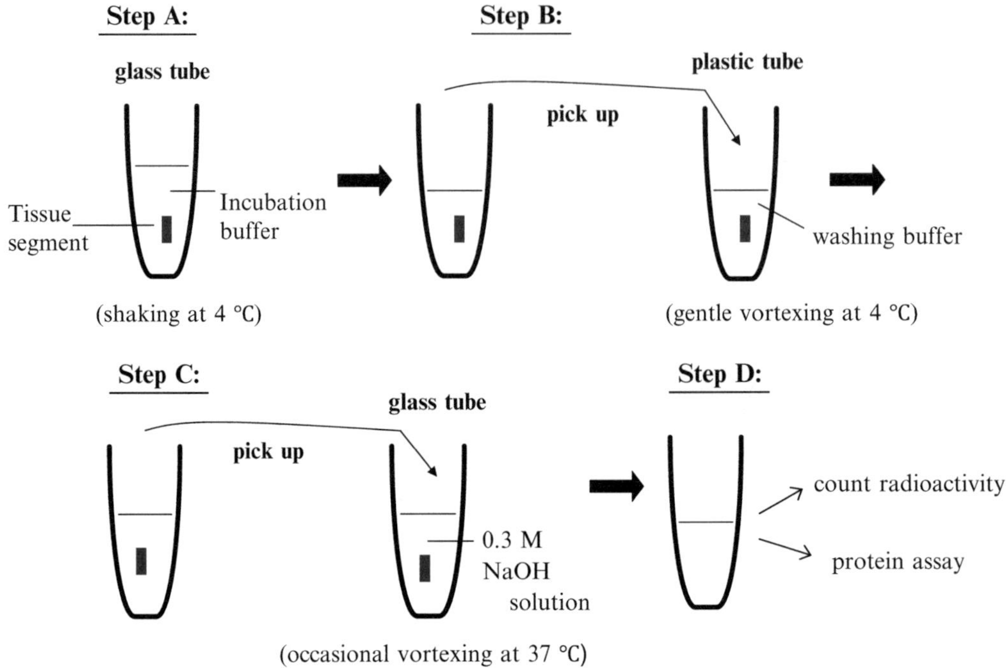

Fig. 1 Experimental protocol for the tissue segment binding method. *Step A*: One tissue segment is incubated with [^{3}H]NMS or [^{3}H]QNB in the absence or presence of competitor in a glass tube at 4 °C. The incubation volume is generally 0.5 ml and the glass tube is shaken 110–120 times per min during incubation. Incubation periods (7–8 or 16 h) and combination of radioligand and competitor depend on the research purpose. *Step B*: Thereafter, the tissue segment is picked up by forceps, and then gently washed in a plastic tube containing a washing buffer at 4 °C for 40–60 s. *Step C*: The tissue segment is again picked up with another forceps, and then solubilized in 1 ml of 0.3 M NaOH solution at 37 °C for ~24 h. The test tube is strongly vortexed several times in order to facilitate the solubilization. *Step D*: After solubilization, the radioactivity and protein content are measured. In general, 500 μl and 20–50 μl of the solubilized solution are used for the measurements of radioactivity and protein content, respectively. *Thick horizontal arrows* between steps A and B, steps B and C, or steps C and D represent that the tubes described in both panels are the same, respectively

and connective tissues are removed. Muscle and mucosal layers can be separated in some tissues such as urinary bladder and stomach. Then the tissue is carefully cut into small pieces with ophthalmic fine scissors under a stereoscopic microscope. In order to allow diffusion of drug into tissue it is necessary to cut tissue in segments of small size. However, segments have to be big enough to keep receptor environment intact. Thus compromised size of sections has to be found. For example, the best size of rat cerebral cortex segments is approximately 1.5 mm in length, 1 mm in width, and 0.5 mm in thickness for the measurement of mAChRs. In rat detrusor and stomach muscles, the best size is 1 × 1.5 mm although the thickness depends on the muscle layer.

3.2 Incubation Volume and Temperature

As shown in Fig. 1, step A, one segment is incubated in one glass tube at 4 °C. The incubation volume is usually 0.5 or 1.0 ml, where a segment is incubated together with different concentrations of competitor and radioligand. In the tissues expressing receptor at high density, larger incubation volume (together with smaller size of tissue segments) is better to avoid a significant reduction of effective concentrations of radioligand and/or competitor added to the incubation solution during incubation. Incubation starts immediately after addition of radioligand. Incubation tubes are usually shaken 110–120 times per min. Some investigators consider 37 °C might be preferable, because this temperature is more physiological and because the resulting binding data might correspond better to the functional data measured at 37 °C. However, it must be noted that the intact tissue segments are incubated in an isotonic solution without bubbling air or oxygen. In order to rule out possible changes in natural states under anoxia at 37 °C, we have therefore used low temperature (generally 4 °C). Receptor trafficking seems to be neglected at this low temperature.

3.3 Incubation Time

In the conventional binding method with cell-free preparations, 2 h incubation is enough to reach equilibrium binding even at low temperature (4 °C). In contrast, binding process to intact tissue segments should be markedly influenced by tissue architecture and physicochemical properties of tested drugs. Figure 2 shows representative time course of specific binding of [^{3}H]QNB in rat urinary bladder and cerebral cortex segments. At 1.5 nM of [^{3}H]QNB, the binding to rat detrusor muscle is monophasic and reaches a plateau in incubation for approximately 8 h (Fig. 2a).

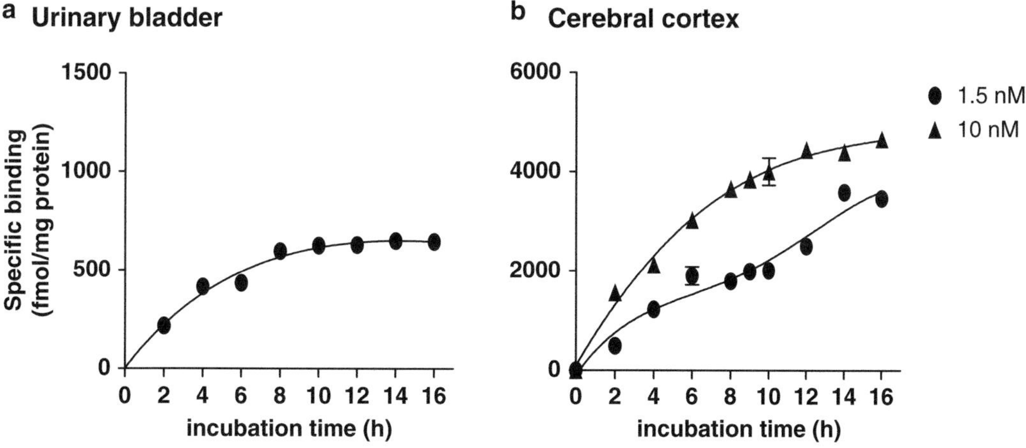

Fig. 2 Time course of [^{3}H]QNB binding to rat urinary bladder segments (**a**, detrusor muscle) and cerebral cortex (**b**) segments at 4 °C. [^{3}H]QNB (1.5 nM in **a**, and 1.5 and 10 nM in **b**) was added at time 0 and then incubated for indicated periods. Specific binding was determined by subtraction of [^{3}H]QNB binding in the presence of 1 μM atropine from total binding. Each point represents the mean of duplicate determinations and SEM in a representative experiment

However, the binding of 1.5 nM [^{3}H]QNB to cerebral cortex is biphasic; the first plateau at approximately 8 h is followed by the second plateau at 16 h incubation (Fig. 2b). On the other hand, at a higher concentration (10 nM) of [^{3}H]QNB, the binding increases monophasically in rat cerebral cortex segments (Fig. 2b). Recently, intracellular distribution of mAChRs in addition to cell surface has been demonstrated in rat, mouse and human brain and in neuroblastoma cells [17, 18], while mAChRs usually occur on the cell surface in the peripheral tissues such as urinary bladder when the receptors are not stimulated by agonists [21, 23–25]. Therefore, it is likely that monophasic binding of [^{3}H]QNB at 10 nM reflects its faster penetration through plasma membrane and rapid association with mAChRs than those at low concentrations (1.5 nM) of [^{3}H]QNB. Although different binding kinetics of QNB among mAChR subtypes cannot be also ruled out, the present and recent additional evidence suggests that the ability of hydrophobic [^{3}H]QNB to bind to cell surface and intracellular mAChRs strongly depends on the plasma membrane permeability (physicochemical property) and concentrations of radioligand, and also on the incubation time. At present, we have used two different incubation periods (8 and 16 h) in [^{3}H]QNB binding experiments in order to detect surface and total mAChRs. In this case, total number of mAChRs is estimated by subtracting nonspecific binding defined with membrane-permeable atropine (1 μM) from total binding at 8 or 16 h incubation, while amount of surface mAChRs is calculated by subtracting the nonspecific binding defined with hydrophilic (membrane-impermeable) NMS (1 μM) at short incubation (8 h). Figure 3 shows the representative saturation curves for [^{3}H]QNB binding to total and surface mAChRs in rat cerebral cortex segments, where similar total number of mAChRs can be estimated by atropine regardless of 8 or 16 h incubation (see more details in Sect. 4.2). Alternatively, surface mAChRs may be more specifically estimated from [^{3}H]NMS binding to the segments, where nonspecific binding should be defined with the use of more hydrophilic N-methylatropine than atropine [21]. Combination of chemically same compounds (e.g., [^{3}H]NMS vs. NMS) must be avoided in order to contaminate non-mAChR sites.

3.4 Washing

After incubation, each segment is carefully picked up by forceps from each incubation tube and quickly moved into a *plastic tube* containing ice-cold washing buffer (2 ml). Then, the plastic tubes are gently vortexed for 40–60 s (Fig. 1, step B). The washing buffer is the same as the incubation buffer used. Figure 4 shows the residual radioactivity remained in the tissue segments after washing for various times, where rat cerebral cortex segments were incubated with 2 nM [^{3}H]NMS (a) or 2 and 10 nM [^{3}H]QNB (b and c) in the absence or presence of 1 μM atropine for 8 h beforehand. The residual radioactivity in the segments rapidly reduced after

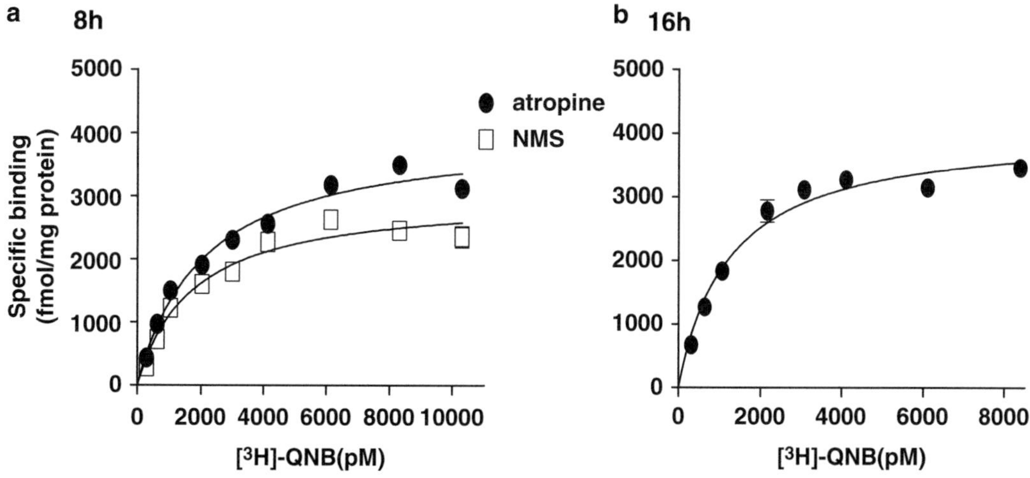

Fig. 3 Representative saturation curves for [^{3}H]QNB binding to intact segments of rat cerebral cortex. (**a**) The segments were incubated for 8 h and the specific binding (*circles* and *squares*) was determined by subtraction of [^{3}H]QNB binding in the presence of 1 μM atropine or 1 μM NMS, respectively. (**b**) The segments were incubated for 16 h and specific binding (*circles*) was determined by subtraction of [^{3}H]QNB binding in the presence of 1 μM atropine from total binding. Each point represents the mean of duplicate determinations and SEM in a representative experiment

washing and was maintained at relatively constant level during 15–120 s. It is interesting to note that tissue radioactivity before washing (at 0 s) is not so high as compared with the residual count after washing, suggesting that contamination of radioligand in extracellular/interstitial spaces is minor in the used segments. This conclusion is also supported by extremely low levels of nonspecific [^{3}H]NMS binding in the segments incubated with atropine (Fig. 4a). In contrast, nonspecific binding of lipophilic [^{3}H]QNB is higher in proportion than that of [^{3}H]NMS, suggesting intracellular accumulation of [^{3}H]QNB and its persistent retention during washing (Fig. 4b, c). After washing, the segment is picked up and moved into a *glass tube* for tissue solubilization (Fig. 1, step C). Different forceps must be used at steps B and C, in order to avoid possible contamination of radioligand. In the case of hard segments like muscle, the blotting on paper may be applied after picking up the segment.

3.5 Tissue Solubilization, and Measurement of Protein and Radioactivity

The washed segments are solubilized in 0.3 M NaOH solution (1 ml) to measure the bound radioactivity and the protein content (Fig. 1, steps C and D). Most of tissue segments are solubilized at 37 °C within 1 day. It must be noted that segment size varies among segments, that results in different protein concentrations between tubes. Therefore, protein content must be measured in every tube. In general, 10–50 μl of the solubilized tissue solution is used for protein assay and 500 μl for measurement of radioactivity of bound radioligand, respectively.

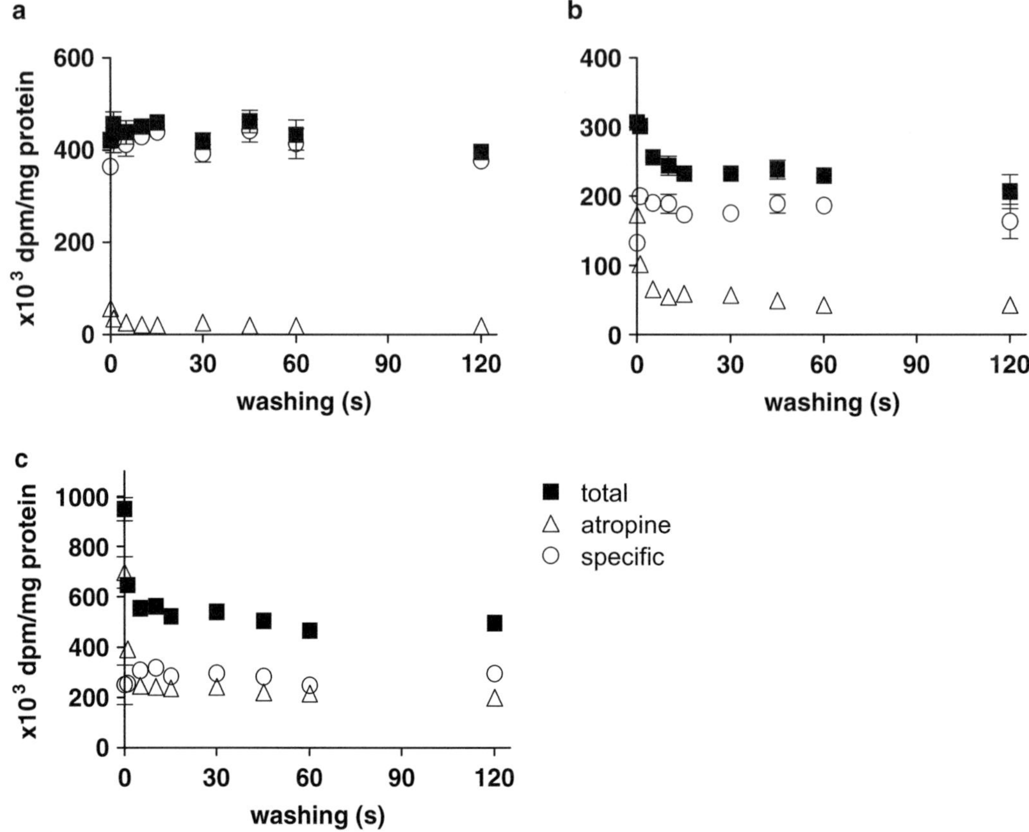

Fig. 4 Residual radioactivity remained in the segments of rat cerebral cortex after washing. The segments were incubated with 2 nM [^{3}H]NMS (**a**) or 2 and 10 nM [^{3}H]QNB (**b** and **c**, respectively) for 8 h beforehand, and then picked up and washed for various times. In abscissa, time 0 means that the radioactivity of the segments was directly measured without washing. *Squares*: total radioactivity. *Circles*: specific radioactivity. *Triangles*: nonspecific radioactivity in atropine-treated segments. Each point represents the mean of duplicate determinations and SEM in a representative experiment

3.6 Data Analysis

Since each tissue segment varies in size, the solubilized tissue solution in each tube has different protein content. For normalization of the data, the radioactivity measured must be adjusted to the counts (dpm) bound per mg of protein. Then, binding data are analyzed using commercially available software (PRISM version 5.01; GraphPad Software, La Jolla, CA, USA) [7].

4 Notes

4.1 Binding Density and Affinity

The most important discrepancy between the tissue segment binding method and the conventional binding method using homogenates or microsomal fraction is a difference in binding density. Recent data of saturation binding experiments with [^{3}H]NMS in

Table 1
The dissociation constants (K_D) and maximal binding capacities (B_{max}) of [^{3}H]NMS in the intact segments and homogenates of various rat tissues

	Segments		Homogenates	
Tissue	K_D (pM)	B_{max} (fmol/mg protein)	K_D (pM)	B_{max} (fmol/mg protein)
Cerebral cortex	1550	3000	178 (134)	2050 (1990)
Striatum	2500	3700	158	2800
Hippocampus	1800	2500	197	1950
Cerebellum	350	280	213	204
Submaxillary grand	280	380	185	153
Left ventricle	1050	420	334 (163)	86 (110)
Gastric muscle	970	1580	245	550
Detrusor muscle	800	810	212	226

Saturation binding experiments were carried out at 4 °C. Incubation periods were 8 h in the segments and 2 h in the homogenates, respectively. Tissues were homogenized with the same buffer as the Krebs' incubation buffer used in the segment binding method, but cerebral cortex and left ventricle were also homogenized with Tris buffer (50 mM Tris, 2 mM EDTA, pH = 7.4) and the data are shown in parentheses. Data represent mean value of 2–3 experiments

the segments and homogenates of various rat tissues are summarized in Table 1. Here, the tissues were homogenized either with the same Krebs' solution as the incubation buffer in the segment binding experiments or with sodium-free Tris-EDTA buffer in order to test the effects of distinct homogenizing buffers. When the binding capacities are compared using the same denominator (that is, per mg of total tissue protein in the segments and the homogenates), lower density of mAChRs can be estimated in the homogenates than in the segments, regardless of distinct homogenizing buffers. Similar differences in binding capacity have been reported in the mAChRs of other tissues (rat gastric mucosa [26]; human airways [27]; mouse epithelial cells [28]). Since the specific binding observed in segments is completely inhibited by not only lipophilic but also hydrophilic antagonists and agonists, higher density in the segments is not an overestimate due to nonspecific accumulation of radioligand into the tissue segments (see Sect. 3.4, Fig. 4). Rather, it is likely that the tissue segment binding method can avoid a yield loss of receptors which may be resulted from homogenization and/or membrane fractionation. This type of yield loss after homogenization has been reported in other receptors (nicotinic receptor [29]; α_1-adrenoceptor [19, 30–32]; β-adrenoceptor [26, 33]). The improved yield of receptor in intact segments is of particular value for dealing with limited amount of tissues and/or small animals. Recently, Ikeda et al. [27] have applied the segment binding method to human airway tissues in order to compare the distribution of mAChR and β-adrenoceptor subtypes between segmental and subsegmental bronchi (approximately 10 mm and 1–4 mm in outer diameter, respectively).

The second significant difference between both binding methods may be observed in binding affinities for radioligand or several drugs. Like [^{3}H]QNB, [^{3}H]NMS has been classically recognized as a non-selective but specific radioligand of mAChRs showing a relatively constant and high (subnanomolar) affinity in various tissues and recombinant receptors [4]; for example, 270 pM in rat cerebral cortex, 290 pM in rat hippocampus, 230 pM in rat corpus striatum, and 547 pM in rat gastric muscle [4, 14, 34, 35]. A similar high affinity (approximately 200 pM) for [^{3}H]NMS was obtained in the present homogenate binding (Table 1). However, the dissociation constants for [^{3}H]NMS estimated in the segments are relatively low and varied among the tested tissues (280–2500 pM). Besides these results, it has been reported that M_3-mAChRs in rat cerebral cortex cannot be identified as high affinity sites for M_3 selective antagonists (darifenacin and solifenacin) under segmental conditions, while M_3-sites have been recognized as their high affinity sites in the homogenates or the membrane preparations [14]. Therefore, it is likely that tissue homogenization may cause a change in receptor profile in addition to yield loss, and it would in part explain well known discrepancy between functional bioassay and conventional binding assay. Although the mechanisms for yield loss and profile change after homogenization are not yet settled, it must be again emphasized that the segment-binding approach may shed light on distinct native phenotypes of cholinergic and probably other receptors observed in functional approach [7, 12, 13, 36].

4.2 Identification of Surface and Intracellular mAChRs

In contrast to peripheral tissues, the mAChRs in the cerebral cortex, striatum, and hippocampus exist not only on the cell membranes but also in the intracellular sites (mainly Golgi apparatus) [17]. The intracellular sites can be accessed by hydrophobic [^{3}H]QNB but not by hydrophilic [^{3}H]NMS [20–22]. Figure 3a shows representative saturation curves for [^{3}H]QNB binding in rat cerebral cortex segments under 8 h incubation, where two distinct binding capacities are estimated using hydrophilic (membrane-impermeable) unlabeled NMS and membrane-permeable atropine, respectively. The density estimated with hydrophilic NMS is significantly lower than that estimated with atropine. Such a high density of mAChRs is also obtained under long incubation (16 h) in the absence or presence of atropine (Fig. 3b). These differences in binding capacities are also observed in striatum and hippocampus of rats and mice, but not in the cerebellum and peripheral tissues.

These results suggest that mAChRs are localized not only on the cell surface but also at intracellular sites in some brain areas, and that such distinct subcellular distribution of mAChRs can be identified by different combinations of [^{3}H]QNB with hydrophobic and/or hydrophilic ligands in the segments. In reference to this point, it is important to emphasize repeatedly that cell surface and intracellular receptors cannot be discriminated after homogenization. Detection of distinct subcellular distribution of receptors would be influenced by physicochemical property and concentrations of ligands, and incubation times.

4.3 Summary and Perspective

Under natural/physiological environment, mAChRs and probably other receptors may exist and function as pharmacologically distinct phenotypes which are different from relatively constant and uniform profile observed in homogenized tissues or recombinant system. The tissue segment binding method is a powerful tool for detecting the native properties of receptors occurring in tissues and cells without altering their environment, and would provide important information of pharmacokinetic analysis, positron emission tomography (PET) analysis and in drug development. Recently, it has been suggested that in vivo distribution of receptor ligands may be related to the distinct binding affinities estimated in the segments of various tissues but not to a uniform affinity estimated in the homogenates [13].

Acknowledgments

This work was supported in part by a Grant-in-Aid for Scientific Research from Japan Society of the Promotion of Science (JSPS), a grant from the Smoking Research Foundation of Japan, and Organization for Life Science Advancement Programs, University of Fukui.

References

1. Yamamura HI, Enna SJ, Kuhar MJ (eds) (1985) Neurotransmitter receptor binding, 2nd edn. Raven Press, New York
2. Bylund DB, Deupree JD, Toews ML (2004) Radioligand-binding methods for membrane preparations and intact cells. Methods Mol Biol 259:1–28
3. Paton WDM, Rang HP (1965) The uptake of atropine and related drugs by intestinal smooth muscle of the guinea-pig in relation to acetylcholine receptors. Proc R Soc Lond B Biol Sci 163:1–44
4. Alexander SPH, Mathie A, Peters JA (2009) Guide to Receptors and Channels (GRAC), 4th edition. Br J Pharmacol 158(suppl):S1–S254
5. Kenakin T (1995) On the importance of the "antagonist assumption" to how receptors express themselves. Biochem Pharmacol 55:17–26
6. Kenakin T, Miller LJ (2010) Seven transmembrane receptors as shape shifting proteins: the impact of allosteric modulation and functional selectivity on new drug discovery. Pharmacol Rev 62:265–304

7. Muramatsu I, Tanaka T, Suzuki F, Li Z, Hiraizumi-Hiraoka Y, Anisuzzaman ASM, Yamamoto H, Horinouchi T, Morishima S (2005) Quantifying receptor properties: the tissue segment binding method—a powerful tool for the pharmacome analysis of native receptors. J Pharmacol Sci 98:331–339
8. Baker JG, Hill SJ (2007) Multiple GPCR conformations and signaling pathways: implications for antagonist affinity estimates. Trends Pharmacol Sci 28:374–381
9. Nelson CP, Challiss RAJ (2007) "Phenotypic" pharmacology: the influence of cellular environment on G protein-coupled receptor antagonist and inverse agonist pharmacology. Biochem Pharmacol 73:737–751
10. Lane JR, Sexton PM, Christopoulos A (2013) Bridging the gap: bitopic ligands of G-protein-coupled receptors. Trends Pharmacol Sci 34:59–66
11. Toung MDT, Garbarg M, Schwartz JC (1980) Pharmacological specificity of brain histamine H2-receptors differs in intact cells and cell-free preparations. Nature 287:548–551
12. Nishimune A, Yoshiki H, Uwada J, Anisuzzaman AS, Umada H, Muramatsu I (2012) Phenotype Pharmacology of lower urinary tract α_1-adrenoceptors. Br J Pharmacol 165:1226–1234
13. Yoshiki H, Uwada J, Anisuzzaman ASM, Umada H, Hayashi R, Kainoh M, Masuoka T, Nishio M, Muramatsu I (2014) Pharmacologically distinct phenotypes of α_{1B}-adrenoceptors: variation in binding and functional affinities for antagonists. Br J Pharmacol 171:4890–4901
14. Anisuzzaman ASM, Nishimune A, Yoshiki H, Uwada J, Muramatsu I (2011) Influence of tissue integrity of pharmacological phenotypes of muscarinic acetylcholine receptors in the rat cerebral cortex. J Pharmacol Exp Ther 339: 186–193
15. Boxall DK, Ford AP, Choppin A, Nahorski SR, Challis RA, Eglen RM (1998) Characterization of an atypical muscarinic cholinoceptor mediating contraction of the guinea-pig isolated uterus. Br J Pharmacol 124:1615–1622
16. Munns M, Pennefather JN (1998) Pharmacological characterization of muscarinic receptors in the uterus of oestrogen-primed and pregnant rats. Br J Pharmacol 123:1639–1644
17. Uwada J, Anisuzzaman AMS, Yoshiki H, Nishimune A, Muramatsu I (2011) Intracellular distribution of functional M1-muscarinic acetylcholine receptors in N1E-115 neuroblastoma cells. J Neurochem 118:958–967
18. Anisuzzaman AMS, Uwada J, Masuoka T, Yoshiki H, Nishio M, Ikegaya Y, Takahashi N, Matsuki N, Fujibayashi Y, Yonekura Y, Momiyama T, Muramatsu I (2013) Novel contribution of cell surface and intracellular M1-muscarinic acetylcholine receptors to synaptic plasticity in hippocampus. J Neurochem 126:360–371
19. Tanaka T, Zhang L, Suzuki F, Muramatsu I (2004) Alpha-1 adrenoceptors: evaluation of receptor subtype-binding kinetics in intact arterial tissues and comparison with membrane binding. Br J Pharmacol 141:468–476
20. Galper JB, Dziekan LC, O'Hara DS, Smith TW (1982) The biphasic response of muscarinic receptors in cultured heart cells to agonists. Effects on receptor number and affinity in intact cells and homogenates. J Biol Chem 257:10344–10356
21. Koenig JA, Edwardson JM (1996) Intracellular trafficking of the muscarinic acetylcholine receptor: importance of subtype and cell type. Mol Pharmacol 49:351–359
22. Dessy C, Kelly RA, Balligand JL, Feron O (2000) Dynamic mediates caveolar sequestration of muscarinic cholinergic receptors and alteration in NO signaling. EMBO J 19:4272–4280
23. Anisuzzaman ASM, Morishima S, Suzuki F, Tanaka T, Yoshiki H, Sathi ZS, Akino H, Yokoyama O, Muramatsu I (2008) Assessment of muscarinic receptor subtypes in human and rat lower urinary tract by tissue segment binding assay. J Pharmacol Sci 106:271–279
24. Uwada J, Yoshiki H, Masuoka T, Nishio M, Muramatsu I (2014) Intracellular localization of M1 muscarinic acetylcholine receptor through clathrin-dependent constitutive internalization via a C-terminal tryptophan-based motif. J Cell Sci 127:3131–3140
25. Van Koppen CJ, Kaiser B (2003) Regulation of muscarinic acetylcholine receptor signaling. Pharmacol Ther 98:197–220
26. Anisuzzaman ASM, Morishima S, Suzuki F, Tanaka T, Muramatsu I (2008) Identification of M1 muscarinic receptor subtype in rat stomach using a tissue segment binding method, and the effects of immobilization stress on the muscarinic receptors. Eur J Pharmacol 599:146–151
27. Ikeda T, Anisuzzaman AMS, Yoshiki H, Sasaki M, Koshiji T, Nishimune A, Muramatsu I (2012) Regional quantification of muscarinic acetylcholine receptors and β-adrenoceptors in human airway. Br J Pharmacol 166:1804–1814
28. Khan RIM, Anisuzzaman AMS, Semba S, Ma Y, Uwada J et al (2013) M1 is a major subtype of muscarinic acetylcholine receptors on mouse epithelial cells. J Gastroenterol 48:885–896
29. Wang MH, Yoshiki H, Anisuzzaman ASM, Uwada J, Nishimune A, Lee KS, Taniguchi T, Muramatsu I (2011) Re-evaluation of nicotinic

acetylcholine receptors in rat brain by a tissue-segment binding assay. Front Pharmacol 2:65. doi:10.3389/fphar.2011.00065

30. Colucci WS, Gimbrone JRMA, Alexander RW (1981) Regulation of the postsynaptic alpha-adrenergic receptor in rat mesenteric artery. Effects of chemical sympathectomy and epinephrine treatment. Circ Res 48:104–111

31. Faber JE, Yang N, Xin XH (2001) Expression of a-adrenoceptor subtypes by smooth muscle cell and adventitial fibroblast in rat aorta and in cell culture. J Pharmacol Exp Ther 298:441–452

32. Hiraizumi-Hiraoka Y, Tanaka T, Yamamoto H, Suzuki F, Muramatsu I (2004) Identification of alpha-1L adrenoceptor in rabbit ear artery. J Pharmacol Exp Ther 310:995–1002

33. Horinouchi T, Morishima S, Tanaka T, Suzuki F, Tanaka Y, Kioke K, Muramatsu I (2006) Pharmacological evaluation of plasma membrane β-adrenoceptors in rat hearts using the tissue segment binding method. Life Sci 79:941–948

34. Delmendo RE, Michel AD, Whiting RL (1989) Affinity of muscarinic receptor antagonists for three putative muscarinic receptor binding sites. Br J Pharmacol 96:457–464

35. Ehlert FJ, Tran LP (1990) Regional distribution of M1, M2 and non-M1, non-M2 subtypes of muscarinic binding sites in rat brain. J Pharmacol Exp Ther 255:1148–1157

36. Sathi ZS, Anisuzzaman ASM, Morishima S, Suzuki F, Tanaka T, Yoshiki H, Muramatsu I (2008) Different affinities of native α1B-adrenoceptors for ketanserin between intact tissue segments and membrane preparation. Eur J Pharmacol 584:22–228

Chapter 5

Use of Antibodies in the Research on Muscarinic Receptor Subtypes

Wisuit Pradidarcheep and Martin C. Michel

Abstract

Antibodies can be a powerful tool to detect receptor expression at the protein level. Their main advantage is the potential of good spatial resolution in immunohistochemistry, whereas their main limitation is that they yield less quantitative results as compared to radioligand binding. However, most available antibodies against muscarinic acetylcholine receptor subtypes have shown poor target selectivity when tested stringently, e.g., often yielded similar staining patterns in wild-type and knockout animals or in cells transfected with the target as compared to a closely related receptor subtype. On the other hand, a small number of antibodies have been validated to some degree for selectivity for a muscarinic receptor subtype. Protocols for their use in immunohistochemistry are discussed. However, it remains a key learning that each investigator should carefully establish whether the intended antibody is indeed selective for the target under investigation under the assay conditions being applied.

Key words Muscarinic receptor, Antibody, Validation, Immunoblot, Immunohistochemistry

1 Introduction

Determination of the number and/or subtype distribution of muscarinic receptors is relevant for the understanding of physiology and pathophysiology. Radioligands are a good tool for the quantification of total muscarinic receptor density in a tissue and its possible regulation by gender, ageing, or pathophysiology [1]. However, the use of radioligands has two limitations in the research field of muscarinic receptors. Firstly, standard muscarinic receptor radioligands such as N-methylscopolamine or quinuclidinylbenzylate are available only in tritiated forms; the associated low specific radioactivity causes a limited sensitivity, i.e. requires large samples and/or a high expression density. As a consequence of this, morphological studies based on autoradiography with the radioligands require long exposure times. Second, N-methylscopolamine or quinuclidinylbenzylate binds with similar affinity to all five muscarinic receptor subtypes. Accordingly, the

Jaromir Myslivecek and Jan Jakubik (eds.), *Muscarinic Receptor: From Structure to Animal Models*, Neuromethods, vol. 107, DOI 10.1007/978-1-4939-2858-3_5, © Springer Science+Business Media New York 2016

relative contribution of any subtype can only be derived from experiments with subtype-selective competitors. However, most muscarinic receptor ligands exhibit only moderate subtype selectivity, which makes robust quantitative analysis of subtypes difficult, particularly within the $M_1/M_3/M_5$ or the M_2/M_4 subfamily of muscarinic receptors [2].

Receptor subtype-selective antibodies could potentially address several of these challenges. Due to their high affinity they can be very sensitive and in immunohistochemical experiments can be an excellent tool for morphological studies. On the other hand, they have the intrinsic disadvantage that the results are difficult to quantify. Most importantly, however, in practical experience most antibodies against (individual) subtypes of muscarinic receptors have proven to lack selectivity for their cognate receptor. Against this background, this chapter initially discusses the selectivity problems with commonly available antibodies against muscarinic receptor subtypes. Thereafter, we discuss protocols which can be used for immunohistochemical detection of muscarinic receptors subtypes for the limited number of cases where antibodies possess the required specificity.

2 Selectivity Problems with Muscarinic Receptor Antibodies

It had been widely assumed that presence of a single band in an immunoblot could be considered as a proof of antibody selectivity. However, the number of bands in an immunoblot can be tweaked in various ways. For example it is influenced by the choice of exposure time and image contrast, which may enhance the visibility of some bands relative to others. This becomes particularly relevant, if the validation immunoblot is generated with a cell line or tissue expressing a very high density of the target protein, for instance with a cell line transfected with the cognate receptor. Such overexpression can enhance target over background signal and may lead to false positive estimates of selectivity when applied to native tissues with a lower expression density.

Another potentially misleading criterion for target selectivity of an antibody can be the disappearance of signal upon co-incubation with a blocking peptide, mostly identical in amino acid sequence to the peptide which had been used to generate the antibody. While it appears obvious that the peptide used for immunization will absorb the antibody, the reasoning for accepting this as specificity evidence ignores the fact that a small peptide in solution may be much more flexible and hence present a very different three-dimensional epitope than a receptor with multiple membrane-spanning domains. Both of these potential problems are worsened by the fact that many commercial suppliers of receptor antibodies provide only limited technical information on the specific experimental conditions which had been used in their validation experiments. Moreover, the frequent absence of cautionary notes that the "representative" immunohistochemical image or immunoblot in a catalog is limited to very specific receptor sources and/or experimental conditions raise doubts about how representative it really is.

Meanwhile evidence from numerous types of G-protein-coupled receptors has shown that even with the best of intentions, selectivity claims on the presence of a single band in an immunoblot and/or signal disappearance in the presence of blocking peptide in many cases provides misleading information on antibody selectivity [3, 4]. Thus, more vigorous approaches to testing antibody selectivity have more often than not failed to confirm selectivity claims based on a single immunoblot band or blocking peptide. For example, when target receptor and a closely related receptor, i.e., another subtype from the same receptor family, were expressed in the same cell line at a comparable density, a given antibody often produced almost identical band patterns in immunoblots with β-adrenoceptor [5] or dopamine receptor subtype antibodies [6]. Similarly, several galanin receptor antibodies produced similar staining patterns in both immunoblots and immunohistochemistry when comparing tissues from wild-type and knockout mice lacking the target receptor [7]. Some investigators lack access to recombinant receptors or knockout animals of the required species; in such cases use of receptor knockdown by small interfering RNA or use of tissues known to lack the receptor of interest may be alternative acceptable validation techniques [8]. The shocking finding from validation approaches using any of these hard criteria was that the vast majority of receptor antibodies failed to exhibit the promised selectivity [3].

Additional potential causes of misleading antibody-based results have been identified. These include the observation that some antibodies may have acceptable specificity in one application, e.g., immunocytochemistry, but not in another, e.g., immunoblotting [9]. Another potential cause of misleading results is the observation that a given antibody may yield acceptable target specificity in one species but not in another [8]. Finally, fixation conditions may also affect the apparent specificity of some antibodies [10]. The sum of these issues has led some investigators to refer to receptor antibodies as "reagents of mass distraction" [11]. In the following we discuss specific evidence in this regard for antibodies against muscarinic receptor subtypes.

In an heroic effort Jositsch et al. [10] have explored the target selectivity of 24 antibodies against muscarinic receptors (1–9 per subtype), with four dilutions and up to 21 different conditions tested for each antibody yielding a total of 1824 conditions being evaluated. In this study staining with several antibodies was abolished by preincubation with blocking peptide.

However, the immunohistochemical signal from M_1 receptor antibodies ABS5164, AMR-001, AS-3701S, GP20a, Rabbit 001, Rabbit 002, and sc-7471 was unaffected in dorsal root ganglia, urinary bladder, and thoracic viscera from M_1 knockout mice. Using a different validation approach, i.e., immunoblotting with membranes from human embryonic kidney (HEK) cells transfected with M_1, M_2, M_3, and M_4 receptors to yield comparable expression levels, we found that AMR-001 exhibited a similar band pattern in immunoblots from all four cell lines [12]. Actually, in these experiments we did not identify a single band that was more prominent in M_1-expressing than in other cells, and one of the bands was actually most prominent in M_2-expressing cells.

Among M_2 receptor antibodies, Jositsch et al. [10] found that immunohistochemical staining was unaffected in dorsal root ganglia, urinary bladder, and thoracic viscera from M_2 knockout mice for AS3721S and AMR-002. For the former, labeling in airways was also unaffected by M_2/M_3 double knockout, indicating that the nonspecific labeling was not due to staining of another muscarinic receptor subtype. For the latter, we have reported a similar band pattern in immunoblots with membranes from cells transfected with M_1, M_2, M_3, and M_4 receptors [12]. However, two M_2 receptor antibodies have shown at least some promise based on the work by Jositsch et al. [10]. The monoclonal antibody mAB367 labelled airway smooth muscle and the cell membrane of a subpopulation of dorsal root ganglion neurons and atrial and pulmonary vein cardiomyocytes in wild-type but not in M_2 knockout mice, particularly when a specific protocol was applied; however, signals from ciliated epithelial cells of the oviduct obtained with the same antibody were not affect in the knock-out mice (Fig. 1). The rabbit polyclonal M_2 receptor antiserum AB5166-50ULa illustrated another problem: while some batches of this antiserum produced labeling specific for wild-type vs. M_2 knockout mice, other batches from the same supplier resulted in identical labeling patterns in both strains, i.e., the producer of this batch may have been unable to deliver a consistent product across batches.

Among antibodies targeted at M_3 receptors, AB9453, AMR-006, AS-3741S, GP19b, R66136, R66431, Rabbit 001, Rabbit 002, and sc-9108 yielded similar staining in immunohistochemical experiments in dorsal root ganglia, urinary bladder, and thoracic viscera from wild-type and M_3 knockout mice (Fig. 1) [10]. AS-3741S also yielded a similar band pattern in immunoblots with membranes from HEK cells transfected with M_1, M_2, M_3, and M_4 receptors [12]. Using a similar approach, another group reported that antibody sc-7474 detected a single band in immunoblots of M_3 receptor-transfected HEK cells, which was absent in non-transfected cells; the apparent molecular weight of this band, 95–100 kDa, was considerably higher that estimated based on the receptor sequence (66 kDa) but might be explained by glycosidation [13]. In a follow-up study

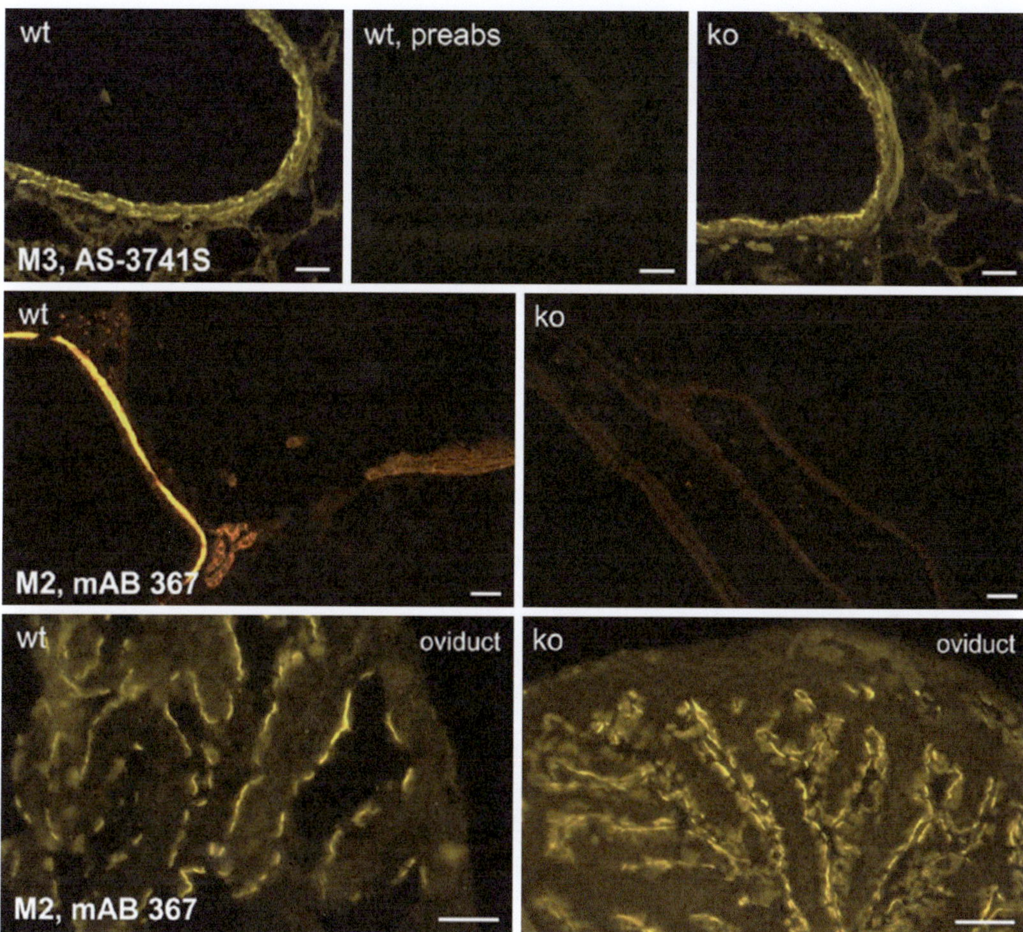

Fig. 1 An example of proven target selectivity and lack of it with antibodies against muscarinic receptor subtypes. The *upper panels* show staining of mouse bronchi with the alleged M_3 receptor antibody AS-37415. Staining appears anatomically selective for smooth muscle in wild-type mice and is prevented by liquid-phase pre-absorption with the corresponding antigen but is not affected in M_3 knockout mice. In contrast, the *middle panels* show that staining of mouse bronchial and pulmonary vein smooth muscle with the M_2 antibody mAB 367 is absent in M_2 receptor knockout mice; however, staining of ciliated epithelial cells of the oviduct was not affected in the knockout mice (*lower panels*). Adapted with permission from [10]

the same group found a similar immunohistochemical staining pattern in the guinea pig urinary bladder for three antibodies, sc-7474, sc-9108, and Ab-13063 [7]. When tested in immunoblots with membranes from HEK cells transfected or non-transfected with M_3 receptors, sc-7474 again yielded a single band of 102 kDa. In contrast, Ab-13063 did not yield any band, and sc-9108 yielded two bands of 45 and 65 kDa. Of note, staining with sc-9108 was abolished by blocking peptide [14] but has not been affected by M_3 receptor knock-out in immunohistochemistry experiments by other investigators [10].

A smaller number of M_4 receptor antibodies have been evaluated by hard criteria. Among these AS-3761S, MAB1576 and sc-9109 yielded similar staining in immunohistochemical experiments in dorsal root ganglia, urinary bladder, and thoracic viscera from wild-type and M_4 knockout mice [10]. MAB1576 also yielded a similar band pattern in immunoblots with membranes from HEK cells transfected with M_1, M_2, M_3, and M_4 receptors [12]. AS-3781S, claimed to be an M_5 receptor antibody, exhibited staining in immunohistochemical experiments in dorsal root ganglia, urinary bladder, and thoracic viscera from wild-type and M_5 knockout mice [10].

Taken together, with the possible exception of few M_2 and M_3 receptor antibodies, most antibodies with claimed selectivity for individual subtypes of muscarinic receptors fail to exhibit target selectivity when tested under stringent conditions. However, even antibodies which exhibit target selectivity under some experimental conditions may not do so under others (Fig. 1). Therefore, the key message from the above is that receptor antibodies need to be validated as carefully as possible and that data based on such antibodies always need to be interpreted very cautiously.

3 Methods for Immunohistochemical Detection of Muscarinic Receptors

Immunostaining is a widely used technique that combines biochemistry and immunology. The concept of immunostaining was developed from the antigen-antibody binding reaction and visualizes the distribution and localization of specific antigens or cellular components (in this case muscarinic receptors) in tissue sections (immunohistochemistry) or isolated cells (immunocytochemistry). Compared to other techniques that are based on the antigen-antibody reaction, such as immunoprecipitation and Western blotting, which provide material for further biochemical analyses and provide information on molecular weight of the antigens (and contaminations), immunostaining provides topographical information.

Immunostaining can be divided into (1) direct staining (one-step staining) and (2) indirect staining (two or more step staining) procedures. The indirect staining technique is more popular, because it allows the amplification of the signals at the site of antigen-antibody binding by different hapten-conjugated antibodies. According to the type of labeling of the antibodies, immunostaining methods can be classified as immunogold, immunofluorescence, and immunoenzyme stainings. In immunogold staining, colloidal gold is bound to the antibodies and visualized. Colloidal gold is the hydrosol form of gold and can bind proteins rapidly and stably. Moreover, colloidal gold has little effect on the biological activity of natural proteins. Therefore, colloidal gold can be conjugated with both primary and secondary antibodies. Due to the high electronic density of colloidal gold, immunogold technique is also suitable for antigen detection with an electron microscope. In immunofluorescence staining methods, the antigens are visualized with fluorescent dyes conjugated to antibodies. Because the exciting and emitted light have to be separated, dedicated fluorescence microscopes are necessary. Due to ease, high sensitivity, and convenience, the immunofluorescence staining method is widely used in biomedical sciences. In immunoenzyme staining, enzymes are coupled to antibodies that are used to bind to specific antigens in tissue samples or cultured cells. After adding substrate, the enzyme generates insoluble or electron-dense particles that can be localized under a light or electron microscope. Compared to immunofluorescence staining, immunoenzyme-stained samples can be stored longer. Two major enzymes covalently linked with secondary antibodies that are commercially available are horse radish peroxidase (HRP) and alkaline phosphatase (AP). These enzymes catalyze reactions that produce stained products that are easily detectable by light microscopy. Binding reaction between the HRP and its substrate yields the products in brown. However, adding metals to HRP changes color. Reaction between AP and its substrate gives rise to products which stain blue (if substrate used is nitroblue tetrazolium chloride/5-bromo-4-chloro-3-indolyl phosphate) or stain red (if aminocarbazol is used as the substrate).

From our own experience with immunostaining of muscarinic receptors in tissue sections and cultured cells, we prefer the AP- to HRP-based method. This is because the staining intensity of properly diluted AP-coupled antisera increases linearly with time for 1–2 h, whereas the product inhibition of peroxidase coupled antisera yields their maximal staining in a few minutes and does, hence, not differentiate very well between locally differing concentrations of antigens. The localization of antigen when stained with AP is adequate but HRP-based staining with diaminobenzidine is superior in this respect. However, the main advantage of staining with AP is that the intensity develops linearly with time for several hours and can also be intensified by developing at a higher temperature [15].

Three major steps in a complete immunoenzyme staining session are the following:

1. Binding of primary antibody to specific antigen (e.g., muscarinic receptors).
2. Forming the antibody-antigen complex by incubation with an (enzyme-conjugated) secondary antibody.
3. Generating a colored precipitate at the sites of antibody-antigen binding by exposing the section to the chromogenic substrate.

3.1 Tips in Immunoenzyme Staining

Tissue preparation or fixation is essential for the preservation of cell morphology and tissue architecture. Inappropriate or prolonged fixation may significantly diminish the accessibility of the antigen to the antibody. Fixatives that are suitable for immunostaining should at least preserve antigenic sites and should not destroy antigenicity by acting as a very strong protein cross-linker. In our experience, 4 % formaldehyde in phosphate-buffered saline (PBS) or an ice-cold mixture of methanol:acetone:water (MAW; 2:2:1 (v/v)) are a proper protein cross-linker and precipitating fixative, respectively. These fixatives are able to successfully preserve antigens in paraffin-embedded tissue sections or in cultured cells. However, some antigens will not survive even moderate amounts of aldehyde fixation. Under such conditions, tissues should be rapidly fresh frozen in liquid nitrogen and cut with a cryostat. The disadvantages of frozen sections include poor morphology, poor resolution at higher magnifications, difficulty in cutting relative to paraffin sections, and the need for a cryotome [16].

If 4 % formaldehyde/PBS is used as fixative, the detection of antigens can be dramatically improved by antigen retrieval. This method breaks up some of the protein cross-links formed by fixation to uncover hidden antigenic sites. This can be accomplished by heating in citrate or EDTA-based solution (e.g., autoclave or microwave) for varying lengths of times [17–19]. However, if MAW is employed as fixative agent, the step of antigen retrieval is not required.

One of the main difficulties with immunostaining is reducing non-specific background. Optimization of fixation methods and times, pre-treatment with blocking agents, incubating antibodies diluted in a high-salt solution (e.g., 500 mM Na-acetate, pH 8), and optimizing post-antibody washing buffers and washing times are all important for obtaining high-quality immunostaining. In addition, the presence of positive and negative controls for staining is essential for determining specificity.

The following and Table 1 reflect the protocols that we have used to successfully reveal antigens in the paraffin-embedded tissue sections [20].

4 Protocol: Indirect Immunoalkaline Phosphatase Staining on Paraffin Sections

1. Deparaffinize sections in three changes of xylene, 3 min each.
2. Hydrate the sections in a graded descending series of ethanol: 100 %, 96 %, 90 %, 80 %, 70 %, and 50 % for 1 min each.
3. Rinse in water.
4. If the sections were fixed with formaldehyde, it is necessary to perform an antigen retrieval step. If the sections were fixed by ice-cold mixture of methanol:acetone:water

Table 1
Chemicals used in [20]

Chemical and material	Manufacturer	Product No.
Paraformaldehyde	VWR	4005
Tween-20	VWR	822184.0500
Normal goat serum (NGS)	Gibco	16210-072
Fetal calf serum (FCS)	Gibco	
NBT/BCIP	Roche	1681 451
GAM-AP	Sigma	A3562
GAR-AP	Dako	D0487
RAG-AP	Sigma	A4187
Super PAP-pen	Beckman	IM3850

(MAW; 2:2:1 (v/v)) (or other alcohol fixative) antigen retrieval is not necessary.

The steps for antigen retrieval are as follows:

(a) Prepare 10 mM sodium citrate from a stock solution of 1 M sodium citrate; adjust the pH to 6.0 with 1 M citric acid.

(b) Put your slides in the solution-containing box and cover with aluminum foil.

(c) Put the slides in the autoclave.

(d) Set the autoclave at 10 min and 120 °C.

(e) Wait until the pressure is off and take the sections out.

(f) Let them cool down to room temperature, wash shortly in distilled water and then continue with the procedure.

5. Wash in PBS, pH 7.4, for a minimum of 5 min on a shaking platform at room temperature.
6. Draw circle around the sections on the glass slides with a Pap-pen to prevent mixing of the different antibodies between adjacent sections.
7. Incubate the sections in 1× TENG-T/10 % serum (normal goat serum (NGS) or fetal calf serum (FCS)) for a minimum of 30 min to reduce nonspecific background staining.
8. Remove the TENG-T + 10 % serum by suction and apply the primary antibody. The appropriate dilution(s) of primary antibody are made in TENG-T/10 % serum. The incubation is done overnight at room temperature in a humidity chamber.

It is noteworthy that the volume applied on each section should not be too big to avoid intermingling with adjacent incubations.

9. Remove the unbound first antibody by gentle suction and drop PBS directly on the sections.
10. Wash the sections in three changes of PBS, 5 min each on a shaking platform at room temperature.
11. Incubate the sections with the AP-conjugated secondary antibody for at least 2 h at room temperature. The optimal dilution of AP-conjugated secondary antibody is dependent on the first antibody used:
    (a) If the first antibody was raised in mouse, use GAM-AP: **GAM-AP**: **G**oat-**a**nti-**M**ouse IgG conjugated with **A**lkaline **P**hosphatase (1:100 in TENG-T/10 % serum).
    (b) If the first antibody was raised in rabbit, use GAR-AP: **GAR-AP**: **G**oat-**a**nti-**R**abbit IgG conjugated with **A**lkaline **P**hosphatase (1:200 in TENG-T/10 % serum).
    (c) If the first antibody was raised in goat, use RAG-AP: **RAG-AP**: **R**abbit-**a**nti-**G**oat IgG conjugated with **A**lkaline **P**hosphatase (1:50 in TENG-T/10 % serum). In this case the serum should *not* be a normal goat serum.
12. Wash the sections in three changes of PBS, 5 min each on a shaking platform at room temperature.
13. Incubate the sections in NBT/BCIP (a substrate of alkaline phosphatase) diluted in NTM at room temperature.

    Note

    (a) NBT/BCIP: nitroblue tetrazolium chloride/5-bromo-4-chloro-3-indolyl phosphate (toluidine salt; Dako).
    (b) NTM contains: (1) 100 mM NaCl, (2) 100 mM Tris pH 9.5, and (3) 50 mM $MgCl_2$.
    (c) $MgCl_2$ should be added to the solution just before use.
    (d) Dilute the NBT/BCIP 1:50 in NTM just before use. Make it fresh, do not store.
14. Stop the reaction in distilled water, after the staining is satisfactory when viewing under light microscope (it can be 30 min, but up to 2 h is permitted).
15. Dehydrate the sections by dipping quickly through a graded ascending series of ethanol (50 %, 70 %, 80 %, 90 %, 96 %, and 100 %). If you do not go quickly through the ethanol, the staining will become faint.
16. Dip the sections in three changes of xylene, 7 min each. This step makes the color in the tissues clearer.
17. Mount the sections in Enthallan (a mounting media).

18. Let the sections dry in fume hood and then overnight in an incubator at 37 °C.
19. Observe and photograph under light microscope.

5 Notes

1. Prepare stock solution of 10× TENG-T containing:
   - 100 mM **T**ris–HCl.
   - 50 mM **E**DTA (pH 8.0).
   - 1.5 M **N**aCl.
   - 2.5 % **G**elatin.
   - 0.5 % v/v **T**ween-20.

   Mix well and store at 4 °C.
2. In order to prepare 1× TENG-T:
   Put 10×TENG-T in warm water and allow the content to melt. Shake gently and dilute it to 1×TENG-T with bidistilled water. Adjust the pH with HCl or NaOH to 8.0.
3. Prepare 1×TENG-T+10 % serum:
   Add 1 ml of serum to 9 ml of 1× TENG-T (NGS, FCS, or another serum can be used, but it must not be the serum from the animal in which the first antibody was raised (e.g., do not use rabbit serum when your first antibody was raised in rabbits)). Ideally primary and secondary antibody should be from different orders of mammals; if this is not feasible, non-specific binding can be tested on a Western blot.

6 Conclusions

The most important conclusion from this data is that investigators must apply great care in their choice of antibody for immunological detection of muscarinic receptor subtypes. As target selectivity may depend on the assay, i.e., immunoblotting vs. immunohistochemistry, each antibody must be carefully validated for the intended use. Choice of fixation protocols and other steps may critically affect signal strength in immunohistochemistry studies.

Acknowledgment

We thank Prof. Wouter H. Lamers, Academisch Medisch Centrum, Universiteit van Amsterdam, for his continued advice and support.

References

1. Schneider T, Hein P, Michel-Reher M et al (2005) Effects of ageing on muscarinic receptor subtypes and function in rat urinary bladder. Naunyn Schmiedebergs Arch Pharmacol 372: 71–78
2. Caulfield MP, Birdsall NJM (1998) International Union of Pharmacology. XVII. Classification of muscarinic acetylcholine receptors. Pharmacol Rev 50:279–290
3. Michel MC, Wieland T, Tsujimoto G (2009) How reliable are G-protein-coupled receptor antibodies? Naunyn Schmiedebergs Arch Pharmacol 377:385–388
4. Kirkpatrick P (2009) Specificity concerns with antibodies for receptor mapping. Nat Rev Drug Discov 8:278
5. Hamdani N, van der Velden J (2009) Lack of specificity of antibodies directed against human beta-adrenergic receptors. Naunyn Schmiedebergs Arch Pharmacol 379:403–407
6. Bodei S, Arrighi N, Sigala S (2009) Should we be cautious on the use of commercially available antibodies to dopamine receptors? Naunyn Schmiedebergs Arch Pharmacol 379:413–415
7. Lu X, Bartfai T (2009) Analyzing the validity of GalR1 and GalR2 antibodies using knockout mice. Naunyn Schmiedebergs Arch Pharmacol 379:417–420
8. Cernecka H, Pradidarcheep W, Lamers WH et al (2014) Rat β_3-adrenoceptor protein expression: antibody validation and distribution in rat gastrointestinal and urogenital tissues. Naunyn Schmiedebergs Arch Pharmacol 387:1117–1127
9. Cernecka H, Ochodnicky P, Lamers WH et al (2012) Specificity evaluation of antibodies against human β_3-adrenoceptors. Naunyn Schmiedebergs Arch Pharmacol 385:875–882
10. Jositsch G, Papadakis T, Haberberger RV et al (2009) Suitability of muscarinic acetylcholine receptor antibodies for immunohistochemistry evaluated on tissue sections of receptor gene-deficient mice. Naunyn Schmiedebergs Arch Pharmacol 379:389–395
11. Rhodes KJ, Trimmer JS (2006) Antibodies as valuable neuroscience research tools versus reagents of mass distraction. J Neurosci 26: 8017–8020
12. Pradidarcheep W, Stallen J, Labruyere WT et al (2009) Lack of specificity of commercially available antisera against muscarinic and adrenergic receptors. Naunyn Schmiedebergs Arch Pharmacol 379:397–402
13. Grol S, Essers PBM, van Koeveringe GA et al (2009) M_3 muscarinic receptor expression on suburothelial interstitial cells. BJU Int 104: 398–405
14. Grol S, Nile CJ, Martinez-Martinez P et al (2011) M_3 muscarinic receptor like immunoreactivity (M_3-IR) in sham-operated and obstructed guinea pig bladders. J Urol 185: 1959–1966
15. van Straaten HWM, He Y, van Duist MM et al (2006) Cellular concentrations of glutamine synthetase in murine organs. Biochem Cell Biol 84:215–231
16. Fischer AH, Jacobson KA, Rose J et al (2008) Cryosectioning tissues. CSH Protoc 2:1–2
17. Yamashita S (2007) Heat-induced antigen retrieval: mechanisms and application to histochemistry. Prog Histochem Cytochem 41: 141–200
18. Syrbu SI, Cohen MB (2011) An enhanced antigen-retrieval protocol for immunohistochemical staining of formalin-fixed, paraffin-embedded tissues. Methods Mol Biol 717: 101–110
19. Gadd VL (2014) Combining immunodetection with histochemical techniques: the effect of heat-induced antigen-retrieval on picro-Sirius red staining. J Histochem Cytochem 62:902–906
20. Pradidarcheep W, Labruyere WT, Dabhoiwala NF et al (2008) Lack of specificity of commercially available antisera: better specifications needed. J Histochem Cytochem 56: 1099–1111

Chapter 6

Allosteric Modulation of Muscarinic Receptors

Jan Jakubik and Esam E. El-Fakahany

Abstract

Allosteric ligands modulate binding and function of muscarinic receptors in a different way than orthosteric ligands. Unlike orthosteric ligands their effects are limited by a cooperativity factor. This imparts them unique properties, including cooperativity-based selectivity, functional selectivity and restoring of physiological-like space and time pattern of signaling under pathological conditions. Therefore, allosteric modulators of muscarinic receptor are intensively studied as possible therapeutics of pathological conditions including Alzheimer's disease and schizophrenia. Research of allosteric modulation has pioneered the way for a whole class A of G-protein coupled receptors and has had an impact beyond its own field. We review principles of allosteric modulations and their implications for proper design of binding as well as functional experiments and for proper data analysis. We demonstrate immense complexity of allosteric modulation of functional responses. Such complexity is reflected in the inability to determine individual microscopic constants in allosterically modulated systems. Therefore, the effects of a given allosteric modulator can be characterized by only two macroscopic parameters, namely a change in the agonist potency and efficacy. We also discuss distinct properties of allosteric interactions that are specific to muscarinic receptors.

Key words Muscarinic receptors, Allosteric modulation, Radioligand binding, Functional response

1 Historic Overview

The concept of allosterism was formally introduced into the field of enzymology by Monod et al. [1, 2] in their description of a generalized model of oligomeric enzymes that contained a number of "stereo-specifically different, non-overlapping receptor sites." The substrate was said to bind at the "primary" or "active" site. On the other hand, an "allosteric effector" (from the Greek word "allo" meaning "other") was defined as a molecule that binds to a site other than the primary binding site. In pharmacology, the term orthosteric ligand denotes a compound that binds to the same binding site as endogenous ligand (neurotransmitter or hormone), while a ligand that binds to other sites on the receptor is termed allosteric. Allosteric ligands influence (modulate) binding and effects of orthosteric ligands in a different way than orthosteric ligands. In the pioneering work by Clark and Mitchelson,

Jaromir Myslivecek and Jan Jakubik (eds.), *Muscarinic Receptor: From Structure to Animal Models*, Neuromethods, vol. 107, DOI 10.1007/978-1-4939-2858-3_6,

gallamine was found to shift the concentration-response curves of acetylcholine in inhibiting the heart atrium to the right but the magnitude of the shifts was smaller than expected for conventional competitive receptor antagonists [3]. They proposed that gallamine interacts with a secondary allosteric site on the receptor. This notion was later confirmed in radioligand binding studies [4]. Subsequently, a wide variety of allosteric modulators of muscarinic receptors was identified including toxiferous alkaloids [5], the L-calcium channel blocker verapamil [6], the potassium channel inhibitor 4-aminopyridine [7], inhibitors of acetylcholinesterase [8–10], strychnos and vinca alkaloids [11], and antibiotics like staurosporine [12]. Numerous site-directed mutagenesis studies located allosteric binding sites for most allosteric modulators to the extracellular domain of the receptor, namely between the second and the third extracellular loops [13–20]. For further details see review by Jakubík and El-Fakahany [21]. This has been recently confirmed by crystallographic studies [22]. Allosteric modulation of muscarinic receptors has been intensively studied for decades for its perspective role in therapy of many pathological conditions including Alzheimer's disease and schizophrenia [23, 24].

2 Principles of Allosteric Modulation

2.1 Allosteric Modulation of Ligand Binding

By definition, allosteric ligands bind to a site on the receptor that is spatially distinct from that of endogenous ligands of the receptor. Consequently, binding of an allosteric ligand (A) and an orthosteric ligand (L) to the receptor (R) is not mutually exclusive, i.e., both ligands may bind to the receptor to form a ternary complex LRA (Scheme 1). Binding of allosteric modulators induces a change in the conformation of the receptor that results in changes in binding and/or effects of the orthosteric ligand. The law of microscopic reversibility of thermodynamics dictates that binding of orthosteric ligand L affects binding of allosteric ligand A in the same way in which the allosteric ligand A affects binding of the orthosteric ligand L. A situation when L and A mutually strengthen each other's binding is called positive cooperativity, i.e., formation of the ternary complex LRA leads to increase in the affinity of both ligands, that is a decrease in the equilibrium dissociation constants of orthosteric ligand (K_D) and allosteric ligand (K_A) (thus $\alpha < 1$) (Scheme 1, Fig. 1 upper graph, green curve). The opposite situation when L and A mutually weaken their binding is called negative cooperativity. Under negative cooperativity formation of the ternary complex LRA leads to a mutual decrease in affinity, that is an increase in equilibrium dissociation constants of orthosteric ligand (K_D) and allosteric ligand (K_A) (thus $\alpha > 1$) (Scheme 1, Fig. 1 upper graph, red curve). In other words, binding of L to R is stronger or weaker in the presence of A in case of positive or negative cooperativity, respectively. In rare situations the two ligands form a ternary complex with the receptor without mutually changing their affinities is called neutral cooperativity ($\alpha = 1$) (Fig. 1 upper graph, blue curve). In this particular case the affinity of L for R or AR is the same. This knowledge is very important in design of binding experiments.

$$\begin{array}{ccc} L + R + A & \underset{}{\overset{K_D}{\rightleftharpoons}} & LR + A \\ K_A \updownarrow & & \updownarrow \alpha K_A \\ L + RA & \underset{}{\overset{\alpha K_D}{\rightleftharpoons}} & LRA \end{array}$$

Scheme 1 Scheme of allosteric interaction. An orthosteric ligand L binds to the receptor R with equilibrium dissociation constant K_D and an allosteric modulator A binds to the receptor R with equilibrium dissociation constant K_A. The orthosteric ligand L and the allosteric modulator A can bind concurrently to the receptor R to form a ternary complex LRA. Binding of one ligand to the receptor changes the equilibrium dissociation constant of the other ligand by factor of cooperativity α

In general ligand association is a fast process that closely parallels ligand diffusion to the receptor. Thus allosteric effects on ligand binding are usually manifested in changes in ligand dissociation; by slowing or accelerating dissociation in case of positive and negative cooperativity, respectively (Fig. 1 lower graph). Because changes in kinetics of ligand binding do not strictly follow changes in ligand affinity, binding of an allosteric agent with neutral cooperativity may be detected by changes in the rate of dissociation of an orthosteric ligand. As changes in ligand kinetics are not possible without formation of the ternary complex they become a hallmark of allosteric interaction and the most straightforward way to identify it.

It should be noted that effects of allosteric modulator on both equilibrium and kinetic binding and on functional effects of orthosteric ligands is limited by a cooperativity factor α. For example, in Scheme 1, with increasing concentrations of the allosteric modulator A the equilibrium dissociation constant of the orthosteric ligand L changes from its original value K_D until it reaches a value $\alpha \times K_D$. Further increase in the concentration of A does not bring further change in K_D. That is in contrast to competition of two orthosteric ligands for the same site where changes in binding of one orthosteric ligand are proportional to the concentration of the other orthosteric ligand without a maximal limit (*see* Chapter 3, Fig. 5). The level of maximum effect of allosteric modulator (also known as "ceiling effect") would confer safety under conditions of

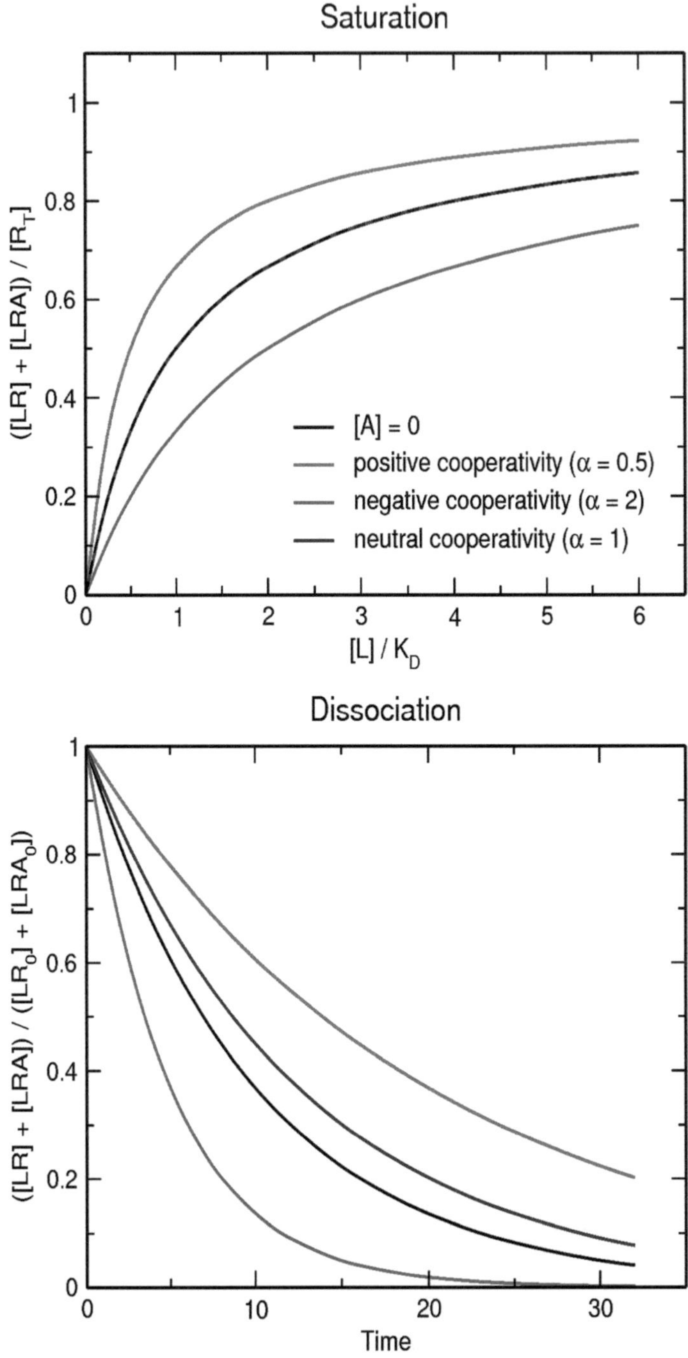

Fig. 1 Allosteric modulation of binding. Theoretical curves of tracer equilibrium saturation binding (*upper graph*) and tracer dissociation (*lower graph*) in the absence (*black*) or in the presence of positive (*green*), negative (*red*) or neutral (*blue*) allosteric modulator. Tracer binding (in binary LR and ternary LRA complexes) is expressed as a fraction of total receptor number R_T (*upper graph*) or fraction of complexes at start of dissociation (*lower graph*). A positive allosteric modulator causes an increase in tracer affinity (decrease in tracer equilibrium dissociation constant K_D) (*upper graph*) and slow down of tracer dissociation (*lower graph*). A negative allosteric modulator causes a decrease in tracer affinity (increase in tracer equilibrium dissociation constant K_D) (*upper graph*) and acceleration of tracer dissociation (*lower graph*)

overdosage. This represents an advantage of development of allosteric receptor modulators for therapeutics purposes.

A conformational change of the receptor induced by an allosteric ligand has different effects on binding of structurally different orthosteric ligands as well as structurally different receptor subtypes. Thus, the factor of cooperativity α depends on combination of all three constituents of allosteric interaction: the receptor and the orthosteric and allosteric ligands. For example eburnamonine decreases affinity of the agonist arecoline at the muscarinic M_2 receptor, but has no effect on the affinity of the agonist arecaidine propargyl ester. Interestingly, it *increases* the affinity of the agonist pilocarpine. In contrast, eburnamonine decreases the affinity of all three agonists at M_1 and M_3 muscarinic receptors [11].

2.2 Anomalies in Allosteric Modulation of Functional Response

Allosteric modulators may affect receptor activation by mechanisms additional to effects on binding of an orthosteric agonist to the receptor. Thus for a given pair of allosteric modulator and orthosteric agonist, positive cooperativity in binding does not necessarily translate into an increase in agonist potency in functional assays and may even lead to "allosteric quenching of agonist efficacy" [25, 26]. Moreover, an allosteric modulator that exerts negative binding cooperativity with an agonist at equilibrium may potentiate agonist-induced activation of the receptor due to acceleration of agonist binding [27].

Another feature of allosteric modulators of muscarinic receptors is their ability to activate the receptors in the absence of agonists. For example, partial stimulation of accumulation of inositol phosphates and inhibition of accumulation of cAMP in response to strychnine-like allosteric modulators has been reported [28]. A whole new class of potent allosteric agonists has been reported recently [29–31].

3 Promises of Allosteric Targeting of Muscarinic Receptors

3.1 Selectivity by Targeting Less Conserved Domains

Muscarinic receptor subtypes share high structural homology in the transmembrane domains where the orthosteric binding site is located. On the other hand, domains out of membrane are less conserved. Targeting allosteric domains allows achieving binding selectivity to an extent that is not possible by orthosteric ligands. Allosteric modulators exhibit a wide range of selectivity for different muscarinic receptor subtypes. For example, while prototypical allosteric modulators like alcuronium and gallamine display selectivity towards M_2 receptors [32], strychnine is M_3 selective [11] and WIN compounds are M_4 selective [33].

3.2 Conservation of Space and Time Pattern of Signaling

Theoretically, a positive allosteric modulator of acetylcholine that has no efficacy on its own would only induce an action when the endogenous acetylcholine is released. Consequently, its action would be restricted in space and time to those regions of the body where signaling is actually taking place. Thus, space and time pattern of signaling could be restored under pathological conditions of diminished acetylcholine release, e.g., degeneration of cholinergic neurons in certain brain areas in Alzheimer's disease.

3.3 Absolute Selectivity

Theoretically, absolute selectivity is achieved by having an allosteric agent with the intended positive or negative cooperativity in combination with the orthosteric ligand at one subtype of the receptor and neutral cooperativity at the rest of the subtypes. Selectivity may be derived from binding and/or activation cooperativity. As proof of concept such cooperativity-based binding selectivity for thiochrome at the M_4 muscarinic receptor has been reported [34].

4 Analyzing Allosteric Modulation of Ligand Binding

The methodology of radioligand binding at muscarinic receptors is described in detail in a previous chapter. Here we only describe the setup and data analysis of experiments with allosteric ligands. The major complication of radioligand binding experiments with allosteric modulators is the lack of suitable allosteric radioligands. So far only a few radiolabeled allosteric ligands are available. These include tritiated ABA-type like compounds (Fig. 2) dimethyl-W84 [35], derivatives of α-truxillic acid anatruxonium and truxillonium [36], and iodinated proteins like the muscarinic toxins MTX2 [37], MTX1 [38, 39], and MTX7 [40]. Tritiated ABA-type compounds have relatively low affinity and extremely high nonspecific binding in comparison to orthosteric antagonists that make their use as tracers difficult. Fluorescent labeling and detection of binding by FRET seems to be the way to reduce nonspecific binding [41, 42]. Muscarinic toxins display very slow kinetics that lead to kinetic artifacts [37, 40]. None of muscarinic radiolabeled ligands are available commercially. Thus almost all binding studies of muscarinic allosteric ligands are conducted indirectly and their binding parameters are inferred from changes in binding of orthosteric tracers.

4.1 Allosteric Modulation of Tracer in Saturation Binding Experiments

Equilibrium dissociation constant of the allosteric ligand K_A and factor of cooperativity α can be determined from a series of experiments of tracer saturation binding. Equilibrium dissociation constants in Scheme 1 are defined as follows:

$$K_D = \frac{[L][R]}{[LR]} \tag{1a}$$

$$K_A = \frac{[R][A]}{[RA]} \tag{1b}$$

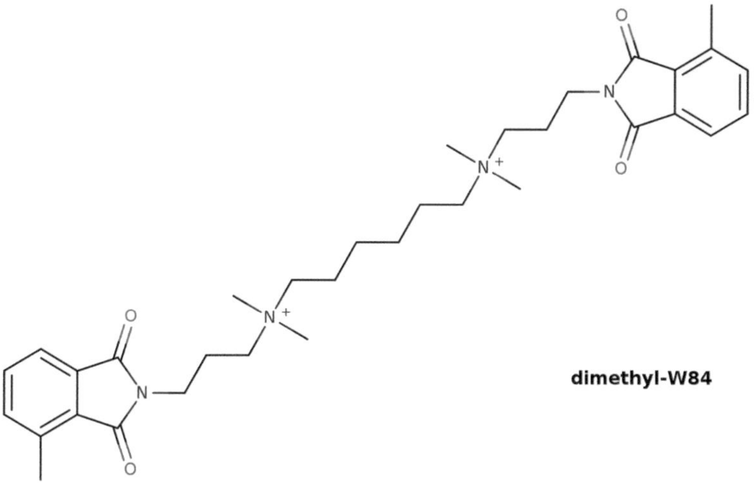

dimethyl-W84

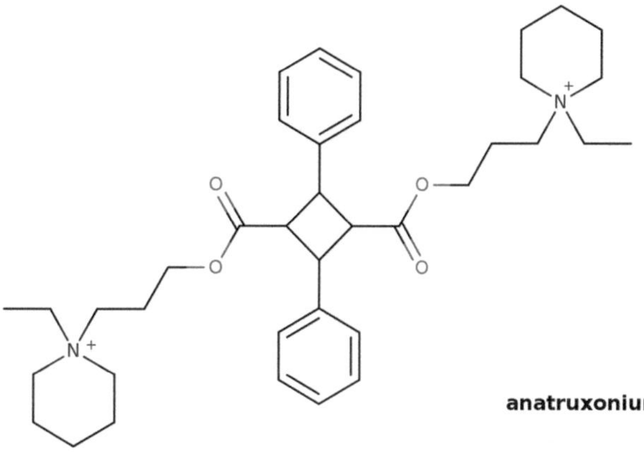

anatruxonium

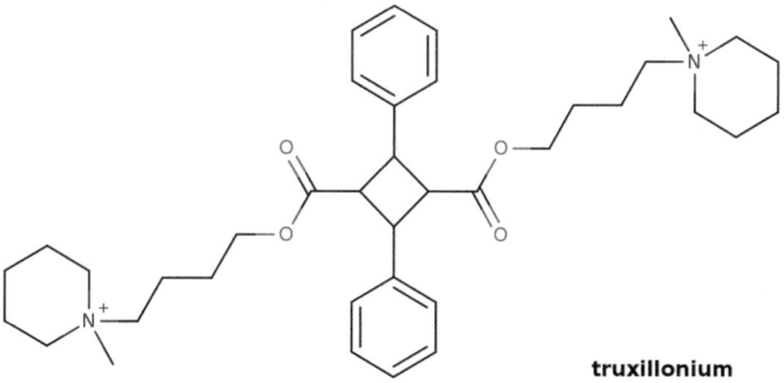

truxillonium

Fig. 2 Radiolabeled allosteric modulators. ABA-type compounds that have been experimentally tritiated. Dimethyl-W84 (6-[dimethyl-[3-(4-methyl-1,3-dioxoisoindol-2-yl)propyl]azaniumyl]hexyl-dimethyl-[3-(4-methyl-1,3-dioxoisoindol-2-yl)propyl]azanium),anatruxonium(1,1′-[(2,4-diphenylcyclobutane-1,3-diyl)bis(carbonyloxypropane-3,1-diyl)]bis(1-ethylpiperidinium)) and truxillonium (bis[4-(1-methylpiperidin-1-ium-1-yl)butyl]2,4-diphenylcyclobutane-1,3-dicarboxylate)

$$\alpha K_D = \frac{[L][RA]}{[LRA]} \tag{1c}$$

$$\alpha K_A = \frac{[LR][A]}{[LRA]} \tag{1d}$$

Total number of receptors is sum of free receptors and binary and ternary complexes:

$$[R_T] = [R] + [LR] + [RA] + [LRA] \tag{2}$$

Fraction of receptors occupied by tracer L:

$$\frac{[LR]+[LRA]}{[R_T]} = \frac{[LR]+[LRA]}{[R]+[LR]+[RA]+[LRA]} \tag{3}$$

Multiplying the numerator and denominator of the fraction on the right side of Eq. (3) by $1/[L][R]$ gives:

$$\frac{[LR]+[LRA]}{[R_T]} = \frac{\frac{[LR]}{[L][R]} + \frac{[LRA]}{[L][R]}}{\frac{[R]}{[L][R]} + \frac{[LR]}{[L][R]} + \frac{[RA]}{[L][R]} + \frac{[LRA]}{[L][R]}} \tag{4}$$

Substituting Eq. (4) by Eq. (1a):

$$\frac{[LR]+[LRA]}{[R_T]} = \frac{\frac{1}{K_D} + \frac{[LRA]}{[L][R]}}{\frac{1}{K_D} + \frac{1}{[L]} + \frac{[RA]}{[L][R]} + \frac{[LRA]}{[L][R]}} \tag{5}$$

Substituting Eq. (5) by Eqs. (1b) and (1c):

$$\frac{[LR]+[LRA]}{[R_T]} = \frac{\frac{1}{K_D} + \frac{[A]}{\alpha K_D K_A}}{\frac{1}{K_D} + \frac{1}{[L]} + \frac{[A]}{\alpha K_D K_A} + \frac{[A]}{[L]K_A}} \tag{6}$$

After simplification:

$$[LR] + [LRA] = \frac{[L][R_T]}{[L] + K_D'} \tag{7a}$$

where K_D' is the apparent equilibrium dissociation constant of the tracer in the presence of allosteric modulator A that is given as:

$$K_D' = K_D \times \frac{K_A + [A]}{K_A + [A]/\alpha} \tag{7b}$$

For saturating concentrations of the allosteric modulator ($[A] \gg K_A$) Eq. (7b) can be reduced to:

$$K_D^{'} = K_D \times \frac{[A]}{[A]/\alpha} = \alpha K_D \tag{8}$$

As can be seen, an allosteric modulator affects the tracer equilibrium dissociation constant (Eq. 7b) without a change in binding capacity (Eq. 7a). Maximum change in the equilibrium dissociation constant is given by the cooperativity factor α (Eq. 8). In case of positive cooperativity the tracer's equilibrium dissociation constant decreases with increasing the concentration of the allosteric modulator (Eq. 7b; Fig. 3, upper graph). Such effects are unique to allosteric interaction and thus positive allosteric modulators are easily identified. In case of negative cooperativity the tracer's equilibrium dissociation constant increases with increasing the concentration of an allosteric modulator (Eq. 7b; Fig. 3, lower graph). At first glance this is similar to competition of orthosteric ligand with the tracer for the same binding site (Fig. 4). However, as stated above, at negative cooperativity the decrease in tracer affinity has its limit given by the cooperativity factor α, while the effects of a classical competitive interaction are directly proportional to the competitor's concentration without a limit. While plotting of tracer equilibrium dissociation constant against concentration of a competitor gives a straight line (with slope equal to 1 and constant equal to 1) (Fig. 5, circles) plotting tracer equilibrium constant against the concentration of a negative allosteric modulator gives a hyperbole with asymptote equal to the cooperativity factor α (Fig. 5, squares). It can be seen that deviations from competitive behavior are more obvious at high concentrations of a negative allosteric modulator. Likewise, plotting tracer equilibrium dissociation constant against the concentration of a positive allosteric modulator gives an inverse hyperbole with asymptote equal to the cooperativity factor α (Fig. 5, triangles).

When equilibrium dissociation constant of the allosteric ligand K_A and factor of cooperativity α are determined from a series of experiments of tracer saturation binding (like in Fig. 3) Eq. (7a) is fitted to data and apparent equilibrium dissociation constant of the tracer $K_D^{'}$ is determined for each concentration of the allosteric modulator A. Then the obtained $K_D^{'}$ values are plotted against the concentration of A (like in Fig. 5) and Eq. (7b) is fitted to data to determine the equilibrium dissociation constant of allosteric modulator K_A and the cooperativity factor α.

4.2 Displacement Binding Experiments

Determination of the equilibrium dissociation constant of the allosteric ligand K_A and factor of cooperativity α from a series of experiments of tracer saturation binding is laborious and expensive. Binding parameters of an allosteric ligand can be determined in a

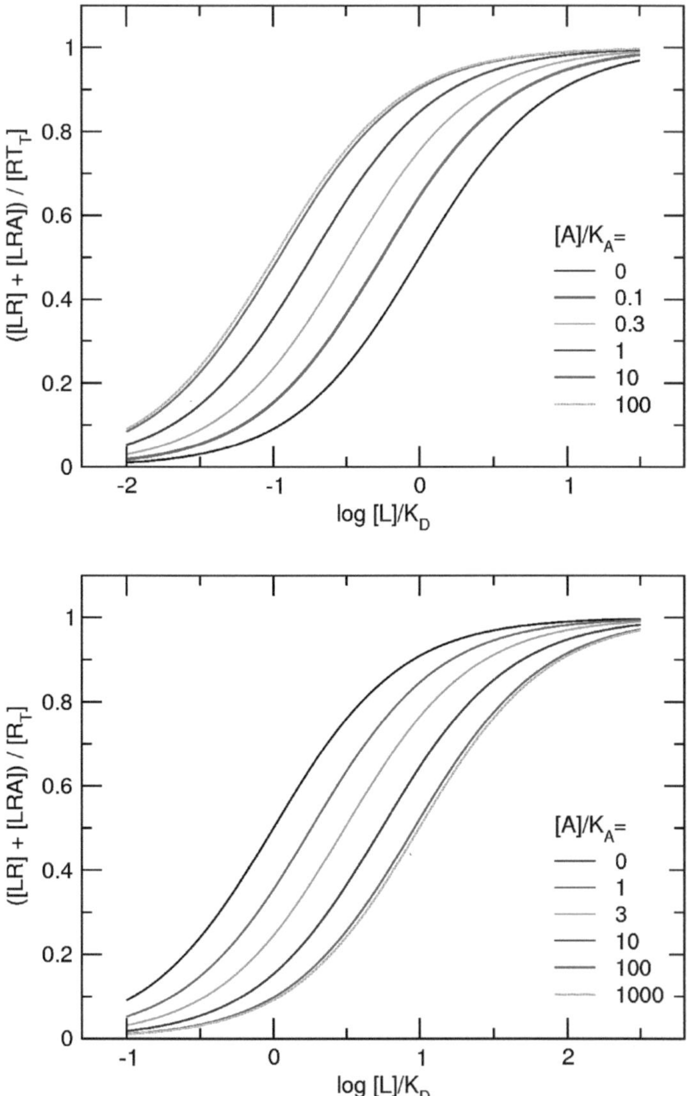

Fig. 3 Allosteric modulation of tracer saturation binding. Effects of positive ($\alpha = 0.1$) (*upper graph*) and negative ($\alpha = 10$) (*lower graph*) allosteric modulators on tracer saturation binding. Abscissa, the concentration of tracer L is expressed as a logarithm of the ratio to its equilibrium dissociation constant K_D. Ordinate, tracer binding is expressed as a fraction of total receptor number R_T. Legend, the concentration of allosteric modulator A is expressed as a ratio to its equilibrium dissociation constant K_A. A positive allosteric modulator concentration dependently decreases tracer K_D, while a negative allosteric modulator concentration dependently increases tracer K_D

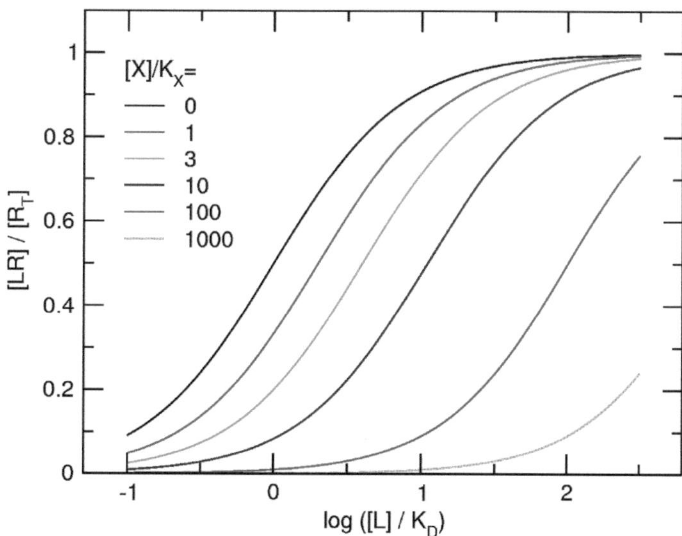

Fig. 4 Competition with tracer saturation binding. Effects of a competitor on tracer saturation binding. Abscissa, concentration of tracer L is expressed as the logarithm of the ratio to its equilibrium dissociation constant K_D. Ordinate, tracer binding is expressed as a fraction of total receptor number R_T. Legend, concentration of competitor X is expressed as ratio to its equilibrium dissociation constant K_X

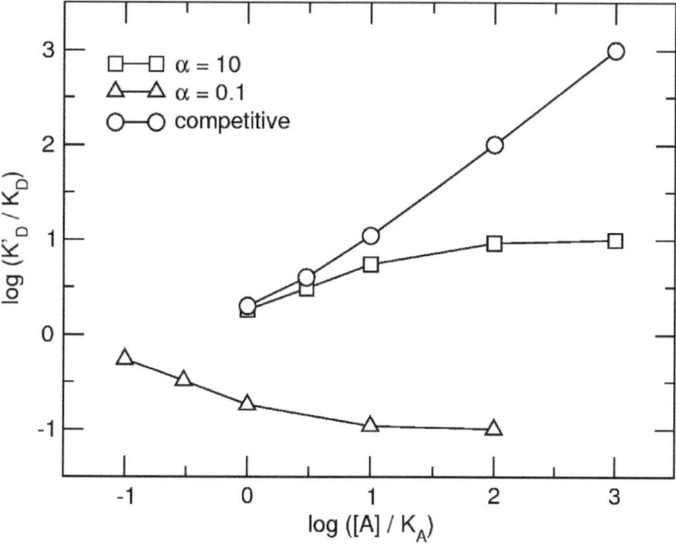

Fig. 5 Shifts in equilibrium dissociation constants. From Figs. 3 and 4 shifts in equilibrium dissociation constant caused by negative allosteric modulator ($\alpha = 10$) (*squares*), positive allosteric modulator ($\alpha = 0.1$) (*triangles*) and competitive ligand (*circles*) are expressed as the logarithm of the ratio of tracer apparent equilibrium dissociation constant K_D' in the presence of the second ligand to tracer equilibrium dissociation constant K_D in its absence. Abscissa, concentration of the second ligand is expressed as logarithm of the ratio of second ligand concentration to its equilibrium dissociation constant

simpler way by measuring the effects of increasing concentrations of an allosteric modulator on binding of a single concentration of the tracer. Binding of tracer L at fixed concentration in the presence of various concentrations of allosteric modulator A is described by Eq. (7) (Fig. 6). It is more convenient to express the data as a fraction of the tracer binding in the presence of the allosteric modulator than its binding in the absence of the allosteric modulator (Fig. 7) that is given:

$$Y = \frac{\dfrac{[L][R_T]}{[L]+K_D'}}{\dfrac{[L][R_T]}{[L]+K_D}} = \frac{[L]+K_D}{[L]+K_D'} \tag{9}$$

After substitution of Eq. (9) with Eq. (7b):

$$Y = \frac{[L]+K_D}{[L]+K_D \times \dfrac{K_A+[A]}{K_A+[A]/\alpha}} \tag{10}$$

When the equilibrium dissociation constant of allosteric modulator K_A and the cooperativity factor α are determined by measuring binding at a single concentration of the tracer and various concentrations of allosteric modulator Eq. (10) is fitted to data expressed as ratio of tracer binding in the presence of A to the tracer binding in the absence of A (like in Fig. 7). Equilibrium dissociation of the tracer K_D has to be determined in separate measurements. Precise concentration of the tracer L used in the assay should be determined by counting total radioactivity added to the sample and division by specific radioactivity of the tracer and sample volume. It can be seen from equations describing allosteric binding that the equilibrium dissociation constant of an allosteric modulator K_A and cooperativity factor α are interdependent parameters. Overestimation of α leads to underestimation of K_A and vice versa. Thus a wide range of concentrations of A has to be used. For proper determination of α a saturating concentration of A has to be used to reach a curve plateau. This plateau defines α according to Eq. (9) where for a saturating concentration of A the apparent equilibrium dissociation constant of tracer K_D' becomes αK_D. The value of αK_A has to be determined properly for accurate determination of K_A. The value of αK_A corresponds to the inflection point of the binding curve. The latter is best determined by measuring binding at several concentrations close to the inflection point (αK_A).

Positive cooperativity is easily spotted as an increase in tracer binding to free receptors (Figs. 6 and 7, upper graphs). The fraction of free receptors decreases with increasing tracer concentration. For positive allosteric modulators with strong cooperativity the fraction of free receptors may be limiting. If a saturating

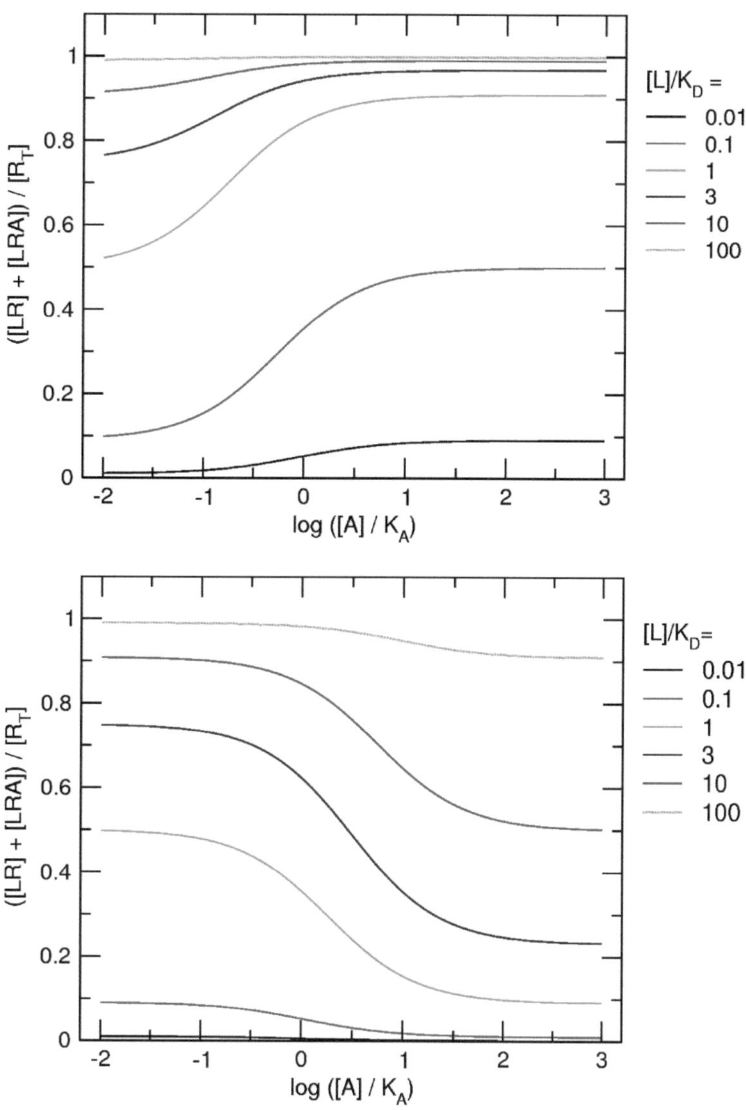

Fig. 6 Effects of an allosteric modulator on binding of the tracer at a fixed concentration. Effects of various concentrations of a positive ($\alpha = 0.1$) (*upper graph*) and negative ($\alpha = 10$) (*lower graph*) allosteric modulators at indicated on the abscissa on binding of the tracer at fixed concentration indicated in legend. Abscissa, concentration of allosteric modulator is expressed as logarithm of ratio to its equilibrium dissociation constant K_A. Ordinate, the tracer binding is expressed as a fraction of total receptor number R_T. Legend, concentration of the tracer L is expressed as ratio to its equilibrium dissociation constant K_D. A positive allosteric modulator concentration dependently increases tracer binding, while a negative allosteric modulator concentration dependently decreases tracer binding. Changes in tracer binding are more obvious at lower concentrations of the tracer

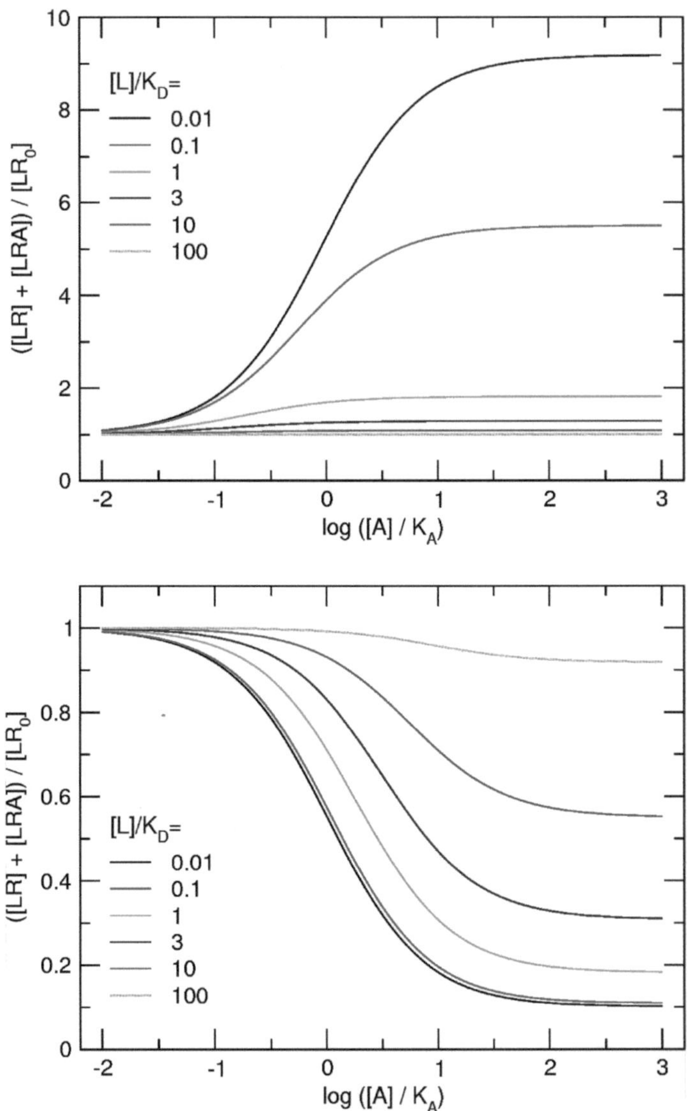

Fig. 7 Effects of an allosteric modulator on binding of the tracer at a fixed concentration. Effects of a positive ($\alpha = 0.1$) (*upper graph*) and negative ($\alpha = 10$) (*lower graph*) allosteric modulators at various concentrations indicated on the abscissa on binding of the tracer at a fixed concentration indicated in the legend. Abscissa, concentration of allosteric modulator is expressed as logarithm of the ratio to its equilibrium dissociation constant K_A. Ordinate, tracer binding is expressed as a fraction of its binding in the absence of allosteric modulator. Legend, concentration of the tracer L is expressed as a ratio to its equilibrium dissociation constant K_D. A positive allosteric modulator concentration dependently increases tracer binding while a negative allosteric modulator concentration dependently decreases tracer binding. Changes in tracer binding are greater at lower concentrations of the tracer

concentration of the tracer is used it is difficult to reliably determine the cooperativity factor α because subtle differences in the level of plateau represent huge differences in α. So in binding experiments with strong positive allosteric modulators low concentrations of the tracer (below its K_D) are desired to prevent full receptor occupancy at saturating concentrations of the allosteric modulator. However too low concentration of the tracer gives low and unreliable control binding in the absence of allosteric modulator. Thus a compromise between the size of allosteric change and the quality of control binding has to be achieved. In the model case of positive cooperativity with $\alpha = 0.1$ a tracer concentration equal to its K_D gives sufficient (>80 %) increase in tracer binding and sufficient proportion (almost 20 %) of the receptors remain free at saturating concentrations of the allosteric ligand (Fig. 7, upper graph).

Negative binding cooperativity can be distinguished from competitive binding by incomplete inhibition of tracer binding, resulting in a plateau in the displacement curve (Figs. 6 and 7, lower graphs). The higher the concentration of tracer, the higher the level of the plateau (less complete inhibition of tracer binding) (Fig. 7). Thus, detection of strong negative cooperativity requires the use of a tracer concentration several times higher than its dissociation constant K_D to get incomplete inhibition of tracer binding. However, the ratio of specific to nonspecific binding decreases with increasing tracer concentration, indicating that extremely high concentrations of the tracer should be avoided. It must be noted that allosteric modulators with very high cooperativity factors ($\alpha > 100$) would cause almost complete inhibition of tracer binding, making them indistinguishable from competitive binding. A more experimental setup in this special case is to construct saturation curves for the tracer in the absence and in the presence of increasing concentrations of the allosteric modulator. The apparent equilibrium dissociation constant of the tracer K_D' is determined for each concentration of the allosteric modulator A (like in Fig. 3, lower graph) and plotted against concentration of A (like in Fig. 5, squares) and Eq. (7b) is fitted to data.

Ligands with weak (either positive or negative) cooperativity induce small changes in tracer binding. As can be seen in Eq. (9) the lower concentration of L the greater the change in fractional binding. Thus low concentrations of the tracer are desired to magnify changes induced by weak allosteric modulators. It can be demonstrated using Eq. (10) that even for ligands with very weak cooperativity ($0.9 < \alpha < 1.1$) lowering tracer concentration below $0.1 \times K_D$ does not bring further increase in allosteric effects. Maximum attainable increase in tracer binding by positive allosteric modulators with $\alpha = 0.9$ is 10 % of the control binding and, analogically, maximum attainable decrease in tracer binding is 10 % by very weak negative allosteric modulators with $\alpha = 1.1$. Such small changes may be problematic to detect and are usually considered as neutral cooperativity.

Protocol A: Determination of K_A and α in equilibrium experiments (96-well plate setup)

1. Determine the equilibrium dissociation constant K_D of [^{3}H]NMS in saturation binding experiments in a buffer of your choice at 25 °C (*see* previous Chapter 3).
2. Add membranes, about 20 fmol of receptors per well.
3. Add [^{3}H]NMS to a final concentration around $0.5 \times K_D$ (when positive cooperativity is expected) or $2 \times K_D$ (when negative cooperativity is expected) for 60 min at 25 °C in a final incubation volume of 0.4 ml.
4. Add tested allosteric modulators to final concentrations ranging from 10 nM to 100 μM (9 concentrations at 0.5 log concentration steps). Make samples of control binding in the absence of allosteric modulator, samples of nonspecific binding in the presence of 1 μM atropine alone and in the presence of 100 μM allosteric modulator (to verify that the allosteric modulator does not change nonspecific binding).
5. Seal the plate and incubate for 20 h at 25 °C (see justification below for the long incubation time).
6. Filter samples though GF/C filters or a filtration plate. Wash with ice-cold deionized water for 6 s.
7. Determine the exact tracer concentration used in the experiment by counting added radioactivity divided by specific radioactivity and incubation volume (0.4 ml).
8. Fit Eq. (10) to specific binding expressed as a fraction of control. Use K_D of [^{3}H]NMS determined in saturation binding experiment and exact radioligand concentration L from step 7.

4.3 Allosteric Modulation of Tracer Binding Kinetics

Usually association of substrates or ligands with enzymes or receptors is a fast process, being controlled by diffusion. Under such conditions a change in the affinity of the orthosteric radioligand by an allosteric modulator is manifested mainly as a change in the dissociation rate of the tracer. Allosteric modulation of the radioligand rate of dissociation in an ideal model system with constant association rate by positive ($\alpha = 0.1$) and negative ($\alpha = 10$) allosteric modulators is shown in Fig. 8. A positive allosteric modulator concentration dependently slows down tracer dissociation. The slowdown is limited by the factor of cooperativity α, in this case 10-times at maximum. Inversely, a negative allosteric modulator concentration dependently speeds up tracer dissociation. Again, the effect is limited by the factor of cooperativity α, reaching a maximum of a tenfold change in this particular case.

However, muscarinic receptors are far from ideal. Almost all muscarinic allosteric ligands, regardless of being positive or negative modulators, slow down both association and dissociation of

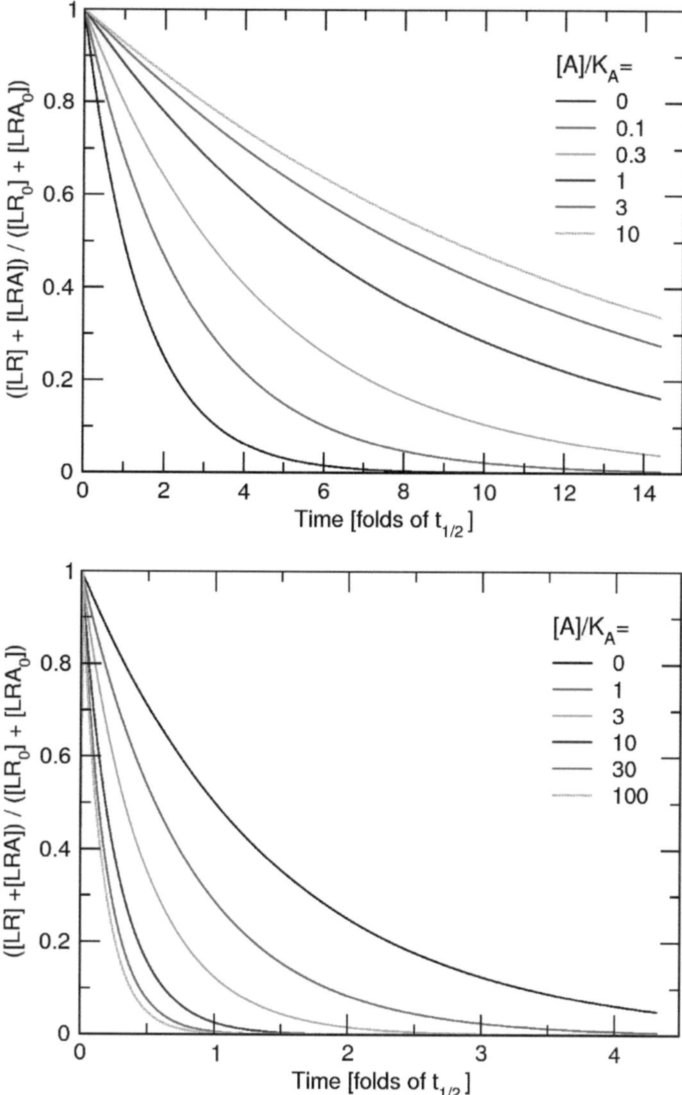

Fig. 8 Allosteric modulation of tracer binding kinetics. Effects of a positive ($\alpha = 0.1$) (*upper graph*) and negative ($\alpha = 10$) (*lower graph*) allosteric modulator on binding kinetics of the tracer. Abscissa, time is expressed as folds of dissociation halftime of tracer-receptor complex LR in the absence of allosteric modulator A. Ordinate, the tracer binding is expressed as a ratio of tracer L binding to its binding at the start of dissociation. Legend, concentration of allosteric modulator A is expressed as a ratio to its equilibrium dissociation constant K_A. A positive allosteric modulator concentration dependently slows down tracer dissociation while a negative allosteric modulator concentration dependently speeds up tracer dissociation

orthosteric ligands. The orthosteric binding site of the receptor is located deep within transmembrane helices so association is relatively slow and may be accelerated by allosteric modulators [27]. The binding site for allosteric ligands is located between the second and the third extracellular loops, being close to the path that an orthosteric ligand takes during association and dissociation. Thus, bound allosteric ligands usually represent a physical obstacle (steric hindrance) for orthosteric ligands on their way to and away from the muscarinic receptor [43]. It is noteworthy that effects on the kinetics of binding of an orthosteric ligand allows for identification of agents with neutral cooperativity, since these agents do not change affinity of the tracer in equilibrium experiments. Furthermore, effects on the kinetics of tracer binding distinguish allosteric modulators with very strong negative cooperativity from competitive agents.

Mechanistically, orthosteric ligand L may bind only to free receptor R and is not able to bind to complex of receptor and allosteric ligand RA. In Scheme 1 reaction L + RA to LRA is not possible (Scheme 2).

The orthosteric ligand L and allosteric modulator A can bind concurrently to receptor R and form a ternary complex. The orthosteric ligand L has to bind first to R followed by binding of A to form the ternary complex LRA. If the receptor is already occupied by an allosteric ligand the orthosteric ligand has to "wait" until the allosteric ligand dissociates. Dissociation of the orthosteric ligand from the ternary complex LRA must take place in the reverse order, i.e., the allosteric ligand has to dissociate first to make way for dissociation of the orthosteric ligand. Thus, this sequential arrangement of binding has profound effects on binding kinetics of an orthosteric ligand as demonstrated in the example of slowing down of binding kinetics of [^{3}H]NMS at M_2 muscarinic receptors by alcuronium. With increasing concentrations of the allosteric modulator alcuronium the proportion of free receptors decreases and thus association of the orthosteric tracer [^{3}H]NMS decelerates, even though alcuronium is a positive allosteric modulator (Fig. 9, upper graph). Concurrently, increasing the concentration of alcuronium is associated with exerting stronger steric

$$\begin{array}{ccc} L + R + A & \overset{K_D}{\rightleftharpoons} & LR + A \\ K_A \updownarrow & & \alpha K_A \updownarrow \\ L + RA & & LRA \end{array}$$

Scheme 2 Scheme of allosteric interaction with allosteric hindrance

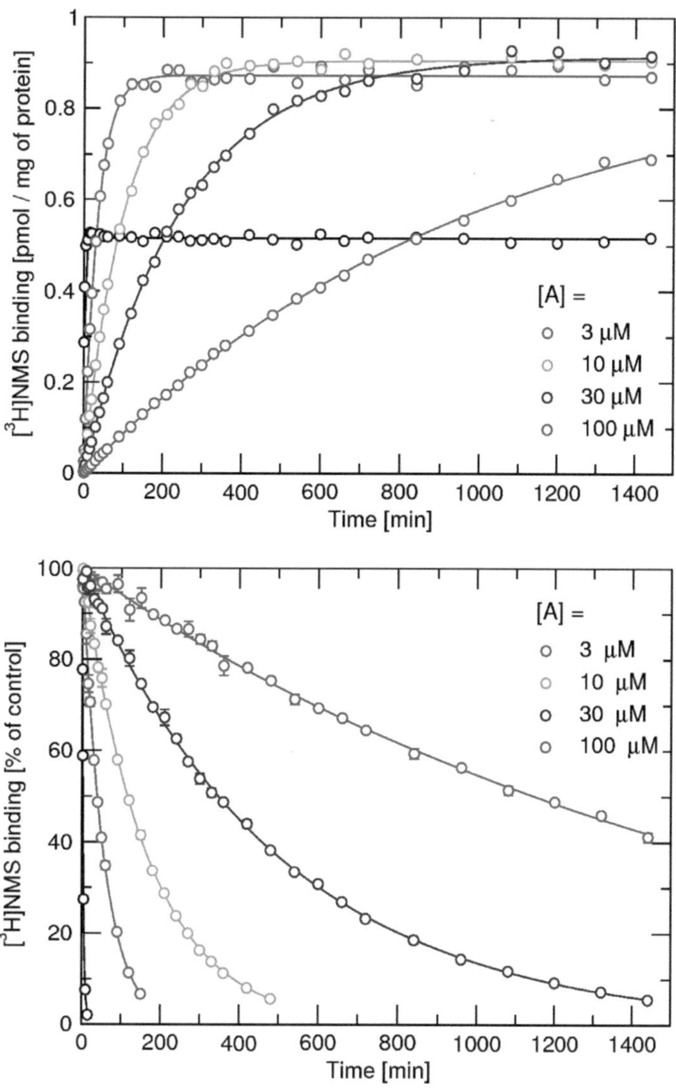

Fig. 9 Slowdown of [^{3}H]NMS binding kinetics by alcuronium. Time courses of association (*upper graph*) and dissociation (*lower graph*) of 100 pM [^{3}H]NMS at M_2 muscarinic receptors in the absence (*black curves*) or in the presence of alcuronium at the concentrations indicated in the legend. *Upper graph*: [^{3}H]NMS binding at the time points indicated on the abscissa is expressed as pmol per mg of protein. Alcuronium concentration dependently increases equilibrium binding and slows down the rate of association of [^{3}H]NMS. *Lower graph*: binding at the time point indicated on the abscissa is expressed as percent of [^{3}H]NMS binding at the start of dissociation. Alcuronium concentration dependently slows down the rate of dissociation

hindrance of [^{3}H]NMS dissociation (Fig. 9, lower graph). Retardation of the on and off biding kinetics of orthosteric ligands by allosteric ligands is proportional to the concentration of the allosteric ligand and is unlimited. Thus extremely long incubation

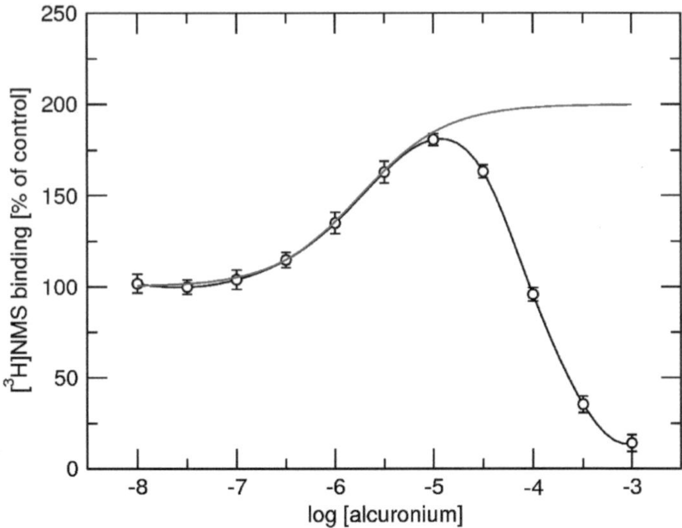

Fig. 10 Artifacts of non-equilibrium binding. Binding of 100 pM [^{3}H]NMS to M_2 receptors after 3 h of incubation at 25 °C in the presence of alcuronium at the concentrations indicated on the asbcissa is expressed as percent of [^{3}H]NMS binding in the absence of alcuronium (*circles*). Equilibrium is not reached in the presence of alcuronium at a concentration of 10 μM and higher. The *red curve* represents binding under equilibrium ($K_D = 250$ pM, $\alpha = 0.32$)

times (hours or even days) are required to reach equilibrium of an orthosteric radioligand at high concentrations (over 100-times of equilibrium dissociation constant) of an allosteric ligand [11].

Non-equilibrium binding leads to kinetic artifacts as shown in Fig. 10 where equilibrium binding is not reached at 10 μM of alcuronium and higher, causing the binding curve to appear bell-shaped (black curve) instead of being sigmoidal (red curve). In such experiment equilibrium may be reached faster by preincubation of receptors with the orthosteric tracer (e.g., 1 h with [^{3}H]NMS) to allow it to bind without slowing-down by the allosteric modulator. Adding a positive allosteric modulator to the preformed [^{3}H]NMS-receptor complex will result in an increase in bound radioactivity. Thus binding lower than control binding (like in Fig. 10) cannot be observed. The situation is more complicated in the case of negative allosteric modulators, where prolonged time may be necessary to allow the tracer to dissociate from the receptor and for binding the allosteric agent to the receptor to reach equilibrium. Lack of equilibrium after the addition of a negative allosteric modulator would lead to underestimation of the factor of cooperativity.

Extremely long incubation times needed to reach equilibrium may be avoided by inferring allosteric modulator binding from changes in tracer kinetics. Because of arrangement of allosteric and orthosteric sites on muscarinic receptors dissociation of the orthosteric ligand from ternary complex is impossible. Thus observed rate of dissociation limits to zero with increase in concentration of allosteric modulator. Observed dissociation rate is inversely proportional to receptor occupancy by allosteric modulator. Receptor occupancy is given by saturation binding isotherm with apparent equilibrium dissociation constant of allosteric modulator $K_A{'}$. Observed dissociation rate $k_{Off}{'}$ is thus given by Eq. (11)

$$k_{Off}^{'} = k_0 \times \frac{K_A^{'}}{[A] + K_A^{'}} \quad (11)$$

where k_0 is dissociation rate constant of the tracer in the absence of allosteric modulator. At saturating concentrations of A dissociation of tracer occurs only from ternary complexes and thus $K_A{'}$ becomes αK_A. For $[A] \gg \alpha K_A$ Eq. (10) simplifies to:

$$k_{Off}^{'} = k_0 \times \frac{\alpha K_A}{[A]} \quad (12)$$

Moreover, at high concentrations of A dissociation of tracer is monophasic and thus observed dissociation rate constant $k_{Off}{'}$ can be determined in a single time-point measurement and the equilibrium dissociation constant of allosteric modulator at the ternary complex (αK_A) determined without prolonged incubation needed to reach equilibrium (Protocol B). An example of determination of apparent equilibrium dissociation constant of the allosteric modulator methoctramine at M_2 muscarinic receptors from changes in [^{3}H]NMS dissociation is shown in Fig. 11. The disadvantage of this approach is that only αK_A can be determined (not α and K_A separately). However, this approach is sufficient for screening purposes (e.g., in structure–function relationship studies where similar α and K_A values are expected for similar compounds). Moreover, in specific conditions (like in the case of methoctramine that binds with high affinity to the orthosteric site and with low affinity to the allosteric site) kinetic experiments are the only way for assessing the apparent equilibrium dissociation constant [42].

Protocol B: Determination of apparent equilibrium dissociation constant in dissociation experiments (96-well plate setup)

1. Determine the equilibrium dissociation constant K_D of [^{3}H] NMS in a buffer of your choice at 25 °C in saturation binding experiment and the dissociation rate constant k_{off} of [^{3}H]NMS in dissociation experiments (*see* previous Chapter 3).
2. Add membranes, about 10 fmol of receptors per well.
3. Add [^{3}H]NMS to a final concentration $3 \times K_D$ for 60 min at 25 °C in a final incubation volume of 0.2 ml.
4. Initiate dissociation by the addition of 0.2 ml of atropine in a final concentration of 1 μM either alone or in combination with the tested allosteric modulator in concentrations ranging

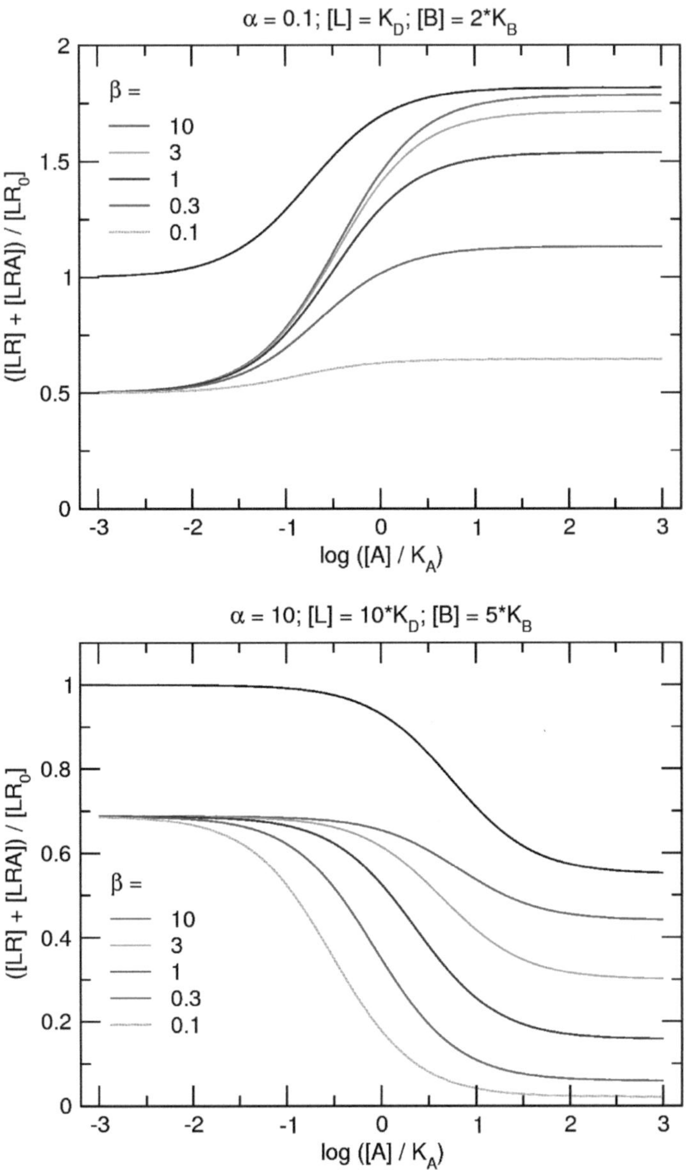

Fig. 11 Dependence of the dissociation rate of [^{3}H]NMS from M_2 receptors on the concentration of methoctramine. Observed rate dissociation constants (k_{off}') are plotted against the concentrations of methoctramine. Fitting Eq. (12) to data yields K_A' around 2.8 μM ($pK_A' = 5.55 \pm 0.5$; means ± SEM, n=4). Data are means ± SEM of 3–4 independent experiments performed in quadruplicates

from 10 μM to 1 mM (5 concentrations at 0.5 log concentration steps). Make samples of control binding (no addition/dilution) and samples of nonspecific binding in the presence of 1 μM (added prior step 3). Sums to 32 samples, one third of 96-well plate, when performed in quadruplicates.

5. Incubate samples for three times of the half-life of ligand dissociation (about 10 min for M_2, 45 min for M_1, M_3, and M_4, and 3 h for M_5).
6. End dissociation by filtering samples through GF/C filters or filtration plates. Wash with ice-cold deionized water for 6 s.
7. Calculate the observed dissociation rates k_{Off}' from decrease in specific binding as negative natural logarithm of fractional binding divided by dissociation time.
8. Plot calculated k_{Off}' values against the concentrations of the tested allosteric modulator and fit Eq. (12) to data.

4.4 Three Ligand System

Many compounds of interest like orthosteric agonists are not candidates as useful tracers because of their low affinity. As explained above, neither are allosteric modulators suitable for radiolabeling. For investigation of allosteric interaction of non-labeled orthosteric ligands and non-labeled allosteric modulators a procedure employing three ligands (orthosteric tracer L, non-labeled orthosteric ligand B, and non-labeled allosteric modulator A) has been devised as depicted in Scheme 3 [11].

Orthosteric tracer L binds to the receptor R with equilibrium dissociation constant K_D, orthosteric non-labeled ligand binds to the receptor R with equilibrium dissociation constant K_B and allosteric modulator binds to the receptor R with equilibrium dissociation constant K_A. The binding of orthosteric ligands L and B is mutually exclusive but the allosteric modulator A can bind concurrently to the receptor R occupied by either of the orthosteric ligands and form a ternary complex LRA or BRA. Alpha and β are factors of binding cooperativity of A with L and A with B, respectively.

In this procedure allosterically induced changes in the affinity for non-labeled orthosteric ligands are reflected in changes in the binding of an orthosteric tracer. The following relations apply besides those described in Eq. (1) in the three ligand system:

$$K_B = \frac{[B][R]}{[BR]} \tag{13a}$$

$$\begin{array}{ccccc}
\mathrm{L + BR + A} & \underset{}{\overset{K_B}{\rightleftharpoons}} & \mathrm{B + L + R + A} & \overset{K_D}{\rightleftharpoons} & \mathrm{LR + A} \\
\beta K_A \updownarrow & & K_A \updownarrow & & \alpha K_A \updownarrow \\
\mathrm{L + BRA} & \overset{\beta K_B}{\rightleftharpoons} & \mathrm{B + L + RA} & \overset{\alpha K_D}{\rightleftharpoons} & \mathrm{LRA}
\end{array}$$

Scheme 3 Interaction of two orthosteric ligands and one allosteric modulator

$$\alpha K_B = \frac{[B][RA]}{[BRA]} \tag{13b}$$

And Eq. (2) becomes to:

$$[R_T] = [R] + [LR] + [RA] + [LRA] + [BR] + [BRA] \tag{14}$$

Analogously to derivation in Eqs. (3)–(7a, b) apparent dissociation constant of the orthosteric tracer L in the presence of the allosteric modulator A and the orthosteric ligand B is derived as:

$$K_D' = K_D \times \frac{[B](K_A + [A]/\beta) + K_B(K_A + [A])}{K_B(K_A + [A]/\alpha)} \tag{15}$$

And the ratio of tracer binding in the presence of A and B to the absence of A and B becomes:

$$Y = \frac{[L] + K_D}{[L] + K_D \times \frac{[B](K_A + [A]/\beta) + K_B(K_A + [A])}{K_B(K_A + [A]/\alpha)}} \tag{16}$$

Experimental setup is similar to measurement at a fixed concentration of the tracer and various concentrations of the allosteric modulator. In this three ligand system two curves are measured. One in the absence of the non-labeled orthosteric ligand and one in the presence of a fixed concentration of the orthosteric ligand (Fig. 12). In control curve (in the absence of B) equilibrium dissociation constant of allosteric modulator K_A and factor of cooperativity α are determined by fitting Eq. (10) to data. These parameters are then used for fitting Eq. (16) to binding data measured in the presence of B. Equilibrium dissociation constants of the tracer K_D and of the orthosteric ligand K_B have to be determined in separate measurements. The precise concentration of the tracer L used in the assay should be determined by counting total radioactivity added to the sample and division by specific radioactivity of the tracer and sample volume. Inhibition of tracer binding by B in the absence of A has to correspond to inhibition calculated using equilibrium dissociation constants K_D and K_B and concentrations of L and B (*see* Chapter 3, Eq. (13)).

A low concentration of the tracer has to be used when studying allosteric modulators that exert positive cooperativity with the tracer (Fig. 12 upper graph) to get a clear increase in tracer binding by allosteric modulators. Vice versa, a high concentration of the tracer has to be used in case of allosteric modulators with negative cooperativity (Fig. 12 lower graph) to get incomplete inhibition of tracer binding by the allosteric modulator. The concentration of B should be chosen based on the factor of cooperativity β. Low concentrations of tracer (in relation to K_B) are suitable in case of positive cooperativity between A and B but leave little room for quantification of negative cooperativity and vice versa. Moreover the concentration of L (in relation to K_D) also affects the range of appropriate concentrations of B. Higher concentration of B is required when a higher concentrations of L (in relation to K_D) is used to reach the same inhibition of tracer binding. In case of negative cooperativity between A and B with increase in concentration of A binding of B becomes weaker and thus inhibition of the tracer binding smaller. As a result the curves of inhibition of tracer binding in the absence and in the presence of B move closer with increasing the concentration of A (Fig. 12, red and green curves). On the other hand, positive cooperativity between A and B leads to strengthening of binding of B and therefore stronger inhibition of tracer binding. As a result the curves of tracer binding in the absence and the presence of B show more diversion with increasing the concentration of A (Fig. 12, magenta and yellow curves).

There are several parameter combinations under which it is very hard or impossible to determine the factor of cooperativity β. One is the case of studying a combination of an allosteric modulator with strong cooperativity with another with weak cooperativity (negative or positive). This is because on the one hand a low concentration of the tracer is a prerequisite for measurements of weak (either positive or negative) cooperativity and on the other hand a high concentration of the tracer is required for measurements of strong cooperativity systems. In such situation a series of tracer saturation binding has to be performed to determine the apparent equilibrium dissociation constant of the tracer K_D' in the presence of various concentrations of A and one concentration of B. Then K_D' has to be plotted against concentration of A and Eq. (15) fitted to data (Fig. 11).

Bitopic ligands that bind both to the orthosteric and allosteric sites at muscarinic receptors have been proposed [44] and subsequently identified [42, 45, 46]. Binding of a bitopic ligand B to the allosteric binding site prevents binding of the allosteric ligand A and vice versa (Scheme 4).

$$
\begin{array}{ccccc}
L + BR + A & \underset{}{\overset{K_B}{\rightleftharpoons}} & B + L + R + A & \overset{K_D}{\rightleftharpoons} & LR + A \\
 & & K_A \updownarrow & & \updownarrow \alpha K_A \\
 & & B + L + RA & \overset{\alpha K_D}{\rightleftharpoons} & LRA
\end{array}
$$

Scheme 4 Interaction of an orthosteric, an allosteric, and a bitopic ligand

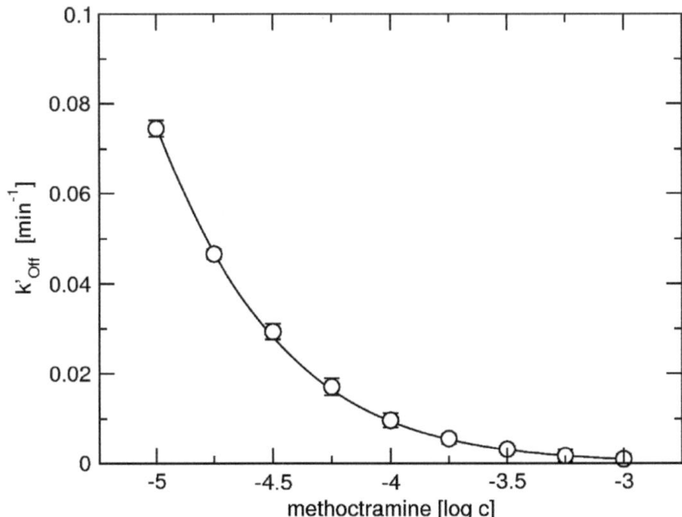

Fig. 12 Effects of an allosteric modulator on binding of a tracer at a fixed concentration in the presence of a non-labeled orthosteric ligand. Effects of a positive (*upper graph*) and negative (*lower graph*) allosteric modulator A at various concentrations indicated on the abscissa on binding of the tracer L at a fixed concentration indicated in the legend, in the absence (*black curves*) or in the presence of orthosteric ligand B at a fixed concentration indicated in the legend. Abscissa, concentration of allosteric modulator is expressed as logarithm of ratio to its equilibrium dissociation constant K_A. Ordinate, the tracer binding is expressed as a fraction of tracer binding in the absence of allosteric modulator. Legend, factors of cooperativity β of A and B binding. Orthosteric ligand B decreases tracer binding (*color curves*). In case of negative cooperativity between A and B (*red and green curves*) A concentration dependently weakens the binding of B that results in smaller inhibition of tracer binding (curves are getting closer to control curve). In case of positive cooperativity between A and B (*magenta and yellow curves*) A concentration dependently strengthens the binding of B that results in greater inhibition of tracer binding (curves are getting apart from control curve)

The orthosteric ligand L and allosteric modulator A can bind concurrently to the receptor R and form a ternary complex LRA. Binding of the bitopic ligand B is mutually exclusive both with binding of L and A. The ternary complex cannot be formed upon binding of B. Equation (16) does not fit the data as fractional binding of the tracer in the presence of A and B and the absence of A and B is:

$$Y = \frac{[L] + K_D}{[L] + K_D \times \frac{[B]K_A + K_B(K_A + [A])}{K_B(K_A + [A]/\alpha)}} \tag{17}$$

5 Analyzing Allosteric Modulation of Functional Responses

5.1 Effects of Allosteric Modulators on Functional Response Under Equilibrium

Muscarinic receptors are spontaneously active that is manifested by activation of second messenger pathways in the absence of agonists [47–49]. Such receptor spontaneous activity implicates that in the absence of the agonist there is a balance between two forms of the receptor (active R_A and inactive R_I) with non-zero R_A number. The thermodynamically complete description of interaction between a agonist and an allosteric modulator on functional receptor is thus described by the cubic ternary complex model (Scheme 5).

The receptor exists in an inactive conformation R_I and an active conformation R_A. The ratio of R_A to R_I is given by the activation constant K_{ACT}. The agonist L and allosteric modulator A bind to inactive receptor R_I with equilibrium dissociation constants K_D and K_A, respectively. Effects of an allosteric modulator on the functional response of the receptor to the agonist are complex. Besides allosteric modulation of agonist binding (factor of cooperativity α) an allosteric modulator affects spontaneous activation of the receptor (factor cooperativity γ) and agonist-induced receptor activation (factor of cooperativity δ). If effects of the allosteric modulator on spontaneous activation of the receptor are positive then the allosteric modulator activates receptors even in the absence of agonists. Such allosteric modulators have been identified [28–31]. The overall effect of an allosteric modulator on the formation of the ternary complex with the active receptor LR_AA is given by multiplication of these three factors of cooperativity (α, γ, δ). If the resultant of multiplication is greater than 1 then the effect of the allosteric modulator on agonist potency is negative even if L and A have positive binding cooperativity.

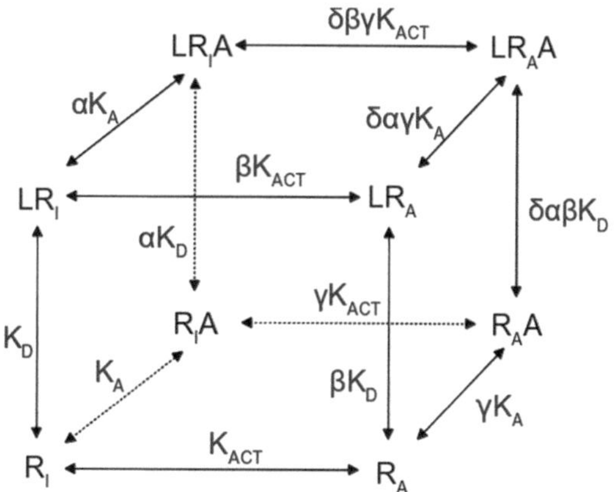

Scheme 5 Cubic ternary complex model

G-proteins bind to the receptor in both activation states (spontaneously active and agonist bound). Effects of agonists on G-protein binding to receptor are described by next cubic ternary complex model (Scheme 6).

G-protein G binds to the receptor in its inactive state R_I with equilibrium dissociation constant K_G. There is mutual allosteric modulation of G-protein binding and receptor activation (factor of cooperativity ε). Agonist L allosterically modulates binding of G to R_I (factor of cooperativity η), receptor activation (factor of cooperativity β) and G-protein binding induced by receptor activity (factor cooperativity ζ). The aggregate effect of agonist on G-protein binding is given by multiplication of these three factors of cooperativity (ζ, η, β). It is obvious that an increase in concentration of G (overexpression of G) leads to activation of receptor. Thus the aggregate effect of agonist is dependent on the receptor to G-protein ratio and is, generally speaking, system dependent.

Besides modulation of receptor activation and agonist binding described in Scheme 5 an allosteric agent can also allosterically modulate G-protein binding to the receptor and receptor agonist complexes (Scheme 7).

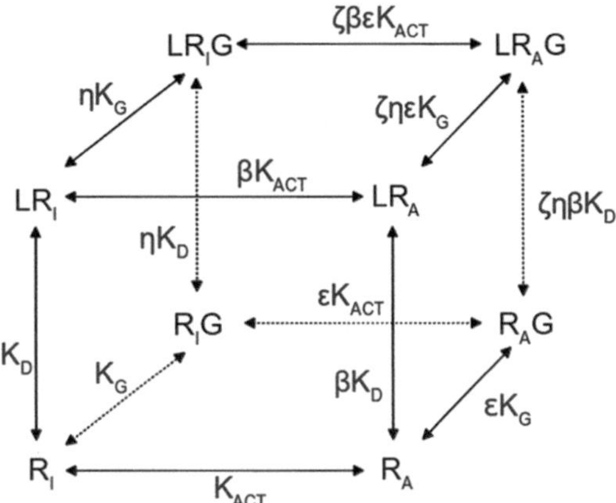

Scheme 6 Agonist effect on G-protein binding and receptor activation

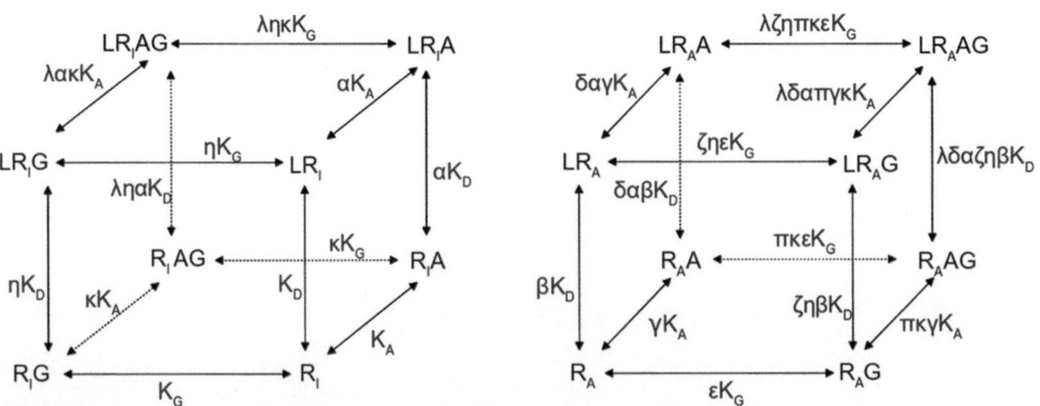

Scheme 7 Effect of an allosteric modulator on G-protein binding

G-proteins can bind both to receptors in inactive (Scheme 7, left cube) and active (Scheme 7, right cube) conformation. As described in Scheme 6 G-protein binds to free receptor in the inactive conformation with equilibrium dissociation constant K_G and to the active conformation with εK_G. In addition to modulation of agonist binding (α), receptor spontaneous activation (γ) and agonist-induced receptor activation (δ) an allosteric modulator affects G-protein binding to R_I (κ) and R_A ($\pi\kappa$) and effects of agonist on G-protein binding (λ). Similar to an agonist, effects of an allosteric modulator depend on the receptor to G-protein ratio. Moreover, effects of an allosteric modulator depend on the direction (activation or inhibition) and magnitude of agonist effects. The overall effect of an allosteric modulator is given by multiplication of all factors of cooperativity involved in transition from state in the absence of A (Scheme 6) to state in the presence of A (Scheme 7) (all factors of cooperativity except ε, β, η and ζ).

Conversions between complexes with bound G-protein have to be added for the scheme describing interactions between receptor, G-protein, agonist and allosteric modulator in order for the model to be thermodynamically complete (Scheme 8).

All eight receptor-G-protein complexes are interchangeable in a step-by-step manner with equilibrium dissociation constants resulting from Scheme 5 through 7. Moreover, G-protein activation is initiated by release of GDP from the G-protein as a result of negative cooperativity between agonist and GDP [50]. There are four receptor-G-protein complexes with bound allosteric modulator in the interaction scheme. Allosteric modulators may affect GDP affinity (and thus activation of G-protein) differently at these four complexes.

Effects of allosteric modulators on functional response depend on the nature of the agonist and system. The ratio of receptor to G proteins affects system basal activity. Systems with high R to G ratio have low basal activity and high receptor reserve. As a result agonists have high efficacy and potency. On the other hand,

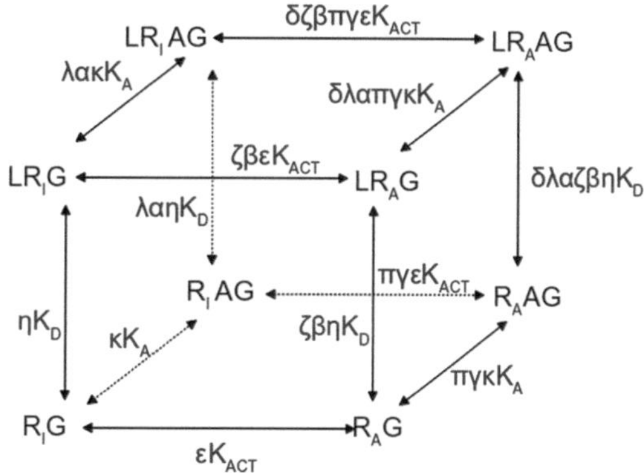

Scheme 8 Conversions between various receptor-G-protein complexes

systems with low R to G ratio have high basal activity and no receptor reserve. As a result agonists have both low efficacy and potency. Effects of both positive and negative allosteric modulators on agonist potency (shift in apparent K_G in the presence versus in the absence of L) (α, γ, δ) are weaker at systems with high R to G ratio due to high receptor reserve. High basal activity of the system decreases agonist efficacy (shift in ratio of active species in the presence versus in the absence of L). At systems with high basal activity effects of positive allosteric modulator on efficacy are weaker. On the other hand, effects of negative allosteric modulators on agonist efficacy may be stronger (if $\delta < \pi\kappa\lambda$).

Full agonists, due to strong positive cooperativity β, act as agonists at all systems (regardless of R to G ratio). Allosteric modulators may act as agonists (activate receptors in the absence of an orthosteric agonist) [28] or inverse agonists depending on the factor of cooperativity γ, activation constant K_{ACT} and receptor to G-protein ratio [51]. Allosteric modulators that have weak positive cooperativity γ act as partial agonists in a system with high R to G ratio and as inverse agonists at a low R to G ratio. As evident, effects of allosteric modulators on functional responses to an agonist are very complex. It is technically unfeasible to experimentally isolate and determine individual constants and factors of cooperativity. From a practical point of view only the overall effect of an allosteric modulator on the potency and efficacy of a given agonist in a given system could be determined.

5.2 Effects of Allosteric Modulators on the Kinetics of Functional Responses

Although a change in agonist potency induced by an allosteric modulator usually follows change in agonist affinity [52, 53] it has been reported that allosteric modulators may have different effects on agonist binding and agonist-mediated functional responses [25]. Moreover, effects of allosteric modulators on functional responses may also differ over time. This is exemplified by the dichotomous effects of the allosteric modulator rapacuronium on acetylcholine equilibrium binding on the one hand and on the kinetics of acetylcholine binding on the other hand [27]. For example, although rapacuronium exerts negative cooperativity with binding of acetylcholine to all muscarinic receptor subtypes at equilibrium it accelerates the rate of acetylcholine binding at odd-numbered subtypes. At low concentrations it transiently increases the potency and efficacy of functional responses to acetylcholine at odd-numbered subtypes (Fig. 13). The time between acetylcholine release and termination of its action by acetylcholinesterase is in the range of a fraction of a second. Therefore, effects of allosteric modulators in the early non-equilibrium stage of receptor signaling are physiologically more relevant than effects on acetylcholine equilibrium binding that does not occur in vivo. Thus fast functional assays that much better simulate physiological conditions are more suitable for screening of potential allosteric modulators of neurotransmission than long-lasting equilibrium binding experiments.

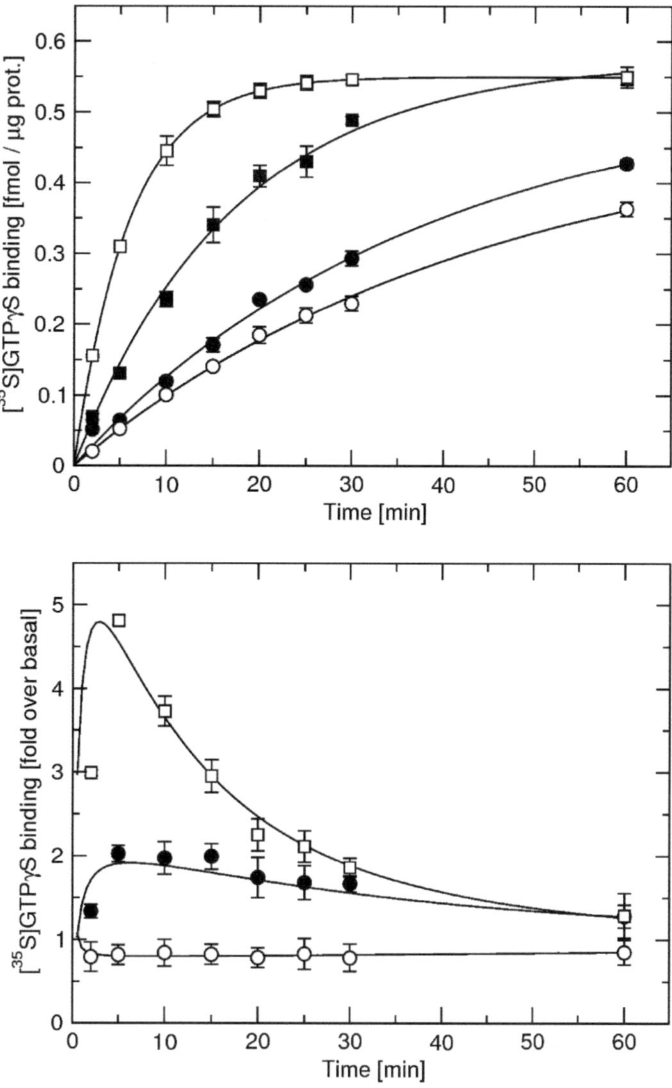

Fig. 13 Effects of rapacuronium on the kinetics of [^{35}S]GTPγS binding. Membranes were preincubated for 60 min in the presence (*open symbols*) or in the absence (*closed symbols*) of 1 μM rapacuronium. Then [^{35}S]GTPγS was added simultaneously with buffer (*circles*) or 10 μM acetylcholine (*squares*). Incubations were terminated at the times indicated on the abscissa. The increase of specific [^{35}S]GTPγS binding is expressed as fmol per μg of protein (*top*) and as fold increase of specific binding under basal conditions (*bottom*). Data are means ± SE of values from three independent experiments performed in quadruplicates

The advantage of [^{35}S]GTPγS binding as a measure of receptor functional response (Fig. 13) is that it can be easily scaled up for high-throughput screening. Its disadvantage is that an agonist must be present during incubation lasting minutes that is still far from physiological conditions. Moreover, the resulting signal is the sum of functional response over the whole time-course of incubation.

Methods that allow real-time measurement of functional responses and transient application of agonist are more appropriate. Such methods include microfluorometry of intracellular calcium release [54] (see Protocol C) or measurement of conformation changes of the receptor by fluorescence resonance energy transfer between two dyes attached to one receptor (*see* Chapter 8).

Protocol C: Measurement of allosteric modulation of a functional response by microfluorometry of intracellular calcium

1. Seed CHO cells stably expressing muscarinic receptors on 24 mm-diameter microscope cover glass in 35 mm-diameter Petri dish and cultivate them until about 80 % confluency.
2. Optional: If CHO cells express M_2 or M_4 receptors transfect cells with $G_{15/16}$ G-protein to couple these subtypes to phospholipase C [55].
3. Prepare DMSO solutions of 2 mM Fura-2AM and 20 % Pluronic F-68 and mix them 1:1.
4. Wash cells with warm Krebs-HEPES buffer (KHB; final concentrations in mM: NaCl 138; KCl 4; $CaCl_2$ 1.3; $MgCl_2$ 1; NaH_2PO_4 1.2; Hepes 20; glucose 10; Probenecid 1; pH adjusted to 7.4).
5. Load cells with Fura-2 by incubating them in 0.5 ml KHB and 10 μl of solution from step 4 for 1 h at 37 °C.
6. Remove KHB, wash cells with fresh KHB.
7. Assemble the cover glass in superfusion chamber and place under a fluorescence microscope. Record Fura-2 emission (>470 nm) at 380 and 340 nm emissions twice a second.
8. Expose cells to increasing concentrations of an agonist (e.g., carbachol or acetylcholine 10 nM to 10 μM) for 5 s. Allow cells to rest for 3–5 min between stimuli.
9. Analyze the ratio of Fura-2 emissions at 380 and 340 nm excitation to determine half-efficient concentration (EC_{50}) and maximum stimulation.
10. Expose cells to the tested allosteric modulator (1 μM to 1 mM) for 5 s to check for possible agonist/inverse agonist effects of the allosteric modulator by itself. Allow cells to rest for 3–5 min between stimuli.
11. To determine the effects of an allosteric modulator on agonist potency expose cells for 5 s to agonist at a concentration around its EC_{50} (for positive cooperativity slightly below and for negative cooperativity slightly above EC_{50}) alone and then in mixture with increasing concentrations of the tested allosteric modulator. Allow cells to rest for 3–5 min between stimuli.
12. To determine effects of allosteric modulator on maximum response to agonist expose cells for 5 s to agonist in saturating

concentration alone and then in mixture with increasing concentrations of the tested allosteric modulator. Allow cells to rest for 3–5 min between stimuli.

6 Conclusions

Allosteric modulation of muscarinic receptors is an interesting phenomenon with great potential for drug discovery and pharmaceutical application. However, detailed studies and understanding are limited to simple systems due to complexity of allosteric interactions. Another limitation is that allosteric modulators of muscarinic receptors generally have affinity that is too low to make them suitable radiolabeled tracers. This necessitates complex experimental arrangements to quantify binding parameters of these agents.

Acknowledgments

This research was supported by Academy of Sciences of the Czech Republic support RVO: 67985823 and Grant Agency of the Czech Republic grant P304/12/G069.

References

1. Monod J, Changeux JP, Jacob F (1963) Allosteric proteins and cellular control systems. J Mol Biol 6:306–329
2. Monod J, Wyman J, Changeux JP (1965) On the nature of allosteric transitions: a plausible model. J Mol Biol 12:88–118
3. Clark AL, Mitchelson F (1976) The inhibitory effect of gallamine on muscarinic receptors. Br J Pharmacol 58:323–331
4. Stockton JM, Birdsall NJ, Burgen AS, Hulme EC (1983) Modification of the binding properties of muscarinic receptors by gallamine. Mol Pharmacol 23:551–557
5. Dalton DW, Tyers MB (1982) A comparison of the muscarinic antagonist actions of pancuronium and alcuronium. J Auton Pharmacol 2:261–266
6. Waelbroeck M, Robberecht P, De Neef P, Christophe J (1984) Effects of verapamil on the binding properties of rat heart muscarinic receptors: evidence for an allosteric site. Biochem Biophys Res Commun 121:340–345
7. Lai WS, Ramkumar V, El-Fakahany EE (1985) Possible allosteric interaction of 4-aminopyridine with rat brain muscarinic acetylcholine receptors. J Neurochem 44:1936–1942
8. Kloog Y, Sokolovsky M (1985) Allosteric interactions between muscarinic agonist binding sites and effector sites demonstrated by the use of bisquaternary pyridinium oximes. Life Sci 36:2127–2136
9. Nedoma J, Tucek S, Danilov AF, Shelkovnikov SA (1986) Stabilization of antagonist binding to cardiac muscarinic acetylcholine receptors by gallamine and other neuromuscular blocking drugs. J Pharmacol Exp Ther 236:219–223
10. Flynn DD, Mash DC (1989) Multiple in vitro interactions with and differential in vivo regulation of muscarinic receptor subtypes by tetrahydroaminoacridine. J Pharmacol Exp Ther 250:573–581
11. Jakubík J, Bačáková L, El-Fakahany EE, Tuček S (1997) Positive cooperativity of acetylcholine and other agonists with allosteric ligands on muscarinic acetylcholine receptors. Mol Pharmacol 52:172–179
12. Lazareno S, Popham A, Birdsall NJ (2000) Allosteric interactions of staurosporine and other indolocarbazoles with N-[methyl-(3)H] scopolamine and acetylcholine at muscarinic receptor subtypes: identification of a second allosteric site. Mol Pharmacol 58:194–207

13. Leppik RA, Miller RC, Eck M, Paquet JL (1994) Role of acidic amino acids in the allosteric modulation by gallamine of antagonist binding at the m2 muscarinic acetylcholine receptor. Mol Pharmacol 45:983–990
14. Matsui H, Lazareno S, Birdsall NJ (1995) Probing of the location of the allosteric site on m1 muscarinic receptors by site-directed mutagenesis. Mol Pharmacol 47:88–98
15. Krejčí A, Tuček S (2001) Changes of cooperativity between N-methylscopolamine and allosteric modulators alcuronium and gallamine induced by mutations of external loops of muscarinic M_3 receptors. Mol Pharmacol 60:761–767
16. Voigtländer U, Jöhren K, Mohr M, Raasch A, Tränkle C, Buller S, Ellis J, Höltje H, Mohr K (2003) Allosteric site on muscarinic acetylcholine receptors: identification of two amino acids in the muscarinic M_2 receptor that account entirely for the M_2/M_5 subtype selectivities of some structurally diverse allosteric ligands in N-methylscopolamine-occupied receptors. Mol Pharmacol 64:21–31
17. Jakubík J, Krejčí A, Doležal V (2005) Asparagine, valine, and threonine in the third extracellular loop of muscarinic receptor have essential roles in the positive cooperativity of strychnine-like allosteric modulators. J Pharmacol Exp Ther 313:688–696
18. Huang X, Prilla S, Mohr K, Ellis J (2005) Critical amino acid residues of the common allosteric site on the M_2 muscarinic acetylcholine receptor: more similarities than differences between the structurally divergent agents gallamine and bis(ammonio)alkane-type hexamethylene-bis-[dimethyl-(3-phthalimidopropyl)ammonium]dibromide. Mol Pharmacol 68:769–778
19. Tränkle C, Dittmann A, Schulz U, Weyand O, Buller S, Jöhren K, Heller E, Birdsall NJM, Holzgrabe U, Ellis J, Höltje HD, Mohr K (2005) Atypical muscarinic allosteric modulation: cooperativity between modulators and their atypical binding topology in muscarinic M_2 and M_2/M_5 chimeric receptors. Mol Pharmacol 68:1597–1610
20. Prilla S, Schrobang J, Ellis J, Höltje H, Mohr K (2006) Allosteric interactions with muscarinic acetylcholine receptors: complex role of the conserved tryptophan M_2 422Trp in a critical cluster of amino acids for baseline affinity, subtype selectivity, and cooperativity. Mol Pharmacol 70:181–193
21. Jakubík J, El-Fakahany EE (2010) Allosteric modulation of muscarinic acetylcholine receptors. Pharmaceuticals 9:2838–2860
22. Kruse AC, Ring AM, Manglik A, Hu J, Hu K, Eitel K, Hübner H, Pardon E, Valant C, Sexton PM, Christopoulos A, Felder CC, Gmeiner P, Steyaert J, Weis WI, Garcia KC, Wess J, Kobilka BK (2013) Activation and allosteric modulation of a muscarinic acetylcholine receptor. Nature 504:101–106
23. Fisher A (2012) Cholinergic modulation of amyloid precursor protein processing with emphasis on M1 muscarinic receptor: perspectives and challenges in treatment of Alzheimer's disease. J Neurochem 120(Suppl 1):22–33
24. Jones CK, Byun N, Bubser M (2012) Muscarinic and nicotinic acetylcholine receptor agonists and allosteric modulators for the treatment of schizophrenia. Neuropsychopharmacology 37:16–42
25. Zahn K, Eckstein N, Tränkle C, Sadée W, Mohr K (2002) Allosteric modulation of muscarinic receptor signaling: alcuronium-induced conversion of pilocarpine from an agonist into an antagonist. J Pharmacol Exp Ther 301: 720–728
26. Jäger D, Schmalenbach C, Prilla S, Schrobang J, Kebig A, Sennwitz M, Heller E, Tränkle C, Holzgrabe U, Höltje H, Mohr K (2007) Allosteric small molecules unveil a role of an extracellular E2/transmembrane helix 7 junction for G protein-coupled receptor activation. J Biol Chem 282:34968–34976
27. Jakubík J, Randaková A, El-Fakahany EE, Doležal V (2009) Divergence of allosteric effects of rapacuronium on binding and function of muscarinic receptors. BMC Pharmacol 9:15
28. Jakubík J, Bačáková L, Lisá V, El-Fakahany EE, Tuček S (1996) Activation of muscarinic acetylcholine receptors via their allosteric binding sites. Proc Natl Acad Sci U S A 93: 8705–8709
29. Lebois EP, Bridges TM, Lewis LM, Dawson ES, Kane AS, Xiang Z, Jadhav SB, Yin H, Kennedy JP, Meiler J, Niswender CM, Jones CK, Conn PJ, Weaver CD, Lindsley CW (2010) Discovery and characterization of novel subtype-selective allosteric agonists for the investigation of M_1 receptor function in the central nervous system. ACS Chem Neurosci 1:104–121
30. Lebois EP, Digby GJ, Sheffler DJ, Melancon BJ, Tarr JC, Cho HP, Miller NR, Morrison R, Bridges TM, Xiang Z, Daniels JS, Wood MR, Conn PJ, Lindsley CW (2011) Development of a highly selective, orally bioavailable and CNS penetrant M_1 agonist derived from the MLPCN probe ML071. Bioorg Med Chem Lett 21:6451–6455
31. Digby GJ, Noetzel MJ, Bubser M, Utley TJ, Walker AG, Byun NE, Lebois EP, Xiang Z, Sheffler DJ, Cho HP, Davis AA, Nemirovsky NE, Mennenga SE, Camp BW, Bimonte-Nelson

HA, Bode J, Italiano K, Morrison R, Daniels JS, Niswender CM, Olive MF, Lindsley CW, Jones CK, Conn PJ (2012) Novel allosteric agonists of M_1 muscarinic acetylcholine receptors induce brain region-specific responses that correspond with behavioral effects in animal models. J Neurosci 32:8532–8544

32. Jakubík J, Bačáková L, El-Fakahany EE, Tuček S (1995) Subtype selectivity of the positive allosteric action of alcuronium at cloned M_1-M_5 muscarinic acetylcholine receptors. J Pharmacol Exp Ther 274:1077–1083
33. Lazareno S, Popham A, Birdsall NJM (2002) Analogs of WIN 62,577 define a second allosteric site on muscarinic receptors. Mol Pharmacol 62:1492–1505
34. Lazareno S, Doležal V, Popham A, Birdsall NJM (2004) Thiochrome enhances acetylcholine affinity at muscarinic M_4 receptors: receptor subtype selectivity via cooperativity rather than affinity. Mol Pharmacol 65:257–266
35. Tränkle C, Mies-Klomfass E, Cid MH, Holzgrabe U, Mohr K (1998) Identification of a [^{3}H]Ligand for the common allosteric site of muscarinic acetylcholine M_2 receptors. Mol Pharmacol 54:139–145
36. Lysíková M, Fuksová K, Elbert T, Jakubík J, Tuček S (1999) Subtype-selective inhibition of [methyl-^{3}H]-N-methylscopolamine binding to muscarinic receptors by alpha-truxillic acid esters. Br J Pharmacol 127:1240–1246
37. Jerusalinsky D, Cerveñasky C, Peña C, Raskovsky S, Dajas F (1992) Two polypeptides from Dendroaspis angusticeps venom selectively inhibit the binding of central muscarinic cholinergic receptor ligands. Neurochem Int 20:237–246
38. Jolkkonen M, Adem A, Hellman U, Wernstedt C, Karlsson E (1995) A snake toxin against muscarinic acetylcholine receptors: amino acid sequence, subtype specificity and effect on guinea-pig ileum. Toxicon 33:399–410
39. Waelbroeck M, De Neef P, Domenach V, Vandermeers-Piret MC, Vandermeers A (1996) Binding of the labelled muscarinic toxin ^{125}I-MT1 to rat brain muscarinic M_1 receptors. Eur J Pharmacol 305:187–192
40. Fruchart-Gaillard C, Mourier G, Marquer C, Ménez A, Servent D (2006) Identification of various allosteric interaction sites on M_1 muscarinic receptor using ^{125}I-Met35-oxidized muscarinic toxin 7. Mol Pharmacol 69: 1641–1651
41. Ilien B, Franchet C, Bernard P, Morisset S, Weill CO, Bourguignon J, Hibert M, Galzi J (2003) Fluorescence resonance energy transfer to probe human M1 muscarinic receptor structure and drug binding properties. J Neurochem 85:768–778
42. Jakubík J, Zimčík P, Randáková A, Fuksová K, El-Fakahany EE, Doležal V (2014) Molecular mechanisms of methoctramine binding and selectivity at muscarinic acetylcholine receptors. Mol Pharmacol 86:180–192
43. Proška J, Tuček S (1994) Mechanisms of steric and cooperative actions of alcuronium on cardiac muscarinic acetylcholine receptors. Mol Pharmacol 45:709–717
44. Melchiorre C, Minarini A, Angeli P, Giardinà D, Gulini U, Quaglia W (1989) Polymethylene tetraamines as muscarinic receptor probes. Trends Pharmacol Sci. Suppl: 55–59
45. Tahtaoui C, Parrot I, Klotz P, Guillier F, Galzi J, Hibert M, Ilien B (2004) Fluorescent pirenzepine derivatives as potential bitopic ligands of the human M_1 muscarinic receptor. J Med Chem 47:4300–4315
46. Daval SB, Valant C, Bonnet D, Kellenberger E, Hibert M, Galzi J, Ilien B (2012) Fluorescent derivatives of AC-42 to probe bitopic orthosteric/allosteric binding mechanisms on muscarinic M_1 receptors. J Med Chem 55:2125–2143
47. Burstein ES, Spalding TA, Braüner-Osborne H, Brann MR (1995) Constitutive activation of muscarinic receptors by the G-protein Gq. FEBS Lett 363:261–263
48. Jakubík J, Bačáková L, El-Fakahany EE, Tuček S (1995) Constitutive activity of the M_1-M_4 subtypes of muscarinic receptors in transfected CHO cells and of muscarinic receptors in the heart cells revealed by negative antagonists. FEBS Lett 377:275–279
49. Burstein ES, Spalding TA, Brann MR (1997) Pharmacology of muscarinic receptor subtypes constitutively activated by G proteins. Mol Pharmacol 51:312–319
50. Jakubík J, Janíčková H, El-Fakahany EE, Doležal V (2011) Negative cooperativity in binding of muscarinic receptor agonists and GDP as a measure of agonist efficacy. Br J Pharmacol 162:1029–1044
51. Jakubík J, Haga T, Tuček S (1998) Effects of an agonist, allosteric modulator, and antagonist on guanosine-gamma-[^{35}S]thiotriphosphate binding to liposomes with varying muscarinic receptor/G_o protein stoichiometry. Mol Pharmacol 54:899–906
52. Birdsall NJ, Farries T, Gharagozloo P, Kobayashi S, Lazareno S, Sugimoto M (1999) Subtype-selective positive cooperative interactions between brucine analogs and acetylcholine at muscarinic receptors: functional studies. Mol Pharmacol 55:778–786
53. Lazareno S, Birdsall B, Fukazawa T, Gharagozloo P, Hashimoto T, Kuwano H, Popham A, Sugimoto M, Birdsall NJ (1999) Allosteric effects of four stereoisomers of

a fused indole ring system with 3H-N-methylscopolamine and acetylcholine at M_1-M_4 muscarinic receptors. Life Sci 64:519–526

54. Santrůčková E, Doležal V, El-Fakahany EE, Jakubík J (2014) Long-term activation upon brief exposure to xanomeline is unique to M_1 and M_4 subtypes of muscarinic acetylcholine receptors. PLoS One 9, e88910

55. Milligan G, Marshall F, Rees S (1996) G_{16} as a universal G protein adapter: implications for agonist screening strategies. Trends Pharmacol Sci 17:235–237

Chapter 7

Subcellular and Synaptic Localization of Muscarinic Receptors in Neurons Using High-Resolution Electron Microscopic Preembedding Immunogold Technique

Véronique Bernard

Abstract

The function of a G protein-coupled receptor in the modulation of neuronal activity is highly dependent on its availability on the cell surface, on its distribution among different subcellular compartments and in relationship with the presynaptic afferents. Therefore, investigation of the precise localization of GPCRs is required to clarify their contribution to neuronal function, and can be achieved only by immunoelectron microscopy. Here, we describe the high-resolution electron microscopic preembedding immunogold technique that we have developed to analyze the subcellular and synaptic distribution of two acetylcholine muscarinic receptors (MR), M_2 and M_4 MRs in neurons in vivo. We have shown that M_2MR and M_4MR are mostly located at the plasma membrane where they are in a right position to interact with acetylcholine to modulate neuronal function. The synaptic and extrasynaptic localization of M_2MR suggests that the effect of acetylcholine might be mediated through a synaptic as well as diffuse type of transmission. The demonstration that M_2MR are present at the postsynaptic membrane beneath glutamatergic terminals provides a direct argument in favor of a co-release of ACh and glutamate. Finally, we have shown that muscarinic receptors are subject to an intraneuronal trafficking when they are stimulated and that this trafficking is different according to the duration of the stimulation (acute versus chronic).

Key words M_2 muscarinic receptor, M_4 muscarinic receptor, Acetylcholine, Immunogold, Synapse, Subcellular localization, Confocal microscopy, Electron microscopy

1 Introduction

Neuronal functions are determined by the highly precise arrangements of presynaptic and postsynaptic elements. G protein-coupled receptor (GPCRs) including MRs, are located to specific presynaptic or postsynaptic sites. The location of MRs within different neuronal compartments has a variety of functional implications, i.e., neurotransmitter release regulation at axonal level [1] or modulation of neuronal excitability at somatodendritic level [2]. Therefore, the precise localization and density of GPCRs on the cell surface seems to be a critical factor for specificity of signaling within and between

Jaromir Myslivecek and Jan Jakubik (eds.), *Muscarinic Receptor: From Structure to Animal Models*, Neuromethods, vol. 107,
DOI 10.1007/978-1-4939-2858-3_7, © Springer Science+Business Media New York 2016

neurons, which can be determined only using high-resolution morphological approaches.

The development of morphological approaches at high-resolution provides a wonderful tool to locate the sites of acetylcholine action, i.e., acetylcholine receptors in normal neurochemical environment and the redistribution of these sites when the cholinergic tone is modified. Especially, this procedure allowed us to identify the localization with a high resolution of two muscarinic receptors, M_2MR and M_4MR and to analyze their trafficking in different subcellular and synaptic neuronal compartments.

We will describe in this chapter how to prepare samples to analyze subcellular and synaptic localization of muscarinic receptors and more generally membrane bound proteins.

2 Materials

2.1 Buffers and Solutions

0.2 M Phosphate Buffer (PB), pH 7.4.

Solution A: 0.2 M Na_2HPO_4-$2H_2O$ (35.6 g/l in distilled water)	800 ml
Solution B: 0.2 M NaH_2PO4-$2H_2O$ (31.2 g/l in distilled water)	200 ml

Adjust at pH = 7.4.
Store at room temperature for up to 2 months.

2.1.1 Phosphate Buffered Saline (PBS)

0.2 M PB	50 ml
NaCl	8.76 g
KCl	0.2 g
Distilled water	up to 1 l

Store at room temperature for up to 2 months.

2.1.2 2 % Paraformaldehyde

Hot distilled water (60 °C, not more)	500 ml
Add paraformaldehyde (Ref: 145.004005.60, Merck)	20 g

Stir until the solution is clear.

Add PB 0.2 M, pH 7.4	500 ml

Store at 4 °C for 1 week.

2.1.3 2 % Paraformaldehyde + 0.2 % Glutaraldehyde

2 % paraformaldehyde	1 l
25 % glutaraldehyde	8 ml

(Glutaraldehyde 25 % for electron microscopy, Ref: 49626, Fluka). Stir and use immediately.

2.2 Cryoprotectant

0.2 M PB pH = 7.4	25 ml
Glycerol (Ref: 24397296, Prolabo)	10 ml
Sucrose (Ref: S7903, Sigma)	25 g
Distilled water	75 ml

Stir to dissolve and store at 4 °C for 1 week.

2.2.1 0.1 M Sodium Acetate Buffer pH = 7

Sodium acetate trihydrate (Ref: S8625, Sigma)	13.6 g
Distilled water	1 l

Adjust at pH = 7 with acetic acid.
Store at 4 °C for up to 1 month.

2.2.2 Tris Buffer

Tris Base (Ref: T1503; Sigma)	6 g
Distilled water	1 l

Adjust at pH = 7.6 with HCl.
Store at room temperature for up to 2 months.

2.3 Pioloform (for Support Film on EM Grids)

Pioloform powder (Ref: R1275; Agar Scientific)	1 g
Chloroform	100 ml

Stir to dissolve.
Store at 4 °C for up to 2 months.

2.4 Resin Durcupan

Single component A (M epoxy resin; Sigma 44611; Sigma-Aldrich)	10 g
Single component B (hardener 964; Sigma 44612; Sigma-Aldrich)	10 g
Single component C (accelerator 960; Sigma 44613; Sigma-Aldrich)	0.3 ml
Single component D (Sigma 44614; Sigma-Aldrich)	0.2 ml

In a fume hood, combine all components in a disposable beaker. Stir until well mixed. Pour in small weighing cup and store until use.

2.5 Materials

- Perfusion stand.
- Vibrating microtome.
- 6 or 12-well plates.
- Mesh baskets fitting with the 6 or 12-well plates.

- Thin paint brush.
- EM copper grids one 2 mm × 1 mm slot (Ref: G2010-Cu; Electron Microscopy Science, Hatfield, USA).
- Thin tweezers.
- Ultramicrotome.
- Antibodies:
  - Anti-Muscarinic Acetylcholine Receptor m2 Antibody (rat), clone M2-2-B3 Ref: MAB367, Millipore.
  - Anti-Muscarinic Acetylcholine Receptor m4 Antibody (mouse), clone 17F10.2, Ref: MAB1576, Millipore.
- Nanogold® Conjugates, Nanoprobes, Yaphank, NY, USA.
  - Nanogold-Anti rat (m2R); Raised in goat, Ref: 2007: IgG molecule.
  - Nanogold-Anti mouse (m4R); Raised in goat, Ref: 2001: IgG molecule.
  - Nanogold-Streptavidin; Ref: 2016: Streptavidin.
- HQ Silver™ Enhancement Kit, Ref: 2012; Nanoprobes, Yaphank, NY, USA.

3 Sequence of Procedures for Detection of mAChRs

3.1 Common Steps for Detection of MR at Light and Electron Microscopic Levels

- Perfuse-fix the animal using a gravity system with 10 ml of NaCl (room temperature) and then with a cold 2 % PFA and 0.2 % glutaraldehyde solution (100 ml for 15 min).
- Remove the brain from the skull and post-fix it in 2 % PFA alone overnight at 4 °C.
- Rinse the brain and store it in PBS at 4 °C until use (may be stored for some weeks).
- Cut sections on a vibrating microtome at 50–70 μm. Use small mesh *baskets* in a 6-well plate and put all the sections from the brain area of interest in a same well. Alternatively, serial sections may be cut. The baskets will allow to easily perform the first steps of freeze–thaw and washes.
- Carry out a freeze–thaw procedure to enhance penetration of immunoreagents:
  - Before freezing, the sections are equilibrated in a cryoprotection solution for 15 min.
  - Then, excess of cryoprotectant is absorbed with a paper towel and the sections are dropped for 5 s into isopentane cooled either in liquid nitrogen or in dry ice.
  - The sections are thawed in the cryoprotectant and then rinsed in PBS 3 times for 5 min.

- Subject sections to immunohistochemistry of muscarinic receptors alone or muscarinic receptors and a marker of cholinergic or non-cholinergic terminals (cholinergic or non-cholinergic [3]) or other markers [4].
  - Blockade of the nonspecific binding with 4 % normal serum for 30 min.
  - Incubate with the primary antibody overnight at RT under shaking. When two targets are co-detected, incubate in a mix of two primary antibodies.

3.2 Detection of mAChRs at Light Microscopic Level

3.2.1 Incubation with the Secondary Fluorescent Antibody

When the receptor only is detected, incubate the secondary antibody coupled to a fluorochrome (Alexa 594 anti-rat for M_2 MR or Alexa 594 anti-mouse for M_4 MR; dilution 1:1000; Life Technologies) for 1 h. Alternatively, if the signal for the receptor is weak, it may be intensified using a biotinylated secondary antibody (dilution 1:200; Vector Laboratories) for 1 h and then a streptavidin coupled to a fluorochrome (Alexa 594 streptavidin; dilution 1:1000; Life Technologies) for 1 h. When two antibodies are used, incubate in a mix of two antibodies coupled to a fluorochrome (Alexa 594 anti-rat for M_2 MR or Alexa 594 anti-mouse for M_4MR; dilution 1:1000; Life Technologies and Alexa 488 anti-species; dilution 1:1000; Life Technologies) for 1 h. Finally, rinse the sections in PBS. Mount the sections in VECTASHIELD (Vector Laboratories, Burlingame, CA, USA).

3.2.2 Observation of the Sections at the Confocal Microscope

The sections were observed using a fully automated upright Leica TCS SP5 fluorescence microscope equipped with a 63x oil immersion lens (numerical aperture, 1.25) and with a Leica SP5 scanning system equipped a Ar white-light laser that allows choosing any excitation wavelength from 470 to 670 nm. We choose 488 nm and 543 nm wavelength to detect Alexa488 and Al594 (Leica Microsystems, Deerfield, IL; USA). Images were treated using ImageJ and Adobe Photoshop softwares.

3.3 Detection of mAChRs at Electron Microscopic Level

3.3.1 Detection of MR

- Incubate with the secondary fluorescent antibody coupled either to a nanogold particle (1.4 nm, Nanoprobes, Yaphank, NY, USA) or to biotin (dilution 1:200; Vector Laboratories). When the secondary antibody is coupled to biotin, incubate the sections with a streptavidin coupled to a nanogold particle (1.4 nm, Nanoprobes, Yaphank, NY, USA).
- Rinse in PBS and post-fix with 1 % glutaraldehyde for 10 min.
- Rinse in PBS and incubate in 0.1 M sodium acetate, pH = 7.0, until silver intensification.
- Increase the diameter of the nanogold particle with silver intensification (HQ Silver™ Enhancement Kit, Nanoprobes, Yaphank, NY, USA; one drop of red, one drop of white, one drop of blue). Incubate in the mix for 3 min.
- Post-fix sections for EM in 1 % osmium tetroxide.

- Dehydrate sections (50 %, 10 min; 70 % + 0.1 % uranyl acetate, 25 min; 95 %, 10 min; 100 %, two times 10 min; propylene oxide, 10 min) and flat-embed in resin between two microscope slides coated with Sigmacote (Sigma SL2, Sigma Aldrich).
- Make the resin polymerize in an oven (60 °C) for 2 days.
- Examine the sections in the light microscope, cut and glue the areas of interest, cut ultrathin sections, and examine in the electron microscope.
- Quantify mAChR localization if needed.

3.3.2 Co-detection of MR and Another Protein of Interest

- When another protein of interest is detected, the secondary antibody against the receptor primary antibody is coupled to a nanogold particle and the secondary Ab against this protein is coupled to biotin. Both antibody are co-incubated.
- After silver intensification of the immunogold signal, incubate the sections in streptavidin coupled to horse radish peroxidase (HRP).
- Reveal HRP with diamino benzidine (DAB).
- Wash in PBS.
- Post-fix sections for EM in osmium tetroxide 1 % diluted in distilled water.
- Dehydrate sections (see above) and flat-embed in resin between two microscope slides.
- Make the resin polymerize in an oven (60 °C) for 2 days.
- Examine the sections in the light microscope, cut and glue the areas of interest (no more than 1 mm^2) at the top of a blank piece of polymerized resin, cut semi-thin and ultrathin sections and examine in the electron microscope.
- Quantify MR localization if needed.

4 Results

4.1 Subcellular Localization of M_2 MR and M_4 MR in Normal Conditions

The M_2 MR and M_4 MR are located mostly at the membrane of cell bodies and dendrites, where they are in a right position to interact with acetylcholine to modulate neuronal excitability. (Figs. 1, 2, 3, and 5a). The M_2 MR is also found at the axonal

Fig. 1 (continued) the striatum of perikarya of control rats using preembedding immunogold method with silver intensification. Proportion of immunoparticles associated with different subcellular neuronal compartments. For each neuron, the number of immunoparticles associated with each compartment was counted, and the proportion in relation to the total number was calculated. The largest portion of immunoparticles are associated with the plasma membrane [1]. In the cytoplasm, the immunoparticles are detected in association primarily with small vesicles [2] and endoplasmic reticulum [5]. A small proportion of immunoparticles are associated with the Golgi apparatus [4] and multivesicular bodies [3]. Some immunoparticles are not seen in association with any identified compartment [6]. *b* bouton, *n* nucleus, *G* Golgi apparatus. From Bernard et al. [11]

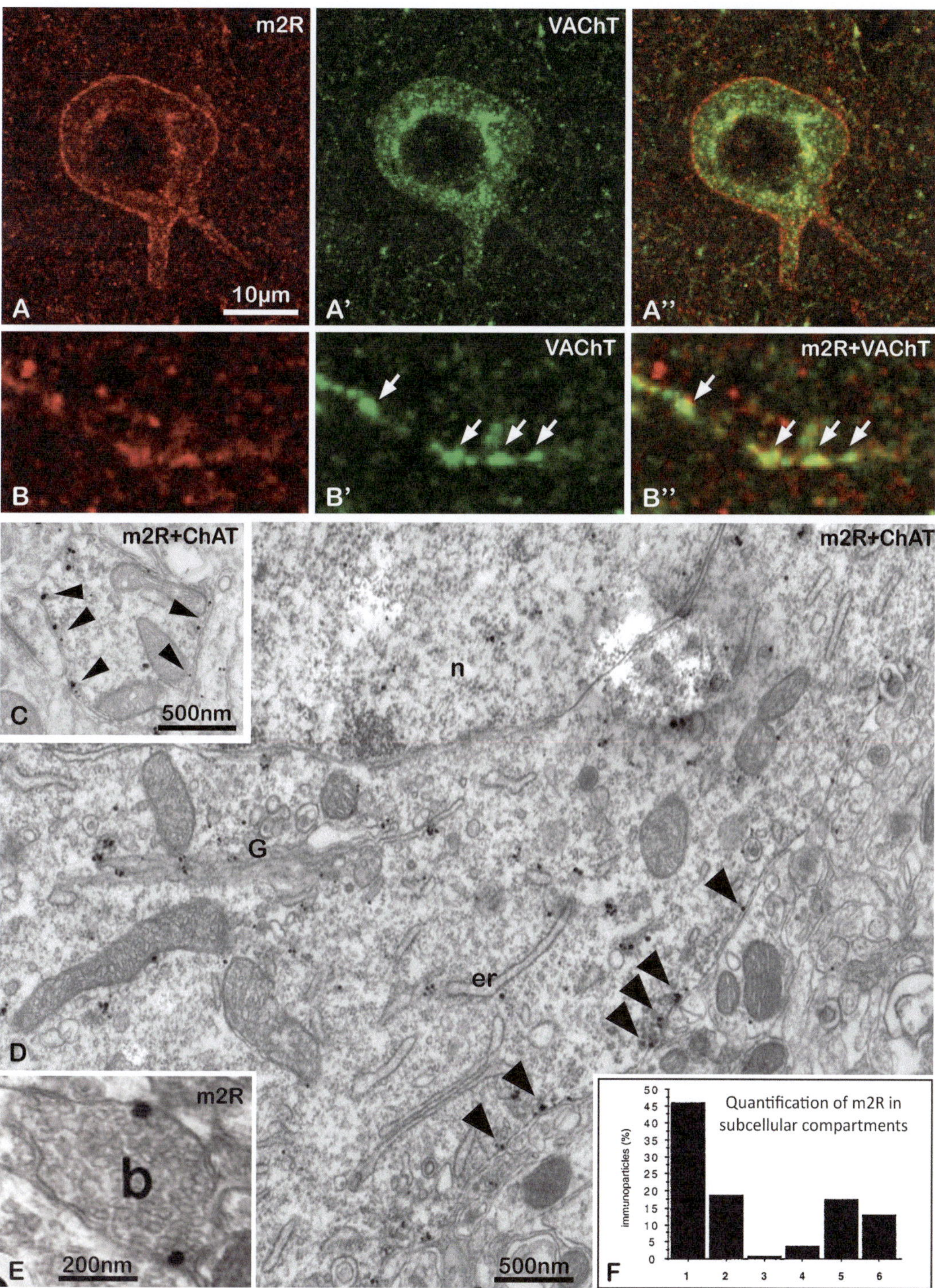

Fig. 1 Cellular and subcellular distribution of M_2 MR in the striatum of a normal animal. (**A–B″**) In a control mouse, the simultaneous detection of M_2 MR and VAChT immunoreactivities at confocal microscopic level shows that m2R is located at the plasma membrane of a cholinergic perikaryon (**A–B″**) and of cholinergic varicosities (**B–B″**). (**C–E**) Subcellular distribution of M_2 MR immunoreactivity in the striatum of rats using preembedding immunogold method with silver intensification. Detail of M_2 MR immunolabeling in the cytoplasm of cell bodies (**D**), a dendrite (**C**) and an axon (**E**). Numerous immunoparticles are associated with the plasma membrane (*arrow heads*) of the perikaryon, dendrite and axon. Some immunoparticles are associated with cytoplasmic compartments like endoplasmic reticulum (er) and Golgi apparatus (G). (**F**) Quantitative analysis of the subcellular distribution of M_2 MR in

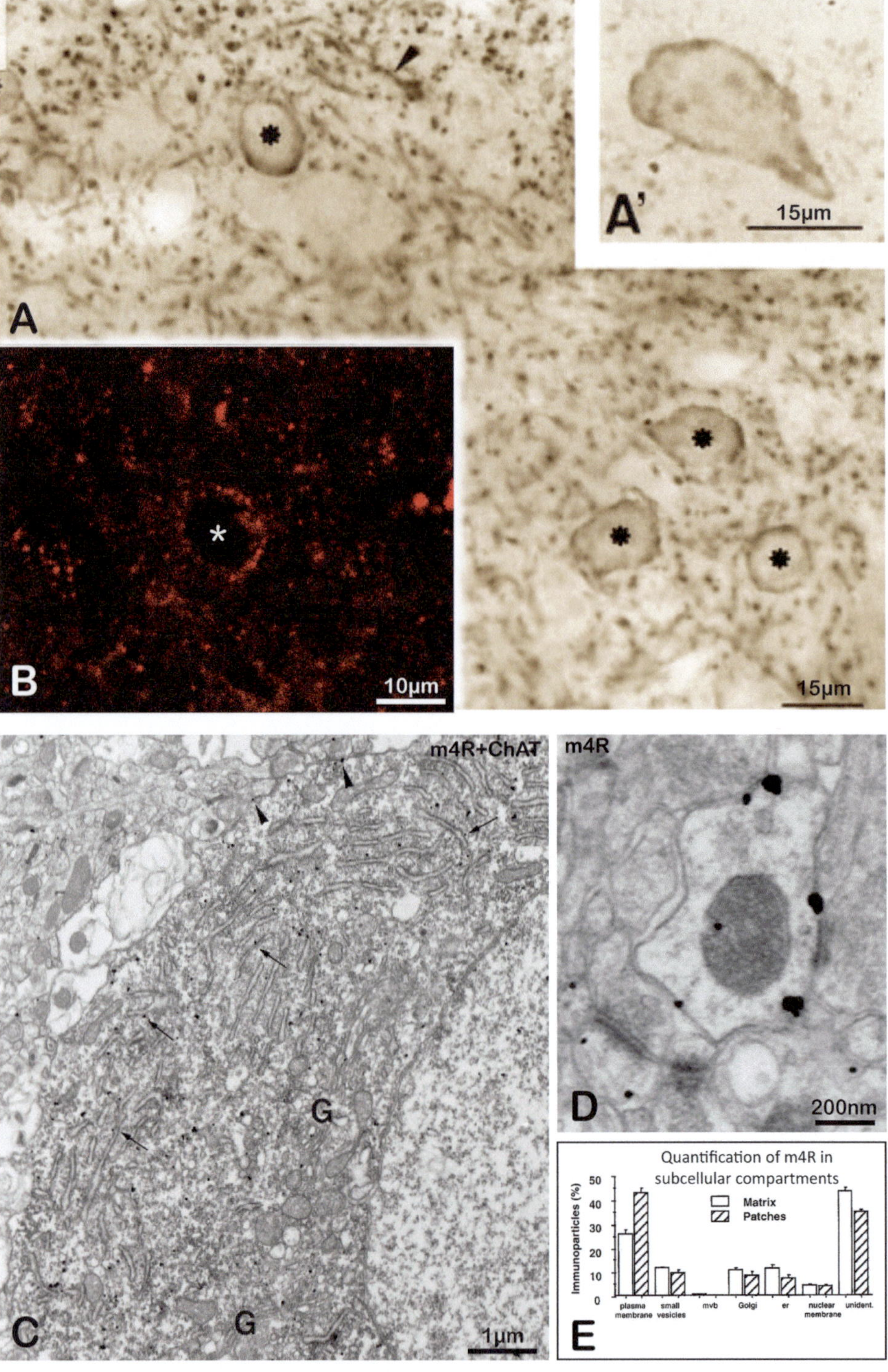

Fig. 2 Cellular and subcellular distribution of M_4 MR. Immunohistochemistry in striatal neurons in control animals (**A**, **A′**, **B**) using the immunoperoxidase (**A**, **A′**) and immunofluorescence (**B**) methods. The M_4 MR immunolabeling is detected at the membrane of some cell bodies of neurons often seen in clusters (*asterisks*). Immunolabeling for m4R is also seen in dendrites (**A**: *arrowheads*), but with reduced immunoreactivity. (**C**, **D**) Subcellular distribution of m4R immunoreactivity in the striatum of normal animals using the preembedding immunogold method with silver intensification. Detail of M_4 MR immunolabeling in the cytoplasm of a cell bodies (**C**) and a dendrite (**D**). Some immunoparticles are associated with the plasma membrane (*arrow heads*) of the perikaryon and dendrite. Immunoparticles are associated also with cytoplasmic compartments like endoplasmic reticulum (*arrows*) and Golgi apparatus (G). (**E**) Quantitative analysis of the subcellular distribution of

membrane where they are supposed to modulate acetylcholine release. Surprisingly, we were not able to detect M_4 MR at axonal varicosities, despite the fact that this receptor was clearly shown to be involved in acetylcholine release [5]. We cannot exclude that this was due to a lack of sensitivity of the immunogold technique.

4.2 Synaptic Localization of M_2 MR

The localization of the M_2 MR was analyzed in correlation with synapses by electron microscopic immunohistochemistry in the mouse trigeminal, facial, and hypoglossal motor nuclei (Fig. 3). In all nuclei, M_2 MR were localized at the membrane of motoneuronal perikarya and dendrites. The M_2 MR were concentrated at cholinergic synapses located on the perikarya and most proximal dendrites. However, M_2 MR at cholinergic synapses represented only a minority (<10 %) of surface M_2 MR. The M_2 MR were also found in abundance at glutamatergic synapses in both motoneuronal perikarya and dendrites. A relatively large proportion (20–30 %) of plasma membrane-associated M_2 MR were located at glutamatergic synapses.

The synaptic and extrasynaptic localization of M_2 MR suggests that the effect of acetylcholine might be mediated through a synaptic as well as diffuse type of transmission. The demonstration that M_2 MR are present at the postsynaptic membrane beneath glutamatergic terminals provides a direct argument in favor of a co-release of ACh and glutamate by the same terminal as suggested by different groups [6, 7].

4.3 Subcellular Redistribution of M_2 MR and M_4 MR After Acute and Chronic Activation

Our data show that muscarinic receptors are subject to an intraneuronal trafficking when they are activated and that this trafficking is different according to the duration of the stimulation (acute versus chronic). When MR are acutely stimulated, few M_2 MR immunoparticles are detected in association with the plasma membrane of the somatodendritic compartment (Figs. 4b, b′, e, e′ and 5c). In parallel, M_2 MR and M_4 MR immunoreactivity is seen in the cytoplasm in association with small endosome-like vesicles (Figs. 4b′, ev and 5c). Muscarinic receptors are thus endocytosed and then either recycled to the plasma membrane or degraded in lysosomes [8]. When ACh receptors are chronically stimulated like in acetylcholinesterase knockout mice (AChE−/−), trafficking of

Fig. 2 (continued) M_4 MR in the striatum of control rats using the pre-embedding immunogold method with silver intensification in two striatal regional compartments, matrix and patches. Proportion of immunoparticles associated with different subcellular neuronal compartments in perikarya of medium spiny neurons. For each cell body, the number of immunoparticles associated with each compartment was counted, and the proportion in relation to the total number was calculated. In medium spiny neurons, of the immunoparticles that are associated with an identified compartment, most of them are preferentially located at the plasma membrane. The proportion of immunoparticles at the membrane is much higher in patches than in matrix. In the cytoplasm, the immunoparticles are mostly detected in association with small vesicles, the Golgi apparatus, and the endoplasmic reticulum. A small proportion of immunoparticles are associated with multivesicular bodies (*mvb*) and the outer nuclear membrane. Some immunoparticles are not seen in association with any of an identified compartment. **(D–E)** Reproduced with permission, from Bernard et al. [4]

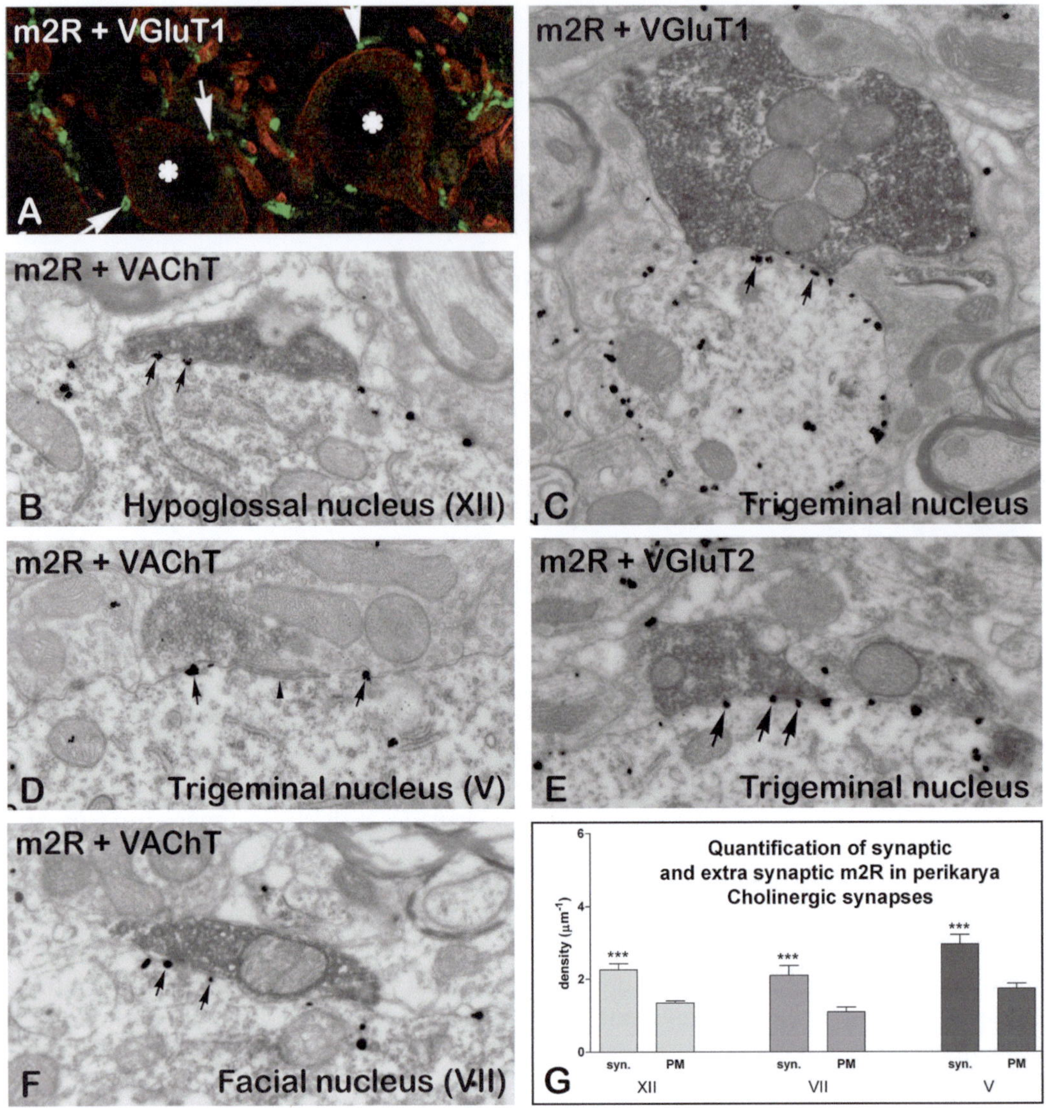

Fig. 3 Synaptic distribution of M_2 MR immunohistochemistry in brainstem motor nuclei in mouse. (**A**) Confocal microscopic illustration of the distribution of M_2 MR and the vesicular transporter of glutamate type 1 (VGLUT1) immunoreactivity in the trigeminal motor nucleus. The surface of motoneuron cell bodies (*asterisks*) is intensely M_2 MR-immunoreactive (*red*). VGLUT1 immunostaining is predominantly confined to large intensely stained varicosities (*green*). VGLUT1-immunoreactive varicosities form close contacts with M_2 MR-immunoreactive perikarya and dendrites (*arrows*). (**B–H**) Electron microscopic qualitative (**B–E**, **H**) and quantitative (**G**) analysis of the subcellular distribution of M_2 MR at cholinergic (**B**, **D**, **F**) and glutamatergic (**C**, **E**) synapses in hypoglossal nucleus (**B**), trigeminal motor nucleus (**C–E**) and facial nucleus (**F**) after preembedding immunogold method with silver intensification. The M_2 MR immunoparticles are detected at the plasma membrane, in association with its inner side. Some immunoparticles are found at the postsynaptic membrane (*arrows*) under VAChT (**B**, **D**, **F**) or VGluT1 (**C**) or vesicular transporter of glutamate type 2 (VGluT2) (**E**) immunoreactive presynaptic boutons (electron-dense peroxidase product). The VAChT-immunoreactive boutons form synapses on perikarya (**B**, **D**, **F**) and the VGluT1 and 2-immunoreactive boutons form synapses on dendrites. (**G**) Statistical analysis (nonparametric Wilcoxon matched pairs test) shows that the density of M_2 MR at VAChT-positive cholinergic synapses (syn.) is significantly higher than the overall density of receptors at the plasma membrane (PM) in motoneurons of the hypoglossal (XII), facial (VII) and trigeminal motor (V) nuclei (e). ***$P < 0.001$. From Csaba et al. [3]

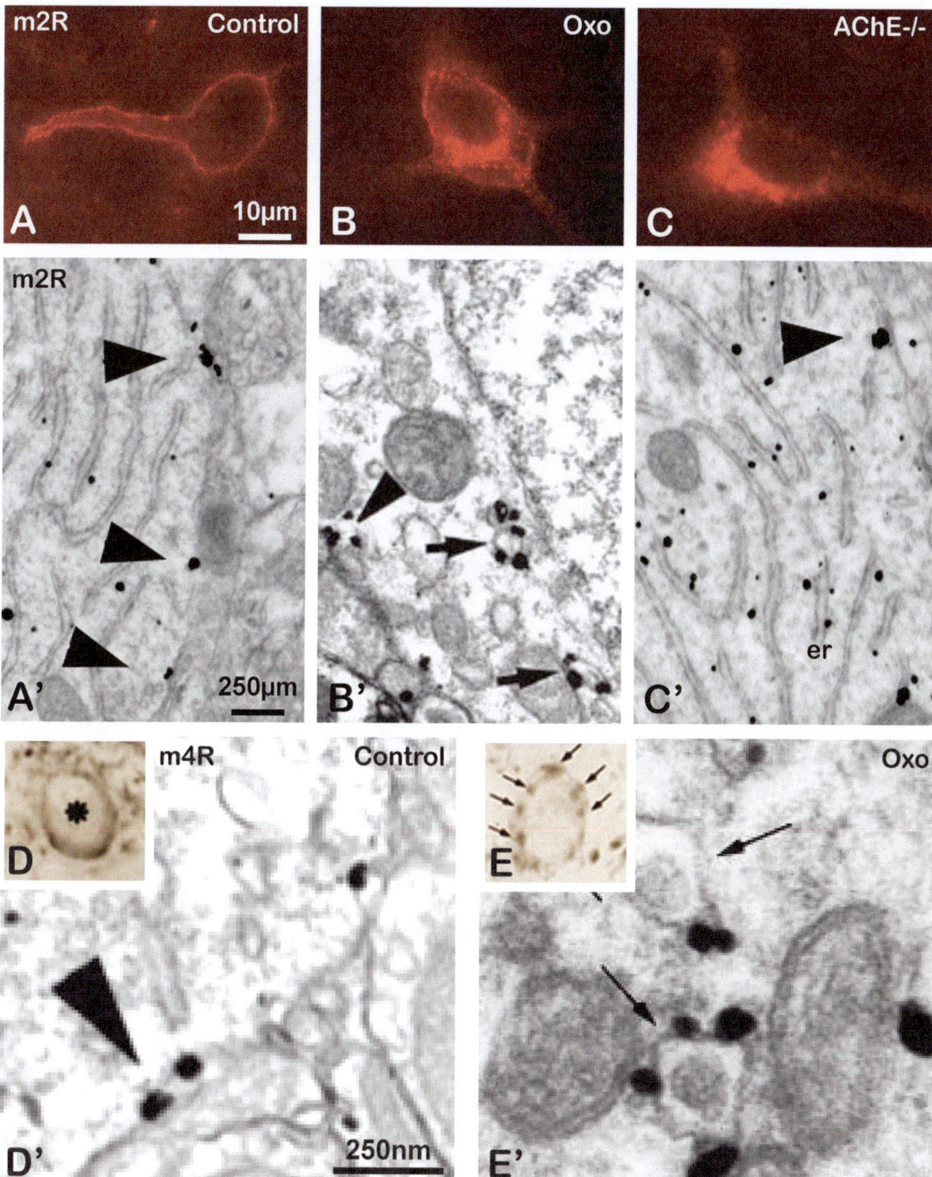

Fig. 4 Effect of acute and chronic modifications of ACh levels on the cellular and subcellular distribution of M_2 MR and M_4 MR in neurons of the striatum in vivo. Images were collected under epifluorescence (**A**–**C**), and electron microscopy (**A**′–**C**′, **D**′, **E**′) using immunofluorescent (**A**–**C**), and peroxidase (**D**, **E**) histochemistry and a pre-embedding immunogold method (**A**′–**C**′, **D**′, **E**′). (**A**, **A**′, **D**, **D**′) In control mice, M_2R and M_4R immunoreactivity are mostly detected at the plasma membrane. Immunoparticles are associated mostly with the internal side of the plasma membrane (*arrowheads*). (**B**, **B**′, **E**, **E**′) After acute treatment with oxotremorine ('Oxo'; 0.5 mg/kg subcutaneously for 1 h), M_2 MR and M_4 MR immunoreactivities are seen in the cytoplasm. (**C**, **C**′) After chronic stimulation of ACh receptors in acetylcholinesterase knockout mice (AChE−/−), no staining is observed at the membrane, whereas strong immunoreactivity is detected in the cytoplasm. Few immunoparticles are detected in association with the plasma membrane (*arrowheads*). By contrast, numerous particles are seen in the cytoplasm associated with the endoplasmic reticulum (er) and Golgi apparatus. This suggests that, when ACh receptors are chronically stimulated, targeting of M_2 receptors is blocked in the intraneuronal compartments of synthesis and maturation, and thus they are no longer targeted to the membrane. Scale bars, 10 mm in (**A**–**E**); 500 nm in (**F**, **G**); 50 nm in (**H**). Reproduced, with permission, from Bernard et al. [8]

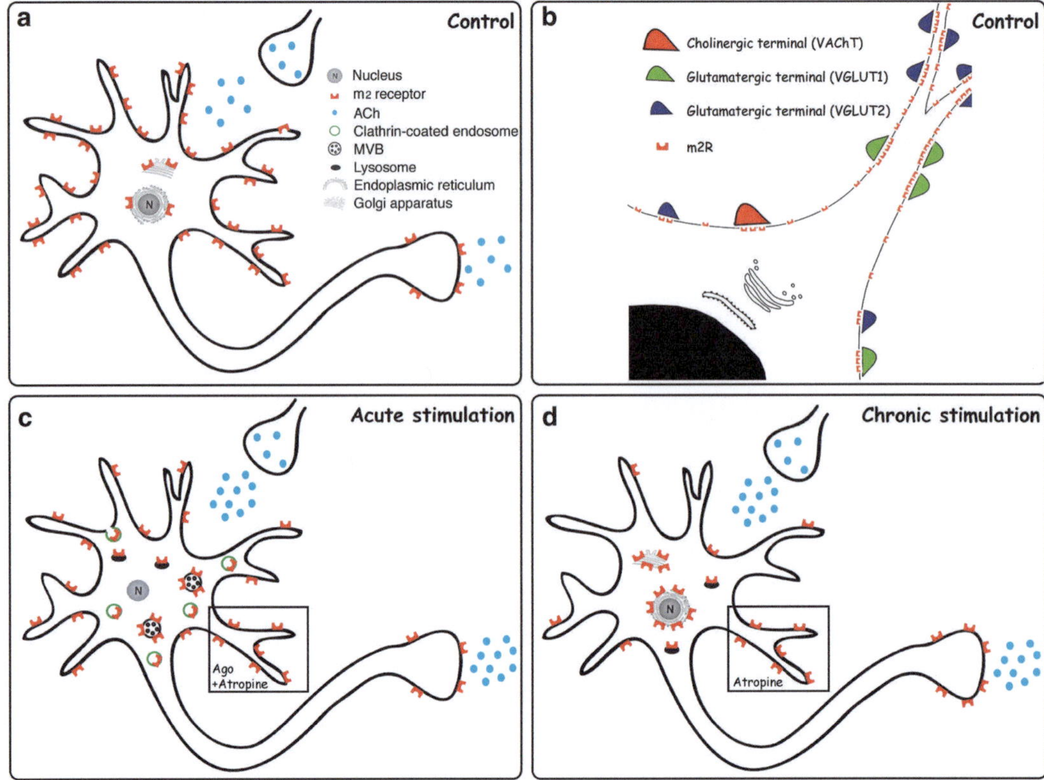

Fig. 5 Schematic representation of the subcellular and synaptic distribution of M_2 MR in neurons. (**a**) In control mice, most M_2 MR are at the plasma membrane of the somatodendritic compartment and at axonal varicosities. (**b**) In motoneurons in the brainstem, the M_2 MR are localized at the plasma membrane of perikarya and dendrites, at higher density in dendrites than in cell bodies. Cholinergic terminals (VAChT-immunopositive) form synapses on perikarya. VGLUT1-labeled glutamatergic terminals form synapses on perikarya and large-caliber dendrites, whereas small VGLUT2-labeled glutamatergic terminals form synapses on perikarya and small-caliber dendrites. The density of both VGLUT1- and VGLUT2-labeled synapses is higher in dendrites than in perikarya. Note the enrichment of M_2 MR at VAChT-labeled cholinergic as well as VGLUT1- and VGLUT2-labeled glutamatergic synapses. (**c**) After acute stimulation of cholinergic neurons, M_2 MR density decreases in the somatodendritic plasma membrane and M_2 MR accumulate in association with endosomes and multivesicular bodies (MVBs). The decrease in M_2 MR density at the membrane is blocked when the muscarinic-receptor antagonist atropine is injected. In cholinergic varicosities, localization of M_2 MR at the plasma membrane is similar to that in control animals. (**d**) After chronic stimulation of ACh receptors, M_2 MR are almost absent from the somatodendritic plasma membrane but accumulate in the cytoplasm in association with the endoplasmic reticulum and Golgi apparatus. The decrease in M_2 MR at the plasma membrane is blocked after atropine injection in a chronically stimulated animal. In cholinergic varicosities, M_2 MR density at the plasma membrane is increased. Intracellular GPCRs are those that have been retrieved from the plasma membrane (acute stimulation) or blocked on their way out to the membrane (chronic stimulation)

muscarinic receptors ends in the intraneuronal compartments of synthesis and maturation, and thus they are no longer targeted to the membrane of the somatodendritic compartment (Figs. 4c, c′ and 5d). Conversely, M_2 MR density increases at axonal varicosities (Fig. 5d).

5 Notes

5.1 Perfusion

5.1.1 Perfusion System

We use a gravity system to perfuse the animals. Though such a system may take longer time to achieve organ perfusion, the results are reproducible and perfusion is thorough. Gravity systems allow consistent pressure and controlled flow rates, providing good perfusion of the major organs, especially brain. Excessive pressure using other methods, like a pump, may cause artifacts in the neuronal ultrastructure.

5.1.2 Fixative

Glutaraldehyde is an essential compound of the fixative solution. The concentration of glutaraldehyde has to be carefully chosen. The percentage has to be high enough to preserve the ultrastructure of the tissue for analysis at EM level and low enough to avoid interference with antibody binding that may cause considerable decrease of intensity of labeling or even nonspecific binding. In our hands, 0.5–0.2 % of glutaraldehyde in addition to 2–4 % paraformaldehyde gave us good results on labeling intensity and ultrastructure preservation for the majority of primary antibodies, including anti-MR antibodies.

5.2 Tissue Permeabilization

The section freezing allows to produce small ice crystals that mechanically disrupt the tissue causing only limited damages and thus enhance the penetration of reagents. Detergents like Triton X-100 must not be used because of the irreversible damages they cause to neuronal ultrastructure.

5.3 Specificity of Primary Muscarinic Receptor Antibodies

We and others have tested different anti-muscarinic receptors antibodies by immunohistochemistry [9]. Only anti-M_2 MR and M_4 MR antibodies commercialized by Millipore (Ref: MAB367 and MAB1576) have been found to selectively label the receptors, i.e., the immunohistochemical M_2 MR and M_4 MR signals were abolished in the corresponding knock-out mice. Anti-M_2 MR antibodies bind rat and mouse M_2 MR. Surprisingly, if we found labeling with anti-M_4 MR antibodies in rats, we were able to detect M_4 MR only in one strain of mouse, e.g., control mice of M_2 KO mice from Taconic farm (Germantown, NY).

5.4 Adjustment of Primary Antibody Concentration

It is important to be aware that the immunogold technique is less sensitive compared to immunofluorescence or immunoperoxidase. Increase of the concentration of the antibody may be necessary to obtain a good signal.

5.5 Gold Coupled Secondary Antibodies or Streptavidin?

When possible, we prefer to use biotinylated secondary antibodies and streptavidin coupled to gold particles instead of secondary antibodies directly coupled to gold beads. First, it allows to use the same gold coupled compound in experiments detecting first antibody produced in different species. Second, we found a more important variability in labeling obtained with secondary antibodies coupled to gold particles compared to the same labeling obtained with biotinylated secondary antibodies and streptavidin coupled to gold particles.

5.6 Coating Grids with Pioloform

Samples for transmission electron microscopy must be supported on a thin electron transparent film, to hold the specimen in place while in the objective lens of the TEM. We use pioloform as a support film. For the detailed procedure, see [10].

5.7 Ultrathin Sections Cutting Procedure

Semi-thin sections (1 μm thick) are first cut and observed under a light microscope. When the labeling correspond to the area of interest, ultrathin sections (5 nm thick) are cut. Since the reagent do not penetrate deeply into the tissue, ultrathin sections have to be cut at the surface of the tissue in the few first micrometers, as parallel as possible to the surface.

The use of single slot grids instead of mesh grids makes the observation more comfortable. It allows for example to analyze the labeling on a whole structure like a cell body and not be hindered by bars of the mesh grid.

5.8 Quantification

Counting immunoparticles at the ultrastructural level is important for comparing the abundance of receptor in each compartment in basal and experimental conditions. For that, sections from control or wild type animals and treated animals or KO mice must be processed for immunohistochemistry in the same time. The analysis are performed on EM at a final magnification of about 4000×, using the Metamorph software (Universal Imaging Corporation, Paris, France) [11] or Image-Pro Plus image analysis software (Media Cybernetics, Bethesda, MD) [3] on a personal computer. In cell bodies, the immunoparticles are identified and counted in association with different subcellular compartments: the plasma membrane, endosome-like vesicles, multivesicular bodies, the Golgi apparatus, endoplasmic reticulum, the nuclear membrane. Some immunoparticles are classified as associated with an unidentified compartment, because they were associated with either no detectable organelles or an organelle that could not be identified as one of the five previous ones. The results are expressed as:

1. The percentage of immunoparticles associated with the different subcellular compartments in normal animals.
2. The number of immunoparticles per membrane length (micrometers), cytoplasmic surface (square micrometers), multivesicular body, or Golgi apparatus in normal and treated rats (mice).

For the analysis of the localization of M_2 MR in relationship with the presynaptic terminals, immunoparticles are identified and counted in association with the plasma membrane and at the postsynaptic membrane beneath the VAChT-, VGLUT1-, and VGLUT2-immunoreactive presynaptic boutons. The latter area included the synaptic complexes and also adjacent plasma membrane in apposition with the VAChT-, VGLUT1-, and VGLUT2-immunoreactive presynaptic boutons. Results are expressed as density of M2 MR at the postsynaptic membrane beneath VAChT-, VGLUT1-, and VGLUT2-labeled boutons and overall at the plasma membrane (number of immunoparticles per 1 μm membrane length). We assume that the number of immunoparticles is proportional to the absolute number of the receptor. The values from control and treated animals are compared using a suitable statistical test.

6 Conclusions

Despite the tricky aspect of the post-embedding immunohistochemistry, this chapter demonstrates the extraordinary power of the high-resolution electron microscopic immunohistochemistry to locate muscarinic receptors in the different neuronal compartments, including synapses and to analyze the trafficking of these receptors in neurons. These data give new keys to understand the mode of transmission of ACh through muscarinic receptors.

Such experimental approaches may be adapted to the analysis of other neurotransmitter receptors localization or other proteins in the brain or in other tissues and may help to better understand the link between localization of GPCRs and their intraneuronal trafficking and neuronal responses induced by GPCR activation. This might enable the development of new strategies for treating neurological diseases associated with altered GPCR signaling, such as Parkinson's disease.

References

1. Slutsky I, Wess J, Gomeza J, Dudel J, Parnas I, Parnas H (2003) Use of knockout mice reveals involvement of m2-muscarinic receptors in control of the kinetics of acetylcholine release. J Neurophysiol 89:1954–1967
2. Santini E, Sepulveda-Orengo M, Porter JT (2012) Muscarinic receptors modulate the intrinsic excitability of infralimbic neurons and consolidation of fear extinction. Neuropsychopharmacology 37:2047–2056
3. Csaba Z, Krejci E, Bernard V (2013) Postsynaptic muscarinic m2 receptors at cholinergic and glutamatergic synapses of mouse brainstem motoneurons. J Comp Neurol 521: 2008–2024
4. Bernard V, Levey AI, Bloch B (1999) Regulation of the subcellular distribution of m4 muscarinic acetylcholine receptors in striatal neurons in vivo by the cholinergic environment: evidence for regulation of cell surface receptors by endogenous and exogenous stimulation. J Neurosci 19:10237–10249
5. Zhang W, Basile AS, Gomeza J, Volpicelli LA, Levey AI, Wess J (2002) Characterization of central inhibitory muscarinic autoreceptors by the use of muscarinic acetylcholine receptor knock-out mice. J Neurosci 22:1709–1717
6. Gras C, Amilhon B, Lepicard EM, Poirel O, Vinatier J, Herbin M, Dumas S, Tzavara ET, Wade MR, Nomikos GG, Hanoun N, Saurini F,

Kemel M, Gasnier B, Giros B, El Mestikawy S (2008) The vesicular glutamate transporter vglut3 synergizes striatal acetylcholine tone. Nat Neurosci 11:292–300

7. Lamotte d'Incamps B, Ascher P (2008) Four excitatory postsynaptic ionotropic receptors coactivated at the motoneuron-renshaw cell synapse. J Neurosci 28:14121–14131
8. Bernard V, Décossas M, Liste I, Bloch B (2006) Intraneuronal trafficking of g-protein-coupled receptors in vivo. Trends Neurosci 29:140–147
9. Jositsch G, Papadakis T, Haberberger RV, Wolff M, Wess J, Kummer W (2009) Suitability of muscarinic acetylcholine receptor antibodies for immunohistochemistry evaluated on tissue sections of receptor gene-deficient mice. Naunyn Schmiedebergs Arch Pharmacol 379: 389–395
10. Kisten M, Harris P (1999) Synapseweb, electron microscopy protocols, coating grids. http://synapses.clm.utexas.edu/lab/howto/protocols/coatinggrids.stm
11. Bernard V, Laribi O, Levey AI, Bloch B (1998) Subcellular redistribution of m2 muscarinic acetylcholine receptors in striatal interneurons in vivo after acute cholinergic stimulation. J Neurosci 18:10207–10218

Chapter 8

Investigation of Muscarinic Receptors by Fluorescent Techniques

Cornelius Krasel, Andreas Rinne, and Moritz Bünemann

Abstract

Since the last decade or so, fluorescent techniques have markedly improved our ability to investigate the localization of proteins within the cell and to measure the kinetics of protein–protein interactions. In this chapter, we discuss how these techniques have been applied to the muscarinic acetylcholine receptor field, with a focus on measuring Förster resonance energy transfer (FRET) by time-resolved sensitized emission.

Key words Sensitized emission, FRET, G-protein-coupled receptors, Muscarinic acetylcholine receptors, Protein–protein interaction

1 Background

Fluorescence-based techniques are well suited to investigate G-protein-coupled receptors on the single-cell level. In most cases, the readout is performed by microscopy, even though it is theoretically possible to perform experiments in other formats. However, the lack of publications suggests that with the exception of time-resolved Förster resonance energy transfer (TR-FRET), it appears that the signal-to-noise ratio is not sufficiently high to reliably detect signals. Fluorescence microscopy has been used to track muscarinic acetylcholine receptors in cells, to investigate conformational changes of muscarinic acetylcholine receptors and to resolve the interaction of muscarinic acetylcholine receptors with other proteins. Basically, the receptor can either be treated with a fluorescent ligand or labeled with a fluorescent dye or protein. A major problem of fluorescent ligands is that ligand modification (to make it fluorescent) may alter its pharmacological properties. For example, fluorescent derivates of the agonist AC-42 are antagonists [1]. Nevertheless, Hern et al. [2] used two fluorescent antagonists, Alexa 488-telenzepine and Cy3B-telenzepine, to monitor dimerization of the M_1 receptor by single-molecule microscopy.

Jaromir Myslivecek and Jan Jakubik (eds.), *Muscarinic Receptor: From Structure to Animal Models*, Neuromethods, vol. 107, DOI 10.1007/978-1-4939-2858-3_8, © Springer Science+Business Media New York 2016

However, the focus of this chapter is the fluorescent labeling of the receptor. This can be achieved either by fusing a fluorescent protein [3] at a suitable position to the receptor of interest, or by inserting a comparatively small targeting sequence that binds specifically to a small organic fluorescent molecule [4]. The advantage of using fluorescent proteins is that labeling is achieved by making appropriate constructs by molecular biology, and no manipulation beyond transfection is required to express fluorescent receptors. In addition, the fluorescent labeling is specific, and the signal-to-noise ratio is quite high. However, fluorescent proteins are bulky (the size of GFP is about 27 kDa) and may therefore sterically interfere with the function of the receptor. This was first demonstrated for a conformational sensor of the adenosine A_{2a} receptor [5]. Labeling with small organic fluorescent molecules will, however, inevitably create background fluorescence within the cell which has frustrated many researchers working with these molecules [6]. It is therefore maybe less suitable for the localization of molecules within cells and more suitable for resonance energy techniques (see below).

Using fluorescent proteins, receptors can be tracked within the cell [7], their pharmacology [8–10] and their quaternary structure [11, 12] may be investigated, and kinetics of their conformational change [13–18] or of their interaction with other proteins [13, 18, 19] can be studied.

Förster resonance energy transfer (FRET), also often called fluorescence resonance energy transfer, is a versatile method to obtain information about the proximity of two suitable fluorescent molecules. In FRET, energy is transferred in a radiation-less way from one fluorescent molecule (the donor) to another molecule (the acceptor). Three conditions are required for FRET to occur [20]: (1) the emission spectrum of the donor and the excitation/absorption spectrum of the acceptor have to overlap, (2) the dipole orientation of the donor and the acceptor must not be orthogonal to each other, and (3) the two fluorescent molecules must be close to one another as FRET efficiency is inversely proportional to the sixth power of the donor–acceptor distance. The spectral overlap is dependent on the two fluorescent molecules used for the experiment. A particularly established donor–acceptor pair for FRET experiments in biology is the combination of cyan and yellow fluorescent protein but FRET can also be performed with other combinations. If the donor or acceptor is a small fluorescent molecule, the other partner of the FRET pair has to be chosen appropriately to ensure spectral overlap. The relative orientation of the dipoles is usually hard to control. If fluorescent proteins are employed, they are often attached to the protein of interest with a relatively long flexible linker to ensure rotational freedom of the fluorophores. This ensures that the relative dipole orientation is variable over the time course of the experiment which should yield at least some FRET. Consequently, changes in FRET over the time course of the experiment should reflect a change in the distance of the two fluorescent molecules which could reflect a conformational change or protein–protein association/dissociation. To investigate whether a fluorescent molecule can indeed rotate freely, fluorescence anisotropy measurements can be employed [21, 22].

Sometimes FRET may fail to report a known protein–protein interaction. The reasons are not entirely clear but may be due to an unsuitable orientation of the two fluorescent molecules. Nevertheless, we and others have successfully investigated the interaction of muscarinic acetylcholine receptors with heterotrimeric G-proteins, G-protein-coupled receptor kinases and arrestins using FRET [13, 18, 19].

2 Creating a Fluorescent Receptor

FRET can be used to monitor the interaction of fluorescent or fluorescence-quenching ligands with receptors in real time. This has been exemplified at the M_1 receptor [9, 10]. In all cases described so far, a fluorescent protein was fused in front of the N-terminus of the receptor. The construction of the EGFP-M_1 fusion protein has been described in some detail [8]. A 31 amino-acid signal sequence from the chicken a7 nicotinic acetylcholine receptor was fused in front of the EGFP to ensure proper translocation. Similarly, a hemagglutinin signal sequence has been used to improve plasma membrane targeting of the β_2-adrenergic receptor [23, 24]. To obtain such constructs, PCR is used to amplify the fluorescent protein of interest. Experimentation with the linker length between the fluorescent protein and the C-terminus may be necessary to obtain good plasma membrane localization. The signal sequence is encoded in the forward primer of the fluorescent protein. For example, the following forward primer encompasses the hemagglutinin signal sequence and should be useful for most GFP derivatives (i.e., YFP, CFP, pHluorin):

aaaaaggatccatgaagacgatcatcgccctgagctacatcttctgcctggtattcgccagta-aaggagaagaacttttcactggagttgtccc

The BamHI restriction site ggatcc can be replaced with any suitable restriction site to facilitate cloning of the PCR product.

In the published work, ligand–receptor interaction was measured by monitoring FRET between a GFP-tagged receptor and a Bodipy-labeled ligand in a cell suspension in a cuvette [8–10].

The M_3 receptor has also been labeled at the N-terminus with a SNAP-tag [11]. The SNAP-tag is derived from O-6-methylguanine-DNA methyltransferase, a protein with a size of 20 kDa (i.e., smaller than GFP), that can be covalently labeled with small fluorescent pseudosubstrates [25]. The advantage of the SNAP tag is that small organic molecules may have advantageous fluorescence properties (e.g., quantum yield, stability against photobleaching) compared to fluorescent proteins. However, in addition to generating the cDNA for the construct of interest, an additional labeling step is required. The introduction of a SNAP tag is performed in the same way as the introduction of a fluorescent protein. Alternatively, the protein of interest can be cloned into a commercial vector already containing the SNAP-tag (available from New England Biolabs). Labeling of proteins with an extracellular SNAP-tag is straightforward, whereas labeling of intracellular SNAP-tags requires different, cell-permeable compounds and should result in a considerable level of background fluorescence as the unbound fluorescent chemical needs to be washed out from the cell.

To look for G-protein-coupled receptor trafficking, receptors are often tagged with fluorescent proteins at the C-terminus. In many cases, this can be achieved by amplification of the open reading frame of the receptor by PCR and replacing the stop codon with a suitable restriction site in this process. This modified receptor can then either be cloned into a commercially or noncommercially available plasmid (pEGFP-N1 and related plasmids) which results in a fairly long linker between the receptor and the fluorescent protein, or the fluorescent protein can be cloned directly behind the restriction site which makes the linker only two amino acids long. However, tagging of the M_4 muscarinic acetylcholine receptor with GFP at the C-terminus reduced the affinity for agonists three- to fourfold and abolished recycling of internalized receptors [7]. This was interpreted as a result of very high overexpression. However, the possibility has to be considered that the extreme C-terminus may contain motifs relevant for protein–protein interactions such as PDZ ligands [26] which may be destroyed by fusing a fluorescent protein to the C-terminus.

FRET can also be used to monitor the conformational change of a G-protein-coupled receptor and has been extensively used for muscarinic acetylcholine receptors. In all cases described so far, either the CFP-YFP pair [13, 16] or the CFP-FlAsH pair [14, 15, 17] was used. Labeling of the intracellular loops is more challenging than labeling of the N- or C-terminus because it may more pronouncedly affect receptor–G protein coupling or trafficking to the plasma membrane. The precise construction of FlAsH-based receptor sensors has been described in a number of papers (e.g., [17, 27]).

3 Fluorescence Microscopy and FRET

3.1 Preparation of Cells for Microscopy

To perform fluorescence microscopy using receptors tagged with fluorescent proteins, the cDNA for these proteins has to be introduced into a suitable cell line. We preferentially use HEK293T cells

(ATCC) because they are easy to transfect, but any type of cell that can be transfected may be used. It is important to use a transfection reagent that does not damage the cells too much. If more than one cDNA is introduced into the cells, it is also important to achieve relatively high transfection efficiencies because otherwise some cells will only be transfected with some of the cDNAs. For HEK293T cells, we use Effectene (Qiagen) which results in high transfection yields (30–50 %) and does not alter cell morphology too much. We also have used polyethylenimine successfully. When working with transiently transfected cells, we perform transfection on day 1, split cells on coverslips on day 2 and do the experiment on day 3.

What cDNAs to transfect depends on the experimental question. When investigating receptor conformational changes or changes in receptor localization, it is often sufficient to transfect just the receptor construct. Some receptor constructs are not targeted very well to the membrane. This may be ameliorated by one of the following:

1. Transfecting a relatively small amount of DNA (e.g., only 20–25 % of what the protocol of the transfection reagents suggests as a maximum) to avoid cellular overload.
2. Cloning a signal sequence in front of the construct (see above).
3. Creating a stable cell line expressing the construct.

When the interaction between receptors and other proteins is studied, at least cDNAs for both fluorescent proteins have to be transfected plus possibly cDNAs for some nonfluorescent proteins. For example, if the interaction between receptors and heterotrimeric G-proteins is to be investigated, one will have to transfect a cDNA for a fluorescent receptor, a cDNA for a fluorescent G-protein subunit (e.g., the β subunit) and two cDNAs for the remaining nonfluorescent G-protein subunits (e.g., the α and γ subunit).

Transfection of HEK293T cells with Effectene:

- On the day before transfection, seed a suitable amount of your cell line of choice on a 6 cm-dish (the manufacturer recommends seeding 200,000–800,000 cells).
- On the day of transfection, mix 200 μl EC buffer, the cDNA (see below) and 16 μl enhancer by pipetting up and down.
- Let the mixture stand for around 5 min at room temperature.
- Add 20 μl Effectene, mix by pipetting up and down.
- Let the mixture incubate for another 5 min at room temperature.
- Add the mixture dropwise to the 6 cm-dish that was prepared the day before.

cDNA:

- To investigate a receptor sensor, use 0.5 μg of the sensor cDNA.
- To investigate receptor–G protein interaction by FRET, use for example 0.5 μg YFP-tagged receptor, 0.5 μg nonfluorescent suitable Gα subunit (e.g., Gαq for M_1, M_3, and M_5 receptors, Gαi1 for M_2 and M_4 receptors), 0.5 μg CFP-tagged Gβ1, and 0.3–0.4 μg nonfluorescent Gγ2.
- To investigate receptor–GRK2 interaction by FRET, use for example 0.5 μg YFP-tagged receptor, 0.5 μg nonfluorescent suitable Gα subunit (e.g., Gαq for M_1, M_3, and M_5 receptors, Gαi1 for M_2 and M_4 receptors), 0.5 μg nonfluorescent Gβ1, 0.3–0.4 μg nonfluorescent Gγ2, and 0.5 μg CFP-tagged GRK2.

On the day after transfection, transfer the cells from the plastic dish onto glass coverslips. One 6 cm-dish of transfected cells will usually be good for six coverslips. Use round coverslips (24–-25 mm diameter, thickness #1, supplier VWR) which will be mounted on the microscope using an Attofluor (Invitrogen) holder. Glass-bottom plates can be used instead but in the long run they will be more expensive than glass coverslips. Glass is preferable to plastic because it has superior optical properties and thus allows more reliable imaging, in particular at higher magnification. Before cell transfer, glass should be coated with polylysine to render it less hydrophobic. Either poly-L-lysine or poly-D-lysine can be used, but poly-D-lysine is less susceptible to degradation by cellular proteases. Coat coverslips as follows:

- Prepare a stock solution of poly-D-lysine by adding 50 ml of sterile water to 5 mg sterile poly-D-lysine hydrobromide (Sigma P-6407). This solution can be stored in the fridge for several weeks.
- The following steps should be performed under a sterile hood.
- Fill each of the six wells of a six-well plate with 2 ml of sterile PBS.
- Use forceps to immerse a round coverslip in a beaker filled with 70 % ethanol for a few seconds, then drop the coverslip into one of the six wells. Repeat with five more coverslips.
- As the coverslips are likely to float on top of the PBS, use a sterile glass Pasteur pipet to fully immerse the coverslip in the PBS. Then remove the PBS by suction. Make sure the glass surface is completely dry.
- Pipet a drop (approx. 200–250 μl) of poly-D-lysine solution in the middle of the glass coverslip. Let it sit there for approx. 1 h.
- Remove the poly-D-lysine by suction using a glass Pasteur pipet. Then add 2 ml of sterile PBS to the well and remove it again using suction, drying the coverslip and the well completely.
- The glass coverslips are now ready for use. They can be prepared in advance and will keep in the fridge for a few days.

3.2 The Microscope Setup

There are several methods to measure FRET, but to obtain information about rapid kinetics we prefer to use a method called sensitized emission in which fluorescence of the FRET donor and FRET acceptor are measured simultaneously. The advantage of this method is that, dependent on the experimental setup, up to 100 values per second can be obtained, making it possible to follow very rapid kinetic changes reliably. The disadvantage is that it is hard to quantify the precise extent of FRET as the concentration of the acceptor is measured only once and not in every frame.

We have several different types of equipment to measure sensitized emission but they all consist of a light source, an inverted microscope and a camera with an additional beam-splitter in front of it. The light paths for observing cells by eye and for measuring sensitized emission are detailed in Fig. 1.

The light source needs a fast triggering possibility as the cells should be illuminated as briefly as possible to avoid photobleaching. We have experience with various Polychrome sources (Till Photonics, now FEI Life Sciences) and with the DG4 (Sutter) but at the moment we like the CoolLED light sources best. They are as bright as the mercury or xenon lamps, relatively inexpensive and, since they are LED-based, should have a longer lifetime than the lamps. Their only disadvantage, compared to a Polychrome, is that if measurements are to be performed at different wavelengths, a new LED must be installed.

The camera should be fairly sensitive (again the more sensitive the camera, the shorter the illumination time) and have a grayscale sensor. We have experience with the Spot Pursuit PR3400 (Spot Imaging Solutions) and Evolve512 (Photometrics) cameras.

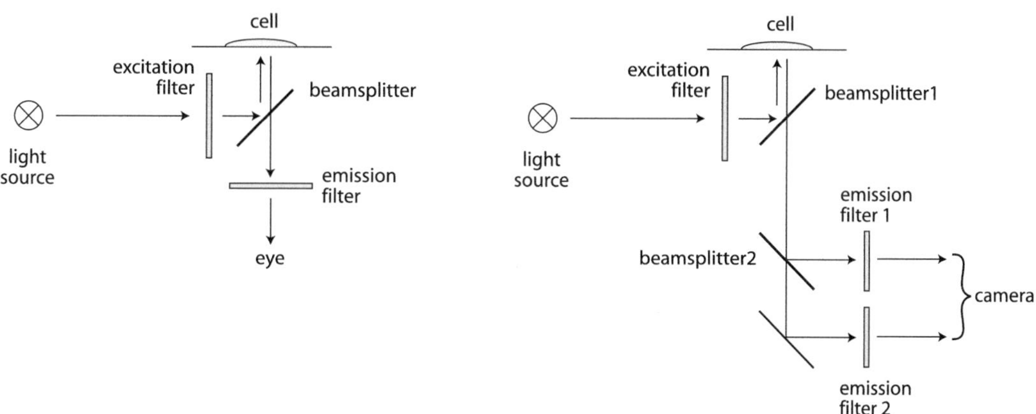

Fig. 1 Scheme of the optical path in the microscope when searching for a cell by eye (*left*) or when measuring sensitized emission (*right*). In the setup on the *right*, beamsplitter 2 and the two emission filters are built into an image splitter that is attached to one of the ports of the microscope. The excitation filter and beamsplitter1 are built into a filter cube within the microscope. Note that this filter cube does not have an emission filter which will increase sensitivity

To measure sensitized emission reliably, fluorescence has to be acquired simultaneously in two channels. While Hamamatsu sells a system that comes with two cameras, one for each channel, it is less expensive to install an additional beamsplitter directly in front of the camera. This way, half of the chip will be illuminated by the fluorescence of the donor and the other half of the chip will be illuminated by the fluorescence of the acceptor. Image splitters offering these possibilities include the Optosplit (Cairn Research) and the DC2 (Photometrics).

For searching a suitable cell by eye, any dual band CFP/YFP filter cube (like the discontinued Chroma 51017 or the Chroma 59017) should work (*see* Fig. 1):

- Excitation filter: dual band exciter (Chroma 59017x).
- Beam splitter: dual band beam splitter (Chroma 69008bs).
- Emission filter: dual band filter (Chroma 59017m).

With such a filter cube, either 430 nm light (to excite CFP) or 500 nm light (to excite YFP) may be used to search for cells expressing the proteins of interest. Once this has been achieved, the filter wheel on the microscope is switched and the following filters are used:

- Excitation filter: ET430/24 (Chroma).
- Beamsplitter 1 (separates emission from excitation light): T455LP (Chroma).
- Beamsplitter 2 (separates CFP from YFP fluorescence): T495lpxr (Chroma).
- Emission filter 1 (CFP): ET480/40m (Chroma).
- Emission filter 2 (YFP): BrightLine HC 534/20 (Semrock).

A number of companies will provide a system consisting of these components, an inverted microscope and the software to run it.

3.3 Measuring FRET

To actually perform the FRET experiment, the glass coverslip is mounted in a suitable holder (e.g., the Attofluor from Invitrogen). If glass-bottom dishes are used, there is no need to mount anything. Throughout the measurement, the cells are continuously superfused with FRET buffer (137 mM NaCl, 5 mM KCl, 2 mM $CaCl_2$, 1 mM $MgCl_2$, 10 mM HEPES pH 7.3–7.4) using a pressurized perfusion system (ALA-VC3-8SP, ALA, Scientific Instruments). For longer measurements, this buffer may be supplemented with 10 mM glucose. If peptides or lipophilic substances are used, 0.1 % BSA should be added to the buffer to reduce absorption of the ligands to the tubes and plastics of the perfusion system. The perfusion system allows fast buffer exchange within less than 10 ms, enabling the measurement of very fast on- or off-rates. Unused channels should be filled with FRET buffer. After each experiment the perfusion system should be washed with water because the very thin tubes tend to become clogged. If BSA or glucose was present in the FRET buffer, the perfusion system should first be washed with water, then with 20 % ethanol to reduce microbial contamination, and finally again with water. Once a month, the system should be cleaned thoroughly by washing it with 5 % acetic acid and 20 % ethanol. Nevertheless, the perfusion system is the part of the setup that is most likely to fail, thus it is useful to have a few spare parts available.

Before the actual kinetic experiment is performed, the dual band filter cube should be used to measure the fluorescence of the YFP-labeled protein upon excitation of YFP (i.e., with 500 nm light). This results in two values, $F_{500,534,\mathrm{cell}}$ and $F_{500,534,\mathrm{BG}}$. After that, the sensitized emission experiment is performed using the 435 nm excitation filter and the T455LP dichroic mirror. In the images obtained in these experiments, two areas of interest should be marked, one containing the cell under investigation, the other containing no cells. This second area of interest is the background. Thus, for each image, four fluorescence intensities are calculated: $F_{430,480,\mathrm{cell}}$, $F_{430,480,\mathrm{BG}}$, $F_{430,534,\mathrm{cell}}$, and $F_{430,534,\mathrm{BG}}$. To compute FRET, the following calculation is performed:

$$\mathrm{FRET} = \frac{\left(F_{430,534,\mathrm{cell}} - F_{430,534,\mathrm{BG}}\right) - \mathrm{BT} \times \left(F_{430,480,\mathrm{cell}} - F_{430,480,\mathrm{BG}}\right)}{F_{430,480,\mathrm{cell}} - F_{430,480,\mathrm{BG}} - \mathrm{DE} \times \left(F_{500,534,\mathrm{cell}} - F_{500,534,\mathrm{BG}}\right)}$$

The two numbers BT and DE are the bleedthrough of CFP into the YFP channel and the direct excitation of YFP by the 430 nm light, respectively. These numbers have to be measured once for each filter combination in cells transfected only with CFP or YFP, respectively, and stay pretty constant over time.

In a cell transfected only with CFP:

$$\mathrm{BT} = \frac{F_{430,534,\mathrm{cell}} - F_{430,534,\mathrm{BG}}}{F_{430,480,\mathrm{cell}} - F_{430,480,\mathrm{BG}}}$$

In a cell transfected only with YFP:

$$\mathrm{DE} = \frac{F_{430,534,\mathrm{cell}} - F_{430,534,\mathrm{BG}}}{F_{500,534,\mathrm{cell}} - F_{500,534,\mathrm{BG}}}$$

Finally, the FRET signal can be plotted against time and evaluated by nonlinear curve fitting.

If investigating protein–protein interactions, both the initial ratio and the amplitude of the FRET signal may be quite sensitive to the precise stoichiometry of fluorescent (and possibly nonfluorescent) proteins in the cell. To average several traces, we subtract a basal ratio (e.g., before the addition of agonist) from each trace and afterwards calculate means and standard error for each time point.

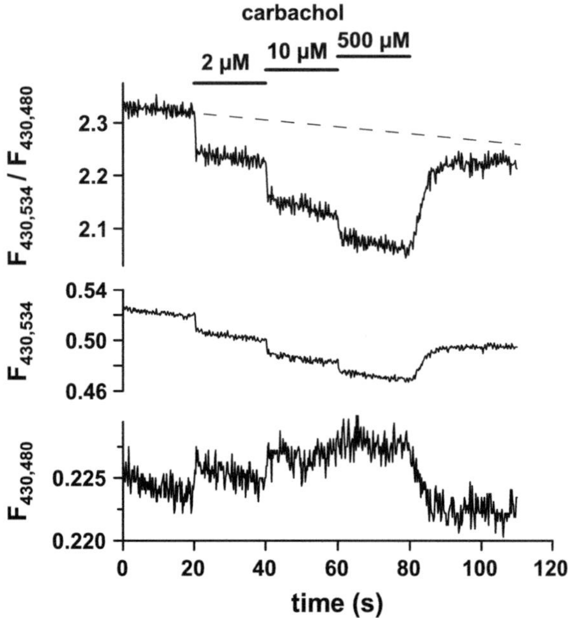

Fig. 2 Example of a FRET experiment. Dual emission FRET recordings were performed with HEK293T cells expressing a conformational sensor of the M_1 receptor [16]. A single cell was imaged as described above and stimulated with three different concentrations of carbachol. Shown are the CFP fluorescence ($F_{CFP} = F_{430,480,cell} - F_{430,480,BG}$, *bottom*), the YFP fluorescence ($F_{YFP} = F_{430,534,cell} - F_{430,534,BG}$, *middle*) and the ratio of the two (*upper trace*). Several such traces could be used to determine a concentration–response curve. To determine the amplitudes correctly, photobleaching has to be taken into account (*dashed line* in the *upper panel*). Because the amplitudes are only compared within the experiment, no further corrections were performed. After agonist washout the conformation of the sensor does not return to the start value. This is partially due to photobleaching but also caused by residual carbachol that cannot be fully removed from the cell by the perfusion system

Kinetics are calculated for each individual trace and the resulting time constants averaged. Sometimes, traces have to be corrected for differential photobleaching of CFP and YFP beforehand—if one of the fluorescent proteins bleaches faster than the other, the ratio will drift which in turn will affect the calculation of the time constants. An example readout for a FRET experiment is shown in Fig. 2.

4 Conclusions

Using the methods described in this chapter we have successfully measured the kinetics of ligand-induced receptor conformation [16, 28] and the interaction of muscarinic acetylcholine receptors with G-proteins and G-protein-coupled receptor kinases [19, 28].

Based on such measurements, others have developed mathematical models of muscarinic acetylcholine receptor signaling [13]. In the future fluorescent techniques may be useful to reveal novel properties of muscarinic acetylcholine receptors.

References

1. Daval SB, Valant C, Bonnet D, Kellenberger E et al (2012) Fluorescent derivatives of AC-42 to probe bitopic orthosteric/allosteric binding mechanisms on muscarinic M1 receptors. J Med Chem 55:2125–2143
2. Hern JA, Baig AH, Mashanov GI, Birdsall B et al (2010) Formation and dissociation of M1 muscarinic receptor dimers seen by total internal reflection fluorescence imaging of single molecules. Proc Natl Acad Sci U S A 107: 2693–2698
3. Giepmans BN, Adams SR, Ellisman MH, Tsien RY (2006) The fluorescent toolbox for assessing protein location and function. Science 312:217–224
4. Griffin BA, Adams SR, Tsien RY (1998) Specific covalent labeling of recombinant protein molecules inside live cells. Science 281: 269–272
5. Hoffmann C, Gaietta G, Bünemann M, Adams SR et al (2005) A FlAsH-based FRET approach to determine G protein-coupled receptor activation in living cells. Nat Methods 2:171–176
6. Stroffekova K, Proenza C, Beam KG (2001) The protein-labeling reagent FLASH-EDT2 binds not only to CCXXCC motifs but also non-specifically to endogenous cysteine-rich proteins. Pflugers Arch 442:859–866
7. Madziva MT, Edwardson JM (2001) Trafficking of green fluorescent protein-tagged muscarinic M_4 receptors in NG108-15 cells. Eur J Pharmacol 428:9–18
8. Ilien B, Franchet C, Bernard P, Morisset S et al (2003) Fluorescence resonance energy transfer to probe human M1 muscarinic receptor structure and drug binding properties. J Neurochem 85:768–778
9. Tahtaoui C, Parrot I, Klotz P, Guillier F et al (2004) Fluorescent pirenzepine derivatives as potential bitopic ligands of the human M1 muscarinic receptor. J Med Chem 47:4300–4315
10. Tahtaoui C, Guillier F, Klotz P, Galzi J-L et al (2005) On the use of nonfluorescent dye labeled ligands in FRET-based receptor binding studies. J Med Chem 48:7847–7859
11. Alvarez-Curto E, Ward RJ, Pediani JD, Milligan G (2010) Ligand regulation of the quaternary organization of cell surface M_3 muscarinic acetylcholine receptors analyzed by fluorescence resonance energy transfer (FRET) imaging and homogeneous time-resolved FRET. J Biol Chem 285:23318–23330
12. Patowary S, Alvarez-Curto E, Xu T-R, Holz JD et al (2013) The muscarinic M_3 acetylcholine receptor exists as two differently sized complexes at the plasma membrane. Biochem J 452:303–312
13. Jensen JB, Lyssand JS, Hague C, Hille B (2009) Fluorescence changes reveal kinetic steps of muscarinic receptor-mediated modulation of phosphoinositides and Kv7.2/7.3 K^+ channels. J Gen Physiol 133:347–359
14. Maier-Peuschel M, Frölich N, Dees C, Hommers LG et al (2010) A fluorescence resonance energy transfer-based M2 muscarinic receptor sensor reveals rapid kinetics of allosteric modulation. J Biol Chem 285:8793–8800
15. Ziegler N, Bätz J, Zabel U, Lohse MJ et al (2011) FRET-based sensors for the human M_1-, M_3-, and M_5-acetylcholine receptors. Bioorg Med Chem 19:1048–1054
16. Markovic D, Holdich J, Al-Sabah S, Mistry R et al (2012) FRET-based detection of M1 muscarinic acetylcholine receptor activation by orthosteric and allosteric agonists. PLoS One 7:e29946
17. Chang S, Ross EM (2012) Activation biosensor for G protein-coupled receptors: a FRET-based m1 muscarinic activation sensor that regulates G_q. PLoS One 7:e45651
18. Tateyama M, Kubo Y (2013) Analyses of the effects of Gq protein on the activated states of the muscarinic M_3 receptor and the purinergic $P2Y_1$ receptor. Physiol Rep 1:e00134
19. Wolters V, Krasel C, Brockmann J, Bünemann M (2015) Influence of $G\alpha_q$ on the dynamics of M3-acetylcholine receptor-G protein-coupled receptor kinase 2 interaction. Mol Pharmacol 87:9–17
20. Stryer L (1978) Fluorescence energy transfer as a spectroscopic ruler. Annu Rev Biochem 47:819–846
21. Vilardaga JP, Steinmeyer R, Harms GS, Lohse MJ (2005) Molecular basis of inverse agonism in a G protein-coupled receptor. Nat Chem Biol 1:25–28
22. Bondar A, Lazar J (2014) Dissociated $G\alpha_{GTP}$ and $G\beta\gamma$ protein subunits are the major activated

form of heterotrimeric Gi/o proteins. J Biol Chem 289:1271–1281

23. Guan XM, Kobilka TS, Kobilka BK (1992) Enhancement of membrane insertion and function in a type IIIb membrane protein following introduction of a cleavable signal peptide. J Biol Chem 267:21995–21998
24. Lampe M, Pierre F, Al-Sabah S, Krasel C et al (2014) Dual single-scission event analysis of constitutive transferrin receptor (TfR) endocytosis and ligand-triggered beta2-adrenergic receptor (beta2AR) or Mu-opioid receptor (MOR) endocytosis. Mol Biol Cell 25:3070–3080
25. Keppler A, Pick H, Arrivoli C, Vogel H et al (2004) Labeling of fusion proteins with synthetic fluorophores in live cells. Proc Natl Acad Sci U S A 101:9955–9959
26. Hall RA, Ostedgaard LS, Premont RT, Blitzer JT et al (1998) A C-terminal motif found in the β_2-adrenergic receptor, P2Y1 receptor and cystic fibrosis transmembrane conductance regulator determines binding to the Na^+/H^+ exchanger regulatory factor family of PDZ proteins. Proc Natl Acad Sci U S A 95: 8496–8501
27. Hoffmann C, Gaietta G, Zurn A, Adams SR et al (2010) Fluorescent labeling of tetracysteine-tagged proteins in intact cells. Nat Protoc 5:1666–1677
28. Hoffmann C, Nuber S, Zabel U, Ziegler N et al (2012) Comparison of the activation kinetics of the M_3 acetylcholine receptor and a constitutively active mutant receptor in living cells. Mol Pharmacol 82:236–245

Chapter 9

Autoradiography Assessment of Muscarinic Receptors in the Central Nervous System

Vladimir Farar and Jaromir Myslivecek

Abstract

The detection of muscarinic receptor binding sites is a crucial step in many experimental conditions. Although in peripheral tissue, the radioligand binding (see appropriate chapter) allows to obtain well-defined receptor characteristic, and also is usable in some central nervous system regions, when trying to determine receptor binding in central nervous system regions with low density or with infinitesimally small receptor changes, the receptor autoradiography is a better method. The development of this method made important progress, and some different modes (phosphor imaging) are used nowadays. Here, we describe muscarinic receptor detection using different radioligands: [^{3}H]-QNB, [^{3}H]-NMS, [^{3}H]-pirenzepine, and [^{3}H]-AFDX-384. Specific attention is paid to the detection of subtypes of muscarinic receptors and the limits of the method are emphasized.

Key words Muscarinic receptors, Autoradiography, [^{3}H]-QNB, [^{3}H]-NMS, [^{3}H]-pirenzepine, [^{3}H]-AFDX-384, Central nervous system

Abbreviations

^{3}H-QNB	^{3}H-quinuclidinyl benzilate, ^{3}H-1-azabicyclo[2.2.2]oct-3-yl 2-hydroxy-2, 2-diphenylacetate
^{3}H-NMS	^{3}H-*N*-methyl-scopolamine, ^{3}H-(1R,2S,4R,5S,7R)-{[(2R)-3-hydroxy-2-phenylpropanoyl]oxy}s[19]-9,9-dimethyl-3-oxa-9-azoniatricyclo [3.3.1.02,4]nonane
^{3}H-pirenzepine	^{3}H-11-[(4-methylpiperazin-1-yl)acetyl]-5,11-dihydro-6H-pyrido[2,3-b][1,4]benzodiazepin-6-one
^{3}H-AFDX-384	^{3}H-*N*-(2-[(2R)-2-[(dipropylamino)methyl]piperidin-1-yl]ethyl)-6-oxo-5H-pyrido[2,3-b][1,4]benzodiazepine-11-carboxamide
RT	Room temperature
MRs	Muscarinic receptors
GPCRs	G protein-coupled receptors

Jaromir Myslivecek and Jan Jakubik (eds.), *Muscarinic Receptor: From Structure to Animal Models*, Neuromethods, vol. 107, DOI 10.1007/978-1-4939-2858-3_9, © Springer Science+Business Media New York 2016

1 Historical Overview

Autoradiography is not a single method, but refers to a general concept shared by a family of experimental techniques. The aim of autoradiography is to visualize and quantify the distribution of radioactive substance within the specimen (e.g., acrylamide gels, agarose gels, nitrocellulose sheets, paper chromatograms, thin layer chromatograms, tissue sections) [1–3]. In principle, the binding of radioactive substance to specific target is not different from radioligand binding (*see* Chapter 3). The radiolabeled compound that is specific to a given receptor is allowed to bind to the receptor and then free radioligand is separated. The first autoradiography screen appeared accidentally when blackening was obtained on silver chloride (and iodide) emulsions (on a photographic plate) by uranium salts (uranium nitrate) [4]. This study and the subsequent work of Henri Becquerel and the Curies (1898) led to the discovery of radioactivity. However, the development of method as a technique to study biological structures became possible only after World War II when photographic emulsions became widely available [5].

Autoradiography technique is able to identify not only proteins but also nucleic acid fragments and can be used in polyacrylamide or agarose gel electrophoresis, in situ hybridization and in subcellular localization of radioactive drug product [1, 3, 6, 7].

One of the first review summarizing pioneering work using acetylcholine receptor imaging was study [8] in which nicotinic acetylcholine receptors were detected using [^{3}H]-bungarotoxin. In fact, autoradiography detection of acetylcholine receptors (nicotinic, i.e., on motor endplate) started in 1970s of the twentieth century. In the same time, Kuhar and Yamamura have published the study on localization of muscarinic receptors (MRs) in the rat brain [9]. Shortly after, autoradiography of muscarinic and their counter-regulatory receptors, adrenoceptors [10] showed its localization on cardiomyocytes in culture. Then, muscarinic receptors were detected in the retina [11], in presynaptic nerve terminals in the heart [12], and repeatedly in the central nervous system (for review *see* [13]). Less work was devoted to characterize muscarinic receptors in peripheral tissue like bladder [14]. Some snake toxins, that are more selective among muscarinic receptor subtypes, were radiolabeled with aim to prepare subtype specific ligand [15]. The main disadvantage of snake toxins is irreversibility of binding and not well-defined allosteric binding [15, 16]. Also, uncertain effects were revealed on adrenoceptors [15]—there is discrepancy between inhibited prazosin binding and abolishment of such binding by atropine: muscarinic toxins can also inhibit the binding of [^{3}H]-prazosin, an antagonist of α-adrenergic receptors. But, binding of radioactive toxins to rat brain was completely abolished by atropine, indicating that the toxins target only muscarinic receptors.

2 Principles of Receptor Autoradiography

Depending on the degree of anatomical resolution, autoradiography can be divided into micro-autoradiography (using exposure and development of photographic emulsion visualized by microscopic technique at cellular and subcellular level) and macro-autoradiography (using X-ray or autoradiography films to produce radioisotope distribution images with resolution at macroscopic level) [6, 7]. The detection and visualization of radioligand tissue distribution throughout the whole body at the level of organs and organ systems is called whole-body macro-autoradiography [7].

In general, there are two ways, how to detect the receptors with radioactive ligands. First is in vivo autoradiography when the receptors are labeled within intact living tissue by systemic or intracerebral administration of radioligand and distribution of radiolabeled receptors within the specimen is then determined ex vivo either by micro or macro-autoradiography [3, 7].

The idea of in vivo labeling of receptors and their visualization gave rise to development of noninvasive in vivo imaging techniques such as positron emission tomography (PET) and single photon emission computed tomography scanning (SPECT, [3]). For detailed description of PET method *see* Chapter 10. Second is in vitro autoradiography [17, 18], that uses slide-mounted tissue sections which are incubated with radioligand what is subject of this chapter. In vitro autoradiography offers several advantages over the in vivo autoradiography. These includes reduced amounts of the radioligand, precise control of the radioligand binding conditions (incubation buffer, pH, incubation temperature and time, exact concentration of the radioligand, duration and number of washings to remove unbound radioligand), use of ligands that do not cross blood–brain barrier or are not metabolically stable, ability to perform competition assays with unlabeled ligands and saturation studies in addition of regional mapping of the ligand binding sites [3, 19].

In principle, the receptor binding of radioactive substance in vitro autoradiography is not different from the radioligand binding to membranes or homogenates (*see* Chapter 3): The radiolabeled compound specific to a given receptor binds to the receptor and then free radioligand is separated.

While receptor autoradiography can provide valuable information about the distribution and density of receptor binding sites it does not evaluate their signaling capacity. In case of distinct G protein-coupled receptors (GPCRs) a specialized in vitro receptor autoradiography was developed to address this issue. While preserving the high degree of anatomical resolution, [^{35}S]GTPγS binding autoradiography, also known as functional receptor autoradiography, was developed [20]. This technique combines the advantages of classical in vitro receptor autoradiography and

[^{35}S]GTPγS binding assay on cell membranes to provide information about signaling capacity – functional state of GPCRs at high anatomical resolution. The aim of receptor autoradiography is visualization and quantification of the binding sites of a given receptor. By contrast functional autoradiography aims to visualize the downstream signaling event upon agonist stimulation. In detail, this method studies CNS distribution of [^{35}S]GTPγS binding under agonist stimulation in the presence of GDP (suppress basal binding—*see* [20] for details) and Mg^{2+} (shifts the equilibrium towards the high-affinity state of receptor). Specificity of binding should be confirmed by antagonizing effects of atropine (in case of MRs). Thus only those potentially active GPCRs (i.e., those receptor-G protein complexes that are still able to function in tissue sections) are recruited and visualized in functional autoradiography. For detailed description of this method, its advantages and limitations *see* [20].

The basic principle of any autoradiography is creation of an image that specifically shows distribution of the radioligand binding within the specimen, the autoradiogram. The radioactive decay of radiolabeled ligand that is bound to a protein or nucleic acid generates changes in detection media that can be autoradiography film exposed directly to the radioactive decay, direct exposure with an intensifying screen [1, 21] and fluorography exposure (fluorography, that is not subject of this chapter) [17]. There are multiple methods how to detect radioactivity in situ: the film autoradiography, the electronic autoradiography, and the biomaging/phosphor imaging. Based on the way how autoradiogram is acquired autoradiography can be divided into direct and indirect autoradiography. A typical example of direct autoradiography is contact film autoradiography when ionizing radiation, that is emitted by the radioligand, is detected by autoradiography film to generate a latent image. Then the autoradiogram is developed by photographic processing of the film. In indirect autoradiography such as phosphor imaging or the use of trans-screens where the radioactive signal is first converted into the light which is subsequently detected with phosphor imager or autoradiography film to generate digitalized autoradiogram or latent image, respectively.

The isotopes usually used in autoradiography determination are listed in Table 1.

For further use we will focus on the in vitro autoradiography of receptors, with specific aim to describe in vitro autoradiography of muscarinic receptors.

2.1 Film Autoradiography

There is a broad range of commercially available films that can be used in autoradiography to detect directly or in conjunction with intensifying screens or trans-screens the radiation within the specimen (e.g., Kodak, Amersham (GE Healthcare)). The selection of autoradiography film type depends on the radioisotope used in the assays, the way of exposure, importance of sensitivity, resolution and speed of detection [2, 21].

Table 1
Radioligands of choice and their half-lives, specific activity (SA), maximal energy, tissue range, and decay

Radionuclide	Half-life	Maximal specific activity (Ci/mol)	Decay mode	Energy (max.) (MeV)	Max. tissue range	Application
^{14}C	5730 years	6.7×10^{1}	β	0.156	0.008	1, 2, 3, 4, 5, 6
^{3}H	12.43 years	2.9×10^{4}	β	0.0186	0.301	4, 5
^{35}S	87.4 days	1.5×10^{6}	β	0.167	0.042	1, 2, 3, 4, 5, 6
^{32}P	14.3 days	9.2×10^{6}	β	1.709	2.750	1, 2, 3
^{125}I	60.0 days	2.2×10^{6}	γ	0.035	0.019	5, 6
^{131}I	8.04 days	1.6×10^{7}	β	0.364	2.300	5

Notes (explanation of Application legend): 1: Southern blots, 2: Northern blots, 3: DNA sequencing, 4: Protein synthesis, 5: In situ hybridization, 6: Western blots
Adapted from refs. [17, 2, 19]

Based on the type of exposure method, film autoradiography can be divided into direct and indirect film autoradiography. In direct exposure procedure, the radioactive specimen is directly apposed to autoradiography film in light-tight autoradiography cassette. In indirect type of exposure such as trans-screen intensifying-screen exposure, the radiolabeled specimen is tightly apposed to a medium that converts ionizing radiation into the light emission that is then detected by the autoradiography film. Converting ionizing radiation to light greatly increases detection efficiency [2, 21].

A typical autoradiography film is composed of a polyester support (clear or colored) that is coated either on one side or on both sides by emulsion of light-sensitive silver halide grains and gelatin. The emulsion layers are then protected by an anti-scratch layer. When the photographic emulsion is exposed to ionizing radiation the silver ions are converted into the silver atoms to produce a stable latent image. In subsequent film processing these few silver atoms catalyze the reduction of the silver halide crystal to metallic silver to produce visible image of the radioisotope distribution within the specimen [2, 21].

Single-coated autoradiography films (e.g., Biomax MR) are optimized for detection of medium-energy radioisotopes (e.g., ^{14}C, ^{35}S, and ^{33}P) in direct exposure mode. When used with high-energy isotopes (e.g., ^{125}I, ^{32}P) they provide maximum resolution. However, because of the anti-scratch protective layer, they are not suitable for detection of ^{3}H (what is the case of muscarinic receptor ligands) [1, 21]. ^{3}H is a β-emitter with low energy, that has only limited permeability through the protective layer to expose the photographic emulsion. Therefore specialized film lacking the overcoat layer and thus allowing direct contact of ^{3}H emission with photographic emulsion has to be used. Then, a special attention has to be paid to handling these films to avoid their damage [19, 21]. The direct exposure is done at room temperature (RT) [21].

Alternatively, ^{3}H can be detected using trans-screen, intensifying screen system. In this type of exposure, the radioactive specimen is apposed to the surface of phosphor layer coated on a thin clear plastic base. The phosphor layer captures the beta particles and converts their energy to photons. Photons are then directed to the autoradiography film placed between the phosphor layer and reflecting layer. Films that are spectrally matched to the trans-screen are used. The optimum exposure temperature range (determined by experience) is –70 to –80 °C [21]. In general, detection of emitted light is significantly improved by reducing the exposure temperature. The latent image center forming in the silver grain becomes more stable, thus reducing latent-image fading (signal loss). The detection sensitivity increases because the latent image accumulates with additional light exposure.

The conversion of ionizing radiation to light emissions is also the principle of intensifying screen exposure. Intensifying screens greatly increase the sensitivity and speed of detection. In this type of exposure, the radiolabeled specimen is placed in direct contact with one side of double-coated autoradiography film and intensifying screen is apposed to the other side of the film. The radioisotope must have sufficient energy (e.g., ^{32}P, ^{125}I) to pass through the autoradiography film before reaching the intensifying phosphors in the screen. The ionizing radiation (γ-rays and high energy β-particles) is then converted by the screen into the light emission that is detected by the film. The optimum exposure temperature range is –70 to –100 °C [1, 21].

Impregnation of the radiolabeled specimen with a scintillator to convert ionizing radiation to light emission and thus enhance detection efficiency by autoradiography film is called fluorography. Fluorographic exposure is used for detection of low- and medium-energy radioisotopes such as ^{3}H, ^{14}C, and ^{35}S. The optimum exposure temperature range is –70 to –100 °C [1, 2, 21].

It is necessary to perform three steps when detecting radioligand distribution within the specimen using film autoradiography: direct or indirect exposure of radiolabeled specimen to the autoradiography film, film development at the end of exposure and evaluation of autoradiograms. The length of exposure for a given radioligand is subject to preliminary experiments. This time depends on the type of radioisotope and radioactivity amount applied to specimen [18, 21]. It is necessary to expose sensitive film to calibrated amount of radioactivity. If the autoradiography resolution is comparable to that of the original chromatogram, the exposition is properly done [17].

Autoradiograms obtained by film autoradiography are analyzed by film densitometry using computer-assisted densitometric systems. Quantification of radioactivity amount bound to the protein (nucleic acid) requires comparison of the measured optical density to a radiation response curve generated using standards on the same media (film, or storage phosphor screen as mentioned below) [17, 19]. Standards can be purchased as radiolabeled plastic polymers. Alternatively, it is possible to prepare standards by labeling tissue paste homogenates or plastics with defined amounts of radioisotope. The apparent disadvantage of autoradiography films is their narrow linear dynamic range of response. The relationship between incident-radiation exposure and optical density is however linear only over limited range. Eventually the media can be saturated and will not be more sensitive to additional exposure. In such cases, the analysis of autoradiograms is not possible. As a result, multiple exposures of the same specimen are often required [19] to determine optimal exposure time. Exposure time depends on the type of isotope and amount of radioactivity applied to the plate. In case of film autoradiography, both the narrow linear dynamic range (from 300 to 1) of the method (when using image analysis) and the lack of sample preservation (in zonal analysis) constitute a definite disadvantage. The linear dynamic ranges of the phosphor image technique are wider (by at least 10^5) than is needed for detection in film autoradiography.

Film development can be made automatically using fully automated benchtop film processors or manually in tanks and trays using film developers and fixators. In principle, developing, rinsing, fixing, washing, and drying are five steps of this process, and the freshness of chemicals is a critical point that can affect the quality of results. For details on manual processing of autoradiography films please *see* ref. 21.

2.2 Bioimaging/ Phosphor Imaging

Storage phosphor screen imaging technology was introduced by Fuji Company in 1980s of the twentieth century. Now, after more than 30 years, other suppliers of this material are on the market (e.g., GE Healthcare, Perkin Elmer, Bio-Rad). The exposition time in phosphor imaging is approximately 1 week, while in classical film autoradiography it is as long as several weeks. This method is similar to film autoradiography but has some advantages, mainly shortening of exposition time (in comparison to working with tritium labeled compounds). In some papers, this type of autoradiography is called "filmless autoradiography" referring to usage of other medium than film.

Phosphor imaging technique uses storage phosphor screens to detect and store energy of ionizing radiation in a stable state to generate latent images of the distribution of radioisotopes within the specimen. Phosphor imaging screens consists of support, photostimulatable phosphor layer (crystals of $BaFBr:Eu^{2+}$) and protective layer. These crystals are able to store the energy in crystal vacancies when irradiated. Then the luminescence occurs, that is evoked by change of Eu^{2+} ions into Eu^{3+} that leads to electron release.

Electrons are trapped in the Br vacancies and color centers are formed [17]. Screens with protective layer are suitable for detection of radioisotopes such as ^{32}P, ^{125}I, ^{35}S, ^{33}P, and ^{14}C, but not for ^{3}H. Similarly to autoradiography film, ^{3}H signal has to be directly accessible to photo-sensitive phosphor layer and imaging plates constructed without protective coat has to be used.

For exposure with phosphor imaging screens, the radioactive specimen is directly apposed to the phosphor screen in light-tight autoradiography cassette. After exposure, latent images are developed and digitized using phosphor imager. The screen is removed from autoradiography cassette under safe-light conditions and placed into the phosphor imager for reading. During scanning the phosphors are activated by a laser beam (633 nm) that leads to the release of stored energy as luminescence. The intensity of released luminescence is proportional to the amount of radioactivity in the specimen. The luminescence is recorded and stored in relation to the position of the laser beam to generate images that are then displayed on video monitor. The stored images are subsequently analyzed with an appropriate software [1].

Phosphor imaging technique offers several advantages over the film autoradiography. These include a linear dynamic range over five orders of magnitude (1.5 orders of magnitude in case of film), higher sensitivity (up to 100 times, depending on sample type and radioisotope used), and reduced exposure times (up to one-tenth, depending on sample type and radioisotope used) [1, 22]. The high sensitivity and wide linear dynamic range allows to detect and analyze weak and strong signals simultaneously. The results come out already in digitized form, which facilitates analysis of autoradiograms. Unlike autoradiography film, storage phosphor screens with protective layer can be used repeatedly if handled carefully. Unfortunately this is not the case for ^{3}H sensitive screens, which lack the protective layer and can be easily contaminated or damaged by exposure to the specimen. However, when ^{3}H sensitive screens are used with thin tissue sections mounted on glass-slides, the surface of the screen that has not been in contact with radiolabeled tissue section can be reused.

The most but not all of stored information is released upon scanning with the phosphor imager. For further use of the screen, the remaining signal must be erased with bright visible light (there are commercially available erasers). This is an important step that cannot be omitted as not properly erased screen retains images from previous exposure that can interfere with the image actually analyzed.

2.3 Electronic Autoradiography

This technique differs from previously mentioned one in the absence of storage medium. The radioactivity is measured directly with imaging detectors. Berthold's digital autoradiograph and Instant-imager (Canberra-Packard) are used for these purposes.

The principle of digital autoradiograph is based on measuring position and intensity of two-dimensional distributions of ionizing radiation on the surface (thin-layer chromatography plate) [17].

Instant imager consists of two sections [17], the microchannel array plate and a multiwire chamber and is continuously flowed by gas (argon, CO_2, and isobutene). The principle is based on gas ionization in one of the microchannels by beta-particle which is emitted from a source of radioactivity. This leads to the production of electrons, that are accelerated by the high electric field in the microchannel. This process leads to the further gas ionization and production of electron cloud, that migrates up an electric field gradient into the multiwire chamber.

3 Advantages and Disadvantages of Method

The principle of ligand binding to receptors is same as in radioligand binding studies. But, there are some advantages of autoradiography method in comparison to binding studies. While radioligand binding studies are restricted to the brain areas that can be precisely dissected, in vitro autoradiography allows explore radioligand binding to MRs in very well-defined brain regions. In case of smaller brain areas or regions with lower MR density it is necessary to pool tissue from more animals. In addition, binding in homogenates/membrane fractions is limited by the density of MR in the sample. By contrast, in vitro autoradiography has high sensitivity allowing explore brain regions even with few MR. The use of very thin tissue sections in vitro autoradiography provides several advantages over the large tissue blocks used in binding studies in homogenates/membrane fractions. The brain sectioning allows analyzing MR density in virtually all brain areas of a single animal greatly reducing the number of experimental animals. Moreover, sectioning of a single brain generates sufficient number of tissue sections to explore the binding of multiple radioligands in a particular brain area of the same animal. This further reduces the number of experimental animals and allows comparing the effect of treatment on multiple targets (receptors, transporters) in a single animal.

The important question, also discussed in the chapter on Radioligand binding, is the selectivity of radioligand used in the experiments. A general problem in identification of muscarinic receptor subtypes present in specific central nervous system area is the lack of highly subtype-selective muscarinic antagonists. The muscarinic receptor subtypes affinities for pirenzepine and AFDX-384 are shown in Table 2. It can be deduced from this table that both pirenzepine and AFDX-384 has high affinity not only for M_1, M_2 muscarinic receptors, respectively, but also for M_4 muscarinic receptor subtype. In radioligand binding studies, it is therefore

Table 2
Pirenzepine and AFDX-384 affinity constants (log affinity or pKi values) for muscarinic receptor subtypes

	Receptor subtype				
Antagonist	M_1	M_2	M_3	M_4	M_5
Pirenzepine	7.8–8.5	6.3–6.7	6.7–7.1	7.1–8.1	6.2–7.1
AF-DX 384	7.3–7.5	8.2–9.0	7.2–7.8	8.0–8.7	6.3

Data were obtained from ref. [35]

necessary to use a combination of various antagonists. However, for autoradiography detection this approach is not suitable because of evaluation limitations of such changed "binding." Thus, the present protocols for M_1 and M_2 muscarinic receptor subtypes identification should be considered as method for detection of M_1 (or M_2) and also small portion of M_4 muscarinic receptors. Unfortunately, only few papers report these binding sites as M_1/M_4 muscarinic receptors (e.g., [23, 24]).

Historically, tritiated pirenzepine was used as ligand that binds to muscarinic receptors with distinct binding in specific brain areas [25]. Further, distinct distribution was found in the central nervous system. [^{3}H]-QNB and [^{3}H]-pirenzepine both label regions of the cerebral cortex, hippocampus, striatum, and dorsal horn of the spinal cord, while sites in the cerebellum, nucleus tractus solitarius, facial nucleus, and ventral horn of the spinal cord are labeled with [^{3}H]-QNB and not by [^{3}H]-pirenzepine [26]. These observations indicated binding to different subtypes of muscarinic receptors. This was further expanded to definition of binding sites as M_1 muscarinic receptors [27] and in the middle of 1980s pirenzepine binding sites were considered as M_1 muscarinic receptors [28, 29]. On the other hand, AFDX-384 was from the beginning considered as M_2 muscarinic receptor specific ligand [30] and some authors became aware of limited selectivity (e.g., [31]). In many cases, however, [^{3}H]-AFDX-384 and [^{3}H]-pirenzepine are considered as selective ligands [32, 33].

4 Equipment, Materials, and Setup

Before the start of work it is necessary to prepare all equipment, surgical instruments, buffers, and all other things that are essential for the method. Appropriate preparation is a basic condition for successful work. Some details about items listed below are discussed in other sections of this chapter.

The items are as follows:

Items for tissue dissection and preparation of glass slide-mounted tissue sections:

1. Target tissue (mice or rat brain). For tissue preparation *see* Section 5.
2. Surgical instruments (e.g., tweezers, forceps, scissors).
3. Gelatine-coated standard microscope glass slides (25 × 5 mm) with frosted ends allowing marking of individual slides or ready to use glass slides treated by manufacturer to attract and firmly bind frozen tissue sections without additional coating of the glass-slide surface (e.g., Superfrost® Plus glass slides).
4. Pencil (for labeling the glass slides).
5. Paintbrush (flat or round with small diameter, for tissue manipulation).
6. Parafilm.
7. Dry ice (better supplied as powdered dry ice).
8. Polystyrene box (for freezing the freshly dissected tissue).
9. Cryostat operating between −10 and −25 °C (e.g., Leica CM3050S).
10. Tissue glue matrix (e.g., Tissue OCT from Labonord).
11. Slide boxes.
12. Desiccant (silica gel with an indicator of moisture content).
13. Isopentane.
14. Brain atlas (e.g., very useful is Paxinos' atlas [34]).

Items for preparation of gelatin-coated microscope glass slides:

15. Gelatin.
16. Chromium(III) potassium sulfate dodecahydrate ($KCr(SO_4)\cdot 12H_2O$.
17. Distilled water.
18. Paper filter.
19. Glass funnel.
20. Slide boxes.
21. Desiccant.

Items for specific labeling of MR in brain sections:

22. Liquid scintillation counter (e.g., Beckman Coulter).
23. Scintillation vials.
24. Scintillation cocktail (e.g., GE Healthcare Life Sciences or homemade (see below)).
25. Pieces of glass fiber filter paper.
26. Hair dryer operating at room temperature.

27. Slide staining set (staining dishes, slide holder).
28. Drain rack.
29. Polystyrene box.
30. Crushed ice.
31. Distilled water.
32. 50 mM sodium/potassium phosphate buffer, pH = 7.4.
33. Atropine sulfate.
34. Radioligands ([^{3}H]-QNB and [^{3}H]-NMS, [^{3}H]-pirenzepine, and [^{3}H]-AFDX-384).

Items for preparation of liquid scintillation cocktail:

35. Naphthalene.
36. 2,5-Diphenyloxazole (PPO).
37. 1,4-Bis(5-phenyl-2-oxazolyl)benzene (POPOP).
38. Methanol.
39. Ethylene glycol.
40. 1,4-Dioxane.

Items for receptor binding evaluation:

41. ^{3}H sensitive storage phosphor screen (e.g., storage phosphor screen BAS IP-TR).
42. Phosphor imager (e.g., Typhoon FLA7000 biomolecular imager).
43. PC with WinXP or Win7 for results evaluation using specific software.
44. Computer-assisted densitometric system (e.g., MCID).
45. Suitable tritium standards on glass slides (e.g., American Radiolabeled Chemicals).
46. Light-tight autoradiography exposure cassette (e.g., Carestream Kodak Biomax).
47. Cartridge paper.
48. Double-sided tape.

5 Procedures

Here we will provide protocols for muscarinic receptors determination using four radioligands: unspecific [^{3}H]-QNB (*see* Fig. 1) and [^{3}H]-NMS (*see* Fig. 2), M_1 specific [^{3}H]-pirenzepine (*see* Fig. 3) and M_2 specific [^{3}H]-AFDX-384 (*see* Fig. 4). In our laboratory we use storage phosphor screen imaging technology and thus this technique is described further.

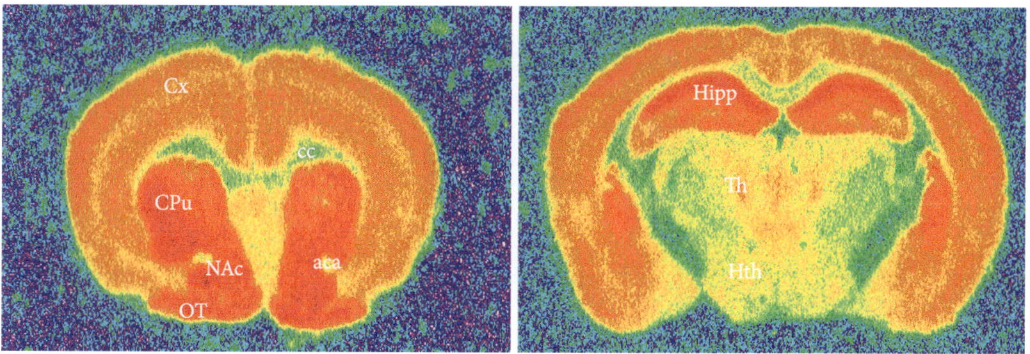

Fig. 1 Illustrative autoradiograms of [^{3}H]-QNB binding in coronal brain sections in wild type mice. *Aca* anterior commissure, anterior part, *cc* corpus callosum, *CPu* caudate putamen, *Cx* cortex, *Hipp* hippocampus, *Hth* hypothalamus, *OT* olfactory tubercle, *Th* thalamus

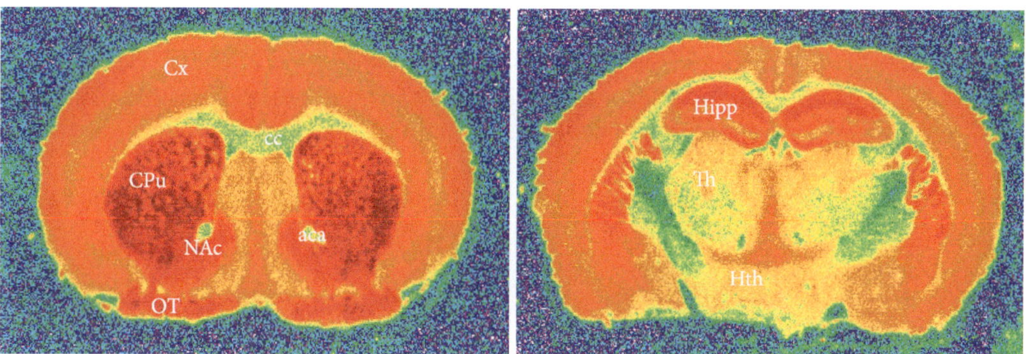

Fig. 2 Illustrative autoradiograms of [^{3}H]-NMS binding in coronal brain sections in wild type mice. *Aca* anterior commissure, anterior part, *cc* corpus callosum, *CPu* caudate putamen, *Cx* cortex, *Hipp* hippocampus, *Hth* hypothalamus, *OT* olfactory tubercle, *Th* thalamus

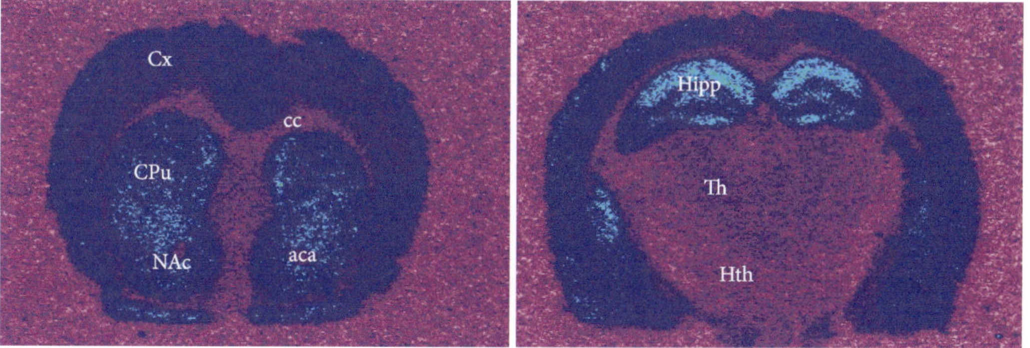

Fig. 3 Illustrative autoradiograms of [^{3}H]-pirenzepine binding in coronal brain sections in wild type mice. Note the barely visible labeling in Th and Hth, brain areas in which M_1 are practically absent. *Aca* anterior commissure, anterior part, *cc* corpus callosum, *CPu* caudate putamen, *Cx* cortex, *Hipp* hippocampus, *Hth* hypothalamus, *OT* olfactory tubercle, *Th* thalamus

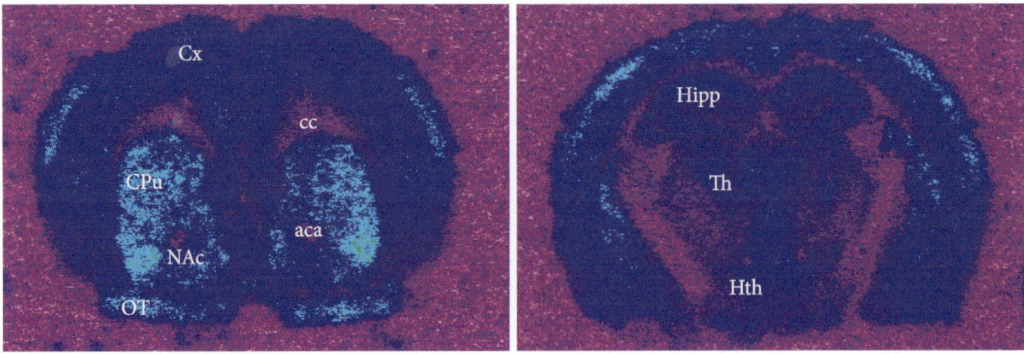

Fig. 4 Illustrative autoradiograms of [^{3}H]-AFDX-384 binding in coronal brain sections in wild type mice. *Aca* anterior commissure, anterior part, *cc* corpus callosum, *CPu* caudate putamen, *Cx* cortex, *Hipp* hippocampus, *Hth* hypothalamus, *OT* olfactory tubercle, *Th* thalamus

In vitro autoradiography of brain muscarinic receptors (MR) is a powerful tool to assess the distribution and MR changes in a variety of experimental models (drug treatment, genetic manipulations, ontogenetic studies) at a high degree of anatomical resolution and reproducibility of results.

Previously the apparent disadvantage of in vitro autoradiography of MR using [^{3}H]-labeled compounds and standard [^{3}H]-sensitive autoradiography films was the time of exposure of radiolabeled sections to the films to detect the bound radioactivity and generate autoradiogram, which could take as long as several weeks. Nowadays, the use of phosphor imaging techniques overcomes this issue and autoradiography of MR can be completed within 1 week.

Similarly to in vitro autoradiography of other receptors, in vitro autoradiography of MR consists of three major steps.

1. Tissue processing and preparation of glass slide-mounted tissue sections.
2. Preincubation, incubation of sections with appropriate radioligand and washing.
3. Detection, visualization and analysis of autoradiographic signal.

5.1 Tissue Processing and Preparation of Glass Slide-Mounted Tissue Sections

1. Sacrifice experimental animal according to the regulations concerning the handle and use of laboratory animals.
2. Rapidly remove the brain from the skull and immediately freeze the brain in isopenthane cooled to −30 °C. Alternatively put the brain on the piece of parafilm and place it on powdered dry ice in a closed polystyrene box. The rapid processing of fresh tissue prevents or minimizes the postmortem loss of receptor-binding sites.

3. If the brains are not directly used for cryosectioning, store the brains at −80 °C. Before cryosectioning of brains stored at −80 °C, transfer brains from −80 to −20 °C for 3 h.
4. Attach the brain to the specimen disc that holds the specimen during sectioning (cryostat chuck).
5. Cover the cryostat chuck with a layer of tissue glue matrix and place it into the quick freeze shelf inside the cryostat chamber.
6. If the cryostat is not equipped with system that secures rapid freezing of the disc, perform the mounting of the brain to the chuck inset into the powdered dry ice.
7. Choose the right orientation of the brain for sectioning of coronal, sagittal or axial brain slices. Always use steel tweezers cooled to the cryostat temperature to manipulate with brain.
8. Once the glue begins to freeze immerse the brain into the glue perpendicular to the chuck. Within few seconds the glue matrix is frozen. Subsequently you can add an additional layer of glue to secure the firm fixation of the brain during the sectioning.
9. Insert the specimen disc with fixed brain into the specimen head. The optimal temperature of the cutting is usually between −15 and −20 °C. The temperature within cryostat chamber and temperature of specimen strongly influences the quality of sections. The appropriate temperature should be set in preliminary experiments.
10. Trim the brain to the anatomical level corresponding to your brain area of interest. Set the appropriate thickness of sections and sectioning speed (in case of motorized devices). When sectioning manually, rotate the hand wheel evenly and at uniform speed. It is recommended to discard first 2–3 sections before tissue section collecting. The most common section thickness is 16 μm. Thinner sections can be cut to collect more sections through a desired brain region.
11. Cut the brain section of appropriate thickness. The tissue section will remain on the knife or knife holder. Use paintbrush to manipulate with section and make it accessible to the glass slide.
12. Thaw and mount the section at the bottom of the glass slide. Place the glass slide of RT over the section as closely as possible. The section will stick to the glass slide. Dry the section at room temperature. Store the dried sections in the slide boxes with desiccant at −80 °C.

Typically four sections are collected on one glass slide. To increase the homogeneity of samples collect every fifth or tenth section on an individual slide (depending on the size of brain area). For brain areas that span several hundred micrometers through the brain (e.g., striatum, dorsal hippocampus) collect brain sections on a set of ten slides. During the collecting sections keep the slides at RT. Thaw mount the first section on the first slide, the second section on the second slide and the tenth section on the last slide. Collect the next ten sections in the same way (thaw mount the 11th section on the first slide, 12th on the second slide, etc.). Collect another two sets of sections. Finally, at individual slide you will have four sections while each section will be 160 μm distant from the previous one.

Recipe 1. Preparation of gelatin-coated microscope glass slides

1. Dissolve 10 g of gelatin in 1 L of distilled water at 60 °C.
2. Dissolve $KCr(SO_4)\cdot 12H_2O$ in gelatin solution chilled to RT.
3. Filter the solution through the paper filter.
4. Store at RT in a dark place.
5. Dip slides into the gelatin solution for 5 min.
6. Dry slides for 24 h at 37 °C in the presence of desiccant
7. Store slides in slide boxes containing desiccant in a dry place.

5.2 Specific Labeling of MR in Brain Sections

Each assay consists of four major steps:

1. Pre-incubation of tissue sections to remove endogenous ligands.
2. Incubation of tissue sections in incubation medium comprising the particular radioligand.
3. Washing of tissue sections to remove unbound radioligand from tissue.
4. Rapid drying of labeled sections to prevent diffusion of radioligand bound to receptors.

Steps 1–3 are described in detail below:

1. Prepare five staining dishes. Fill one with maximal volume of pre-incubation medium, the second one with incubation medium containing radioligand at appropriate concentration at suitable volume. The volume of incubation medium is chosen to following two conditions: First, it has to be sufficient to cover all tissue sections. Second, the concentration of radioligand in incubation medium should be only marginally affected by binding. This can be experimentally addressed by sampling the incubation medium before, during and after incubation and determining concentration of free radioligand by liquid scintillation counting. Concentration of free radioligand should not differ among determinations.
2. Shortly before the end of incubation period fill two dishes with ice-cold washing medium and one with ice-cold water. Place the dishes containing ice-cold solutions into the ice in polystyrene box to maintain their temperature.

Recipe 2. Preparation of liquid scintillation cocktail

1. Add 120 g of naphthalene, 8 g of PPO and 0.4 g of POPOP to a beaker.
2. Add 500 ml of dioxane, 200 ml of methanol and 40 ml of ethylene glycol to a beaker.
3. Stir until naphthalene dissolves (ca. 30 min).
4. Add 1260 ml of dioxane and stir for 2 h.

Detailed protocols for above mentioned radioligands are in further sub-sections.

5.2.1 Labeling of MR with [^{3}H]-QNB

1. Remove the slide-mounted tissue sections from the freezer and allow them to thaw and dry for 20 min at RT. Transfer the slides into the slide holder.
2. Pre-incubate dried sections for 30 min in 50 mM sodium/potassium phosphate buffer (pH 7.4) at RT (pre-incubation medium).
3. Transfer the sections into the fresh 50 mM sodium/potassium phosphate buffer (pH 7.4) containing 2 nM [^{3}H]-QNB (incubation medium) and incubate for 2 h at RT.
4. To determine nonspecific binding of [^{3}H]-QNB, label adjacent sections with incubation medium supplemented with 10 μM atropine (final concentration) sulfate in the medium.
5. Wash the sections for 5 min in ice-cold 50 mM sodium/potassium phosphate buffer (pH 7.4).
6. Wash the sections second time for 5 min in fresh ice-cold 50 mM sodium/potassium phosphate buffer (pH 7.4).
7. Dip the slides for 2 s in ice-cold distilled water.
8. Immediately place the slides upright in the drain rack with tissue sections on the top and dry sections with gentle stream of room temperature air.
9. Store the dried section in slide boxes containing desiccant.

5.2.2 Labeling of MR with [^{3}H]-NMS

1. Remove the slide-mounted tissue sections from the freezer and allow them to thaw and dry for 20 min at RT. Transfer the slides into the slide holder.
2. Pre-incubate dried sections for 30 min in 50 mM sodium/potassium phosphate buffer (pH 7.4) at RT.
3. Transfer the sections into the fresh 50 mM sodium/potassium phosphate buffer (pH 7.4) containing 2.5 nM [^{3}H]-NMS and incubate for 1 h at RT.
4. To determine nonspecific binding of [^{3}H]-NMS, label adjacent sections with incubation medium supplemented with 10 μM atropine (final concentration) in the medium.

5. Wash the sections for 5 min in ice-cold 50 mM sodium/potassium phosphate buffer (pH 7.4).
6. Wash the sections second time for 5 min in fresh ice-cold 50 mM sodium/potassium phosphate buffer (pH 7.4).
7. Dip the slides for 2 s in ice-cold water.
8. Immediately place the slides upright in the drain rack with tissue sections on the top and dry sections with gentle stream of room temperature air.
9. Store the dried section in slide boxes containing desiccant.

5.2.3 Labeling of MR with [^{3}H]-pirenzepine

1. Remove the slide-mounted tissue sections from the freezer and allow them to thaw and dry for 20 min at RT. Transfer the slides into the slide holder.
2. Pre-incubate dried sections for 30 min in 50 mM sodium/potassium phosphate buffer (pH 7.4) at RT.
3. Transfer the sections into the fresh 50 mM sodium/potassium phosphate buffer (pH 7.4) containing 5 nM [^{3}H]-pirenzepine and incubate for 1 h at RT.
4. To determine nonspecific binding of [^{3}H]-pirenzepine, label adjacent sections with incubation medium supplemented with 10 μM atropine sulfate (final concentration) in the medium.
5. Wash the sections for 5 min in ice-cold 50 mM sodium/potassium phosphate buffer (pH 7.4).
6. Wash the sections second time for 5 min in fresh ice-cold 50 mM sodium/potassium phosphate buffer (pH 7.4).
7. Dip the slides for 2 s in ice-cold distilled water.
8. Immediately place the slides upright in the drain rack with tissue sections on the top and dry sections with gentle stream of room temperature air.
9. Store the dried section in slide boxes containing desiccant.

5.2.4 Labeling of MR with [^{3}H]-AFDX-384

1. Remove the slide-mounted tissue sections from the freezer and allow them to thaw and dry for 20 min at RT. Transfer the slides into the slide holder.
2. Pre-incubate dried sections for 30 min in 50 mM sodium/potassium phosphate buffer (pH 7.4) at RT.
3. Transfer the sections into the fresh 50 mM sodium/potassium phosphate buffer (pH = 7.4) containing 2 nM [^{3}H]-AFDX-384 and incubate for 1 h at RT.
4. To determine nonspecific binding of [^{3}H]-AFDX-384, label adjacent sections with incubation medium supplemented with 10 μM atropine sulfate (final concentration) in the medium.
5. Wash the sections for 5 min in ice-cold 50 mM sodium/potassium phosphate buffer (pH 7.4).

6. Wash the sections second time for 5 min in fresh ice-cold 50 mM sodium/potassium phosphate buffer (pH 7.4).
7. Dip the slides for 2 s in ice-cold distilled water.
8. Immediately place the slides upright in the drain rack with tissue sections on the top and dry sections with gentle stream of room temperature air.
9. Store the dried section in slide boxes containing desiccant.

5.3 Generation of Autoradiograms

1. Cut the cartridge paper to fit the autoradiography cassette.
2. Organize the slides and standard in a logical manner. Use double-sided tape to attach slides and standard to cartridge paper. In case that the size of microscope slide layer does not match that of the storage phosphor screen use additional microscope slides for covering of the free surface of the paper sheet to secure uniform contact of the screen.
3. Place the sheet with fixed slides into autoradiography cassette.
4. Appose the storage phosphor screen to the labeled sections and standard.
5. Close autoradiography cassette and note the date and time.
6. Expose storage phosphor screen for the predetermined period of time at RT.
7. After exposure, dim the lights and remove storage phosphor screen from the autoradiography cassette.
8. Scan the storage phosphor screen in a phosphor imager.
9. Save the digitized autoradiograms for further analysis.

5.4 Analysis of Autoradiograms

Proper analysis of autoradiograms requires reliable anatomical identification of brain regions of interest. This can be achieved in several complementary ways. In all cases, a detailed brain atlas is indispensable tool for identification of regional anatomy. The use of brain atlas by Franklin and Paxinos [34] is very useful, as it provides brain sections stained for acetylcholinesterase, which shows distribution pattern similar to that of MR. Autoradiograms in itself can be used to directly identify distinct brain regions. The heterogeneous distribution of MR binding sites within the tissue sections and thus signal intensity within corresponding autoradiogram can provide readily observable details and contrasts to directly identify many brain regions. As can be seen in Fig. 1 the white matter is virtually devoid of labeling and can be clearly distinguished from surrounding grey matter which shows different levels of labeling. Accordingly, the white matter structures and tracts such as internal capsule, corpus callosum, anterior commissure, fornix, and others can be clearly recognized. In addition there are significant differences in MR density between individual brain regions. For instance the density of MR in striatum is about twofold higher than that in cortex.

The anatomical identification of brain regions can be facilitated by conventional histology techniques such as Nissl staining. Either those radiolabeled and used for exposure or adjacent tissue sections can be processed for Nissl staining [19].

There is specialized software for analysis of digitized autoradiograms generated by phosphor imagers (e.g., MCID). The basic procedure for quantifying digitized autoradiograms with MCID Analysis is as follows:

1. Load the autoradiographic image file.
2. Establish a density calibration.
3. Use any Sample tool to gather data from tissue section(s).
4. Repeat steps 1–3 for all remaining image files.
5. Summarize and save the data.

6 Conclusions

In vitro autoradiography of MR is widely recognized method in neuroscience and neuropharmacology that can bring out unique information about the regulation of MR abundance under physiological as well as pathophysiological conditions (drug treatment, genetic manipulations, stress conditions, physiological processes like learning and memory and many others) at a high degree of anatomical resolution and reproducibility of results. The use of radioligand membrane binding techniques has been useful tool to determine pharmacological properties of wide range of receptors in the peripheral tissues as well as in the brain and is the method of choice for kinetic and competition assays. The apparent limitations of these techniques is however the resolution at the anatomical level. In particular, mapping the distribution of receptor binding sites in discrete functionally and/or anatomically defined regions of the brain is not possible by membrane binding techniques. While the use of radioligand membrane binding techniques is restricted to larger brain areas that can be precisely dissected, in vitro autoradiography allows exploring radioligand binding to MR in very discrete brain regions. In vitro autoradiography has high sensitivity allowing to explore brain regions even with few MR. The brain cryosectioning allows analyzing MR density in virtually all brain areas of a single animal greatly reducing the number of experimental animals. Moreover, cryosectioning of a single brain generates sufficient numbers of tissue sections to explore the binding sites of additional radioligands in a particular brain area of the same animal. Alternate sections collected during cryosectioning can be also processed in multiple assays such as in situ hybridization, histochemistry and functional autoradiography. This further reduces the number of experimental animals and allows comparing the effect of treatment on multiple targets (receptors, coupling of receptors, transportes, mRNA, enzymes) in a single animal. There are multiple methods for detection and visualization of autoradiographic signal including sensitive media (film, screen) and media-free (electronic) autoradiography. As recommended method for this purposes, phosphor imaging with [^{3}H]-QNB and [^{3}H]-NMS, [^{3}H]-pirenzepine, and [^{3}H]-AFDX-384 is discussed in this chapter.

Acknowledgments

The research on this topic was supported by grant GAUK 328314 from Grant Agency of Charles University and by projects PRVOUK P25/1LF and PRVOUK P35/1LF.

References

1. Voytas D, Ke N (2001) Detection and quantitation of radiolabeled proteins and DNA in gels and blots. In: Frederick M Ausubel et al (eds.) Current protocols in molecular biology. Appendix 3: appendix 3A. doi:10.1002/0471142727.mba03as48
2. Laskey RA (1993) Efficient detection of biomolecules by autoradiography, fluorography or chemiluminescence. Principles of detection using radiographic film. Amersham Life Sci. Review 23:Part I
3. Chabot JG, Kar S, Quirion R (1996) Autoradiographical and immunohistochemical analysis of receptor localization in the central nervous system. Histochem J 28:729–745
4. de St Victor N (1867) Sur une noublle action de la lumière Sixième Mémoire. Hebdomadaire des Séances de l'Academie des Sciences 65: 505–507
5. Ross R (1966) Electron microscope autoradiography. Adv Tracer Methodol 3:131–137
6. Harvey B (2008) Autoradiography and fluorography. In: Rapley R, Walker J (eds) Molecular biomethods handbook. Humana, Totowa, pp 396–410. doi:10.1007/978-1-60327-375-6_26
7. Stumpf WE (2013) Whole-body and microscopic autoradiography to determine tissue distribution of biopharmaceuticals—target discoveries with receptor micro-autoradiography engendered new concepts and therapies for vitamin D. Adv Drug Deliv Rev 65:1086–1097. doi:10.1016/j.addr.2012.11.008
8. Porter CW, Barnard EA (1976) Ultrastructural studies on the acetylcholine receptor at motor end plates of normal and pathologic muscles. Ann N Y Acad Sci 274(1):85–107. doi:10.1111/j.1749-6632.1976.tb47678.x
9. Kuhar MJ, Yamamura HI (1975) Light autoradiographic localisation of cholinergic muscarinic receptors in rat brain by specific binding of a potent antagonist. Nature 253(5492): 560–561
10. Lane M-A, Sastre A, Law M, Salpeter MM (1977) Cholinergic and adrenergic receptors on mouse cardiocytes in vitro. Dev Biol 57(2): 254–269, doi:http://dx.doi.org/10.1016/0012-1606(77)90213-5
11. Sugiyama H, Daniels MP, Nirenberg M (1977) Muscarinic acetylcholine receptors of the developing retina. Proc Natl Acad Sci U S A 74(12):5524–5528
12. Hartzell HC (1980) Distribution of muscarinic acetylcholine receptors and presynaptic nerve terminals in amphibian heart. J Cell Biol 86(1):6–20
13. Hoss W, Messer W Jr (1986) Multiple muscarinic receptors in the CNS. Significance and prospects for future research. Biochem Pharmacol 35(22):3895–3901
14. Yoshida A, Fujino T, Maruyama S, Ito Y, Taki Y, Yamada S (2010) The forefront for novel therapeutic agents based on the pathophysiology of lower urinary tract dysfunction: bladder selectivity based on in vivo drug-receptor binding characteristics of antimuscarinic agents for treatment of overactive bladder. J Pharmacol Sci 112(2):142–150. doi:10.1254/jphs.09R14FM
15. Karlsson E, Jolkkonen M, Mulugeta E, Onali P, Adem A (2000) Snake toxins with high selectivity for subtypes of muscarinic acetylcholine receptors. Biochimie 82(9–10):793–806,

doi:http://dx.doi.org/10.1016/S0300-9084(00)01176-7

16. Olianas MC, Adem A, Karlsson E, Onali P (2004) Action of the muscarinic toxin MT7 on agonist-bound muscarinic M1 receptors. Eur J Pharmacol 487(1-3):65–72
17. Hazai I, Klebovich I (2003) Thin-layer radiochromatography. In: Sherma J, Fried B (eds) Handbook of thin-layer chromatography. Marcel Dekker, Inc., New York, pp 442–470
18. Kuhar MJ (2001) In vitro autoradiography. In: Enna SJ (editor-in-Chief) et al (eds.) Current protocols in pharmacology, Chapter 8:Unit 8.1. doi:10.1002/0471141755.ph0801s00
19. Frey KA, Albin RL (1997) Receptor binding techniques. In: Jacqueline N Crawley et al (eds.) Current protocols in neuroscience, Chapter1:Unit1.4.doi:10.1002/0471142301.ns0104s00
20. Sóvágó J, Dupuis DS, Gulyás B, Hall H (2001) An overview on functional receptor autoradiography using [35S]GTPgammaS. Brain Res Brain Res Rev 38:149–164
21. Bundy DC (2001) Autoradiography. In: John E Coligan et al (eds.) Current protocols in protein science, Chapter 10:Unit 10.11. doi:10.1002/0471140864.ps1011s10
22. Kanekal S, Sahai A, Jones RE, Brown D (1995) Storage-phosphor autoradiography: a rapid and highly sensitive method for spatial imaging and quantitation of radioisotopes. J Pharmacol Toxicol Methods 33:171–178
23. Zavitsanou K, Katsifis A, Mattner F, Huang X-F (2003) Investigation of M1//M4 muscarinic receptors in the anterior cingulate cortex in schizophrenia, bipolar disorder, and major depression disorder. Neuropsychopharmacology 29(3):619–625
24. Wang Q, Wei X, Gao H, Li J, Liao J, Liu X, Qin B, Yu Y, Deng C, Tang B, Huang XF (2014) Simvastatin reverses the downregulation of M1/4 receptor binding in 6-hydroxydopamine-induced parkinsonian rats: The association with improvements in long-term memory. Neuroscience 267:57–66, doi:http://dx.doi.org/10.1016/j.neuroscience.2014.02.031
25. Yamamura HI, Wamsley JK, Deshmukh P, Roeske WR (1983) Differential light microscopic autoradiographic localization of muscarinic cholinergic receptors in the brainstem and spinal cord of the rat using [3H]pirenzepine. Eur J Pharmacol 91(1):147–149, doi:http://dx.doi.org/10.1016/0014-2999(83)90379-5
26. Wamsley JK, Gehlert DR, Roeske WR, Yamamura HI (1984) Muscarinic antagonist binding site heterogeneity as evidenced by autoradiography after direct labeling with [3H]-QNB and [3H]-pirenzepine. Life Sci 34(14):1395–1402
27. Villiger JW, Faull RLM (1985) Muscarinic cholinergic receptors in the human spinal cord: differential localization of [3H]pirenzepine and [3H]quinuclidinylbenzilate binding sites. Brain Res 345(1):196–199, doi:http://dx.doi.org/10.1016/0006-8993(85)90854-6
28. Cortes R, Palacios JM (1986) Muscarinic cholinergic receptor subtypes in the rat brain. I Quantitative autoradiographic studies. Brain Res 362(2):227–238
29. Buckley NJ, Burnstock G (1986) Autoradiographic localization of peripheral M1 muscarinic receptors using [3H]pirenzepine. Brain Res 375(1):83–91, doi:http://dx.doi.org/10.1016/0006-8993(86)90961-3
30. Aubert I, Cecyre D, Gauthier S, Quirion R (1992) Characterization and autoradiographic distribution of [3H]AF-DX 384 binding to putative muscarinic M2 receptors in the rat brain. Eur J Pharmacol 217(2–3):173–184
31. Mulugeta E, Karlsson E, Islam A, Kalaria R, Mangat H, Winblad B, Adem A (2003) Loss of muscarinic M4 receptors in hippocampus of Alzheimer patients. Brain Res 960(1–2):259–262, doi:http://dx.doi.org/10.1016/S0006-8993(02)03542-4
32. Tien L-T, Fan L-W, Sogawa C, Ma T, Loh HH, Ho I-K (2004) Changes in acetylcholinesterase activity and muscarinic receptor bindings in μ-opioid receptor knockout mice. Molecular Brain Research 126(1):38–44, doi:http://dx.doi.org/10.1016/j.molbrainres.2004.03.011
33. Wolff SC, Hruska Z, Nguyen L, Dohanich GP (2008) Asymmetrical distributions of muscarinic receptor binding in the hippocampus of female rats. Eur J Pharmacol 588(2–3):248–250, doi:http://dx.doi.org/10.1016/j.ejphar.2008.04.002
34. Paxinos G, Franklin KBJ (2008) The mouse brain in stereotaxic coordinates. The coronal plates and diagrams, Compact edn, 3rd edn. Academic, London
35. Caulfield MP, Birdsall NJ (1998) International Union of Pharmacology. XVII. Classification of muscarinic acetylcholine receptors. Pharmacol Rev 50(2):279–290

Chapter 10

Imaging of Muscarinic Receptors in the Central Nervous System

Hideo Tsukada, Shingo Nishiyama, and Kazuhiro Takahashi

Abstract

For the quantitative imaging of muscarinic acetylcholine receptors (mAChR), we developed novel PET probes, (+)*N*-^{11}C-methyl-3-piperidyl benzilate (^{11}C-(+)3-MPB), and its *N*-alkyl substitute analogs, and evaluated them in the brains of conscious monkeys (*Macaca mulatta*) using high-resolution positron emission tomography (PET). Although (+)3-MPB had relatively poor selectivity to the subtypes of mAChR, the regional cortical distribution of ^{11}C-(+)3-MPB was found to be consistent with mAChR density in the living monkey brain as reported in vitro. In contrast, its enantiomeric analog ^{11}C-(−)3-MPB provided homogeneous distribution with no significant specific binding throughout the whole brain. The *N*-alkyl substitution of alkyl moiety from methyl (^{11}C-(+)3-MPB) to ethyl (^{11}C-(+)3-EPB) and propyl (^{11}C-(+)3-PPB) resulted in lower affinities to mAChR in vitro, the faster kinetics in the living brain, and greater sensitivity to increased endogenous ACh level, induced by acetylcholinesterase (AChE) inhibitor, than ^{11}C-(+)3-MPB. Administration of scopolamine, a mAChR antagonist, reduced ^{11}C-(+)3-MPB binding to mAChR in all regions except the cerebellum, and the reduction of ^{11}C-(+)3-MPB uptake was well correlated with the degree of impairment of working memory performance assessed in conscious monkeys. These results demonstrated that PET imaging with ^{11}C-(+)3-MPB could be useful for diagnosis of neurological diseases associated with impaired mAChR function and cognitive function.

Key words Brain, Muscarinic acetylcholine receptor, ^{11}C-(+)3-MPB, PET

1 Introduction

In the cholinergic neuronal system, acetylcholine (ACh) is the primary neurotransmitter released in the central nervous system (CNS). The cholinergic receptor (AChR) population is divided into nicotinic and muscarinic subclasses, and the AChR appears to be predominantly of the muscarinic-type receptor (mAChR) in the CNS. The mAChR belongs to the family of receptors coupled to heterotrimeric GTP-binding proteins (G-proteins), and the CNS mAChR system plays an important role in memory and cognitive functions. Alzheimer-type dementia (AD) has been neuropathologically characterized by the presence of neurofibrillary tangles with the deposition of hyperphosphorylated tau protein inside nerve cells

Jaromir Myslivecek and Jan Jakubik (eds.), *Muscarinic Receptor: From Structure to Animal Models*, Neuromethods, vol. 107, DOI 10.1007/978-1-4939-2858-3_10, © Springer Science+Business Media New York 2016

and senile plaques with extracellular aggregation of amyloid-β (Aβ) protein [1, 2]. Moreover, loss of cholinergic neurons in the forebrain [3], reduced cholinergic activity in the hippocampus and cortical loss of choline acetyltransferase [4], and reduced central mAChR binding have been observed in the brains of AD patients [5–7]. In addition, the severity of these cholinergic abnormalities is closely correlated with the degree of dementia [6, 7].

Positron emission tomography (PET) has been applied for noninvasive investigations of physiological functions as well as neurodegenerative dysfunctions of mAChR in the living brain. Several antagonist-based ^{11}C-labeled-PET probes for imaging of mAChR in the CNS have been developed and evaluated, including ^{11}C-scopolamine [8–10], ^{11}C-dexetimide [11], ^{11}C-quinuclidinyl benzilate (QNB) [12], ^{11}C-benztropine [13], and ^{11}C-tropanyl benzilate (TRB) [14, 15]. These labeled ligands for mAChR, however, show poor subtype selectivity, relatively low uptake to the brain, and also slow dissociation rates from the binding sites which may limit quantitative measurement of the density of mAChR in vivo [15]. A low dissociation rate from mAChR makes analysis difficult, because a true equilibrium state cannot be obtained within the PET scanning time with the short half-lives of positron emitters. To solve these problems, a radiolabeled mAChR probe, *N*-^{11}C-methyl-4-piperidyl benzilate (^{11}C-4-MPB) (Fig. 1), with more favorable kinetic properties than previous compounds was proposed [16]. However, the properties of ^{11}C-4-MPB with

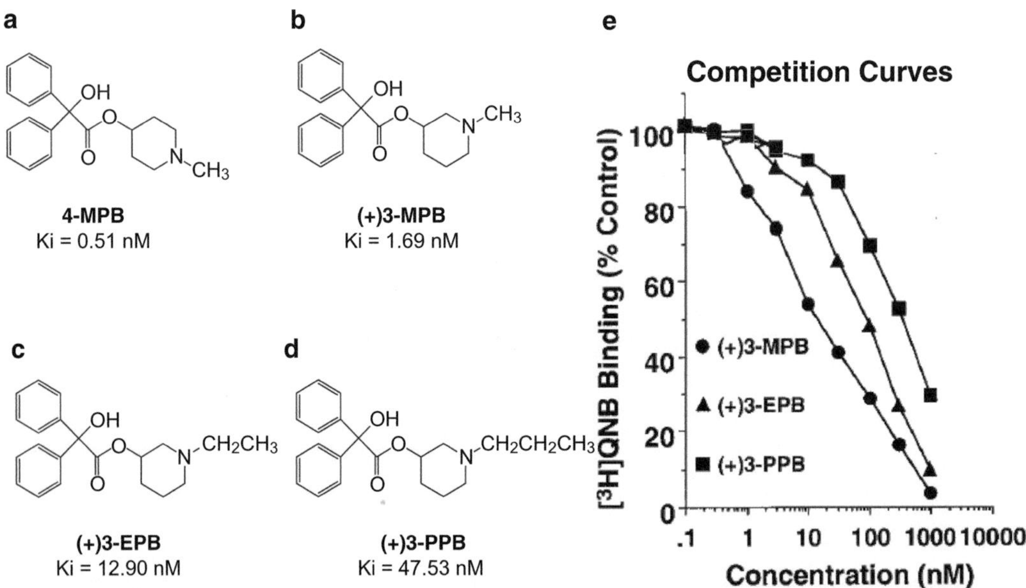

Fig. 1 Chemical structures of 4-MPB (**a**), (+)3-MPB (**b**) and its *N*-alkyl substitution analogs, (+)3-EPB (**c**) and (+)3-PPB (**d**). Affinity (K_i) values of each compound were determined by in vitro competitive assay in rat brain slices with ^{3}H-QNB, as shown in **e**

relatively high affinity to mAChR were still insufficient for quantitative imaging of mAChR in the living brain. In 1997, we proposed a novel mAChR probe, *N*-^{11}C-methyl-3-piperidyl benzilate (^{11}C-3-MPB) [17] (Fig. 1). Since its chemical structure contained a chiral carbon, it provides two stereoisomers, an active form is ^{11}C-(+)3-MPB and an inactive one is ^{11}C-(−)3-MPB. Active form, ^{11}C-(+)3-MPB, revealed relatively low affinity to mAChR as determined by in vitro binding assay [17, 18]. Furthermore, ^{11}C-(+)3-MPB was evaluated for the quantification of cerebral mAChR binding in the brain of conscious monkey (*Macaca mulatta*) [19]. By comparison between young and aged monkeys changes in ^{11}C-(+)3-MPB binding to mAChR were found [20, 21]. The temporal relationship between the occupancy level of central mAChR by scopolamine, as measured by ^{11}C-(+)3-MPB, and cognitive impairment, as assessed by the delayed matching-to-sample, was determined in conscious monkeys [22].

This chapter will provide a detailed overview of the development of a novel PET probe, ^{11}C-3-MPB, for mAChR imaging from design of its chemical structure, synthesis, radiolabeling to its assessments in vitro as well as in vivo on experimental animals.

2 Materials and Methods

2.1 Syntheses of Cold 3-PB, 3-MPB, 4-PB, and 4-MPB

It is well known that stereoisomers exhibit different properties in target-binding affinity/specificity, which results in their different pharmacological properties. To apply the chemical compounds with stereoisomers like 3-MPB as PET probe agents, quality control of enantiomeric purity is a very critical factor in the production of their corresponding precursors to radiolabeling.

As a precursor of ^{11}C-3-MPB, 3-piperidyl benzilate (3-PB) was prepared following a previously reported procedure [17]. A mixture of benzene (30 mg), 3-piperidinol (0.4 g), and methyl benzilate (1.0 g) was stirred and heated to reflux in a flask with a Molecular Sieve 4A column, a reflux condenser, and a soda lime tube. When all methyl benzilate was dissolved, sodium methoxide (20 mg) was added to the flask. After 3-h reaction, the reaction mixture was cooled down to room temperature, and 50 ml of 1 mol/l hydrochloric acid was added for separation into two phases. The aqueous phase was washed twice with 50 ml of ether, and made the phase basic with ammonium hydroxide to precipitate the benzyl ester. Thirty milliliters of ether was added twice to the aqueous phase to dissolve the benzyl ester and separate the ether layer. The ether layer was washed twice with water, dried with anhydrous potassium carbonate, filtered, and then the solvent was removed by distraction in vacuum. The residue was crystallized in ether-hexane solution, and the solid was separated by filtration as enantiomeric 3-PB. (+)3-PB and (−)3-PB were separated using an

HPLC system with a chiral column (Chiralcel OJ column, Daiseru Corporation, Osaka, Japan) eluted with a mixture of hexane/ethanol = 80/20 (v/v) under a flow rate of 0.5 ml/min. The retention times of (+)3-PB and (−)3-PB were 27.0 min and 16.8 min, respectively.

As the standard compound of ^{11}C-3-MPB for quality control analysis, 3-MPB was obtained using *N*-methyl-3-piperidinol (0.45 g) instead of 3-piperidinol by the same reaction as 3-PB synthesis described above. (+)3-MPB and (−)3-MPB were isolated using same HPLC system used in (+) and (−)3-PB isolation. The retention times of (+)3-MPB and (−)3-MPB were 15.8 min and 13.2 min, respectively.

For the comparison, 4-piperidyl benzilate (4-PB) and 4-methylpiperidyl benzilate (4-MPB) were prepared according to a previously reported method [23]. A mixture of benzene (30 mg), 4-piperidinol (0.4 g), and methyl benzilate (1.0 g) was stirred and heated to reflux in a flask with a Molecular Sieve 4A column, a reflux condenser, and a soda lime tube. When all methyl benzilate was dissolved, sodium methoxide (20 mg) was added to the flask. After 3 h reaction, the reaction mixture was cooled down to room temperature, and 50 ml of 1 mol/l hydrochloric acid was added for separation into two phases. The aqueous phase was washed twice with 50 ml of ether, and the phase was made basic with ammonium hydroxide to precipitate the benzyl ester. Thirty milliliters of ether was added twice to the aqueous phase to dissolve the benzyl ester and separate the ether layer. The ether layer was washed twice with water, dried with anhydrous potassium carbonate, filtered, and then the solvent was removed by distraction in vacuum. The residue was crystallized in ether-hexane solution, and the solid was separated by filtration as enantiomeric 4-PB. 4-MPB was obtained using *N*-methyl-4-piperidinol (0.45 g) instead of 4-piperidinol by the same reaction as 4-PB synthesis described above.

2.2 In Vitro Assessment of Novel PET Probes

The binding affinity of each probe was assessed by a competitive binding assay in rat brain slices using ^{3}H-QNB autoradiography. Since it is generally considered that the information derived from autoradiography is more pertinent to the in vivo properties of the receptor than that obtained from "pure" binding studies carried out using cell-free preparations, we applied this method for PET imaging agents.

The frozen brain was sectioned on a Cryostat (CM3000, Leica Microsystems, Nussloch, Germany) into 10 μm coronal tissue sections (bregma −2.3 to −2.8 mm), which were transferred onto cooled, gelatinized glass slides. The cryosections were incubated with ^{3}H-QNB (1.2 nM) and varying amounts of each PET probe for 30 min at 25 °C in Tris–HCl buffer. Nonspecific binding of the radioligand in the brain slice was determined by including a saturating concentration (1 μM) of atropine in the incubation medium. The incubation was terminated by rinsing the sections twice for 2 min each in cold buffer, then dipping briefly in cold distilled water (4 °C), and slides were dried rapidly on a hot-plate (50 °C). The labeled slices were exposed for 3 days to an imaging plate (TR-2040, Fuji Film Co., Tokyo, Japan), and the imaging data were analyzed (BAS-2500, Fuji Film Co., Tokyo, Japan). Specific binding of ^{3}H-QNB was estimated as the difference between total binding and nonspecific binding. Specific binding was plotted against the concentration of each ligand to determine concentration causing 50 % inhibition (IC_{50} values), which were converted to inhibition constants (K_i) using the Cheng and Prusoff equation [24]. The rat brain anatomical structures in autoradiographic images were visually identified according to the atlas of Paxinos and Watson [25].

Subtype specificity of (+)3-MPB and 4-MPB was assessed using human mAChRs (M_1–M_5) transfected in CHO-K1 cells (Receptor Biology, Inc., Beltsville, USA) [26]. For the experiments, the membranes were thawed and diluted in phosphate-buffered saline (PBS, pH 7.4) and homogenized in a glass vessel with the aid of a Teflon pestle. The final concentration of receptors in the assay was in the range of 50–100 pM. For the association study, homogenate was incubated with labeled compound at 22 °C, and the binding was started at different time points in reverse order and terminated simultaneously in all samples at time 0. The incubation was terminated by rapid filtration through Whatman GF/B glass fiber filters (FPB-148, Gaithersburg, USA). For the dissociation study, the association of samples was terminated after 50 min. The dissociation was started by adding unlabeled compound in excess at different time points, and the reactions were terminated simultaneously in all samples at time 0. After termination of the incubation, the filters were rinsed four times with 2 ml of ice-cold incubation buffer each time. Thereafter, the filter rings were collected from the cell harvester (Brandel, Gaithersburg, USA) and were transferred into vials for counting in a γ-counter.

To measure lipophilicity of the labeled ligands, their partition coefficients (log D at pH 7.4) were measured. Octanol (2 ml) and 0.066 M phosphate-buffer, pH 7.4 (2 ml) were mixed for 3 min. Ten milliliters of each labeled compound was mixed, vortexed for 3 min, centrifuged for 5 min, separated, and ^{11}C-radioactivity was counted. The lipophilicity of each labeled PET probe was then calculated as logarithm of the distribution constant between lipophilic organic (octanol) phase and polar aqueous (water) phase.

The buffer solutions can be stored conveniently at 4 °C and are stable for several months. Drug solutions and radioligands are distributed in small aliquots and stored at −20 °C. Atropine solutions are prepared fresh monthly.

2.3 Radiosynthesis of ^{11}C-(+)3-MPB, ^{11}C-(−)3-MPB, and ^{11}C-4-MPB

Carbon-11 (^{11}C), a positron emitter with a 20 min half-life, is a candidate for the isotopic or non-isotopic labeling of any organic compound used for PET imaging. In isotopic labeling, since ^{11}C replaces stable $^{12}C/^{13}C$ in the molecule, the biological properties of the molecule are virtually unchanged. In non-isotopic labeling, a group containing ^{11}C is added to the molecule of interest, producing a new compound with different properties to the original.

The reaction time with ^{11}C that has a short-half life (=20 min) needs to be as short as possible, and the reaction needs to be driven to give useful yields within one physical half-life. Reactions are promoted by (1) using a large excess of precursor to consume labeling agent; (2) using a high precursor concentration in small volumes; (3) using sealed vessels for elevated reaction temperature; and (4) using microwaves or sonication. In addition, since the stability of some ^{11}C-PET probes was low even at the end of synthesis (EOS) because of a radiolysis, this should be taken into account by suppressing radiolysis by adding a selective scavenger for hydroxyl radicals and/or hydrated electrons into the PET probe solution.

The concentration of ^{11}C-labeled probe was calculated from calibration curves by simultaneous monitoring of radioactivity and UV absorbance with authentic substances used as standards. Specific radioactivity of ^{11}C-labeled probe was defined as the ratio of radioactivity (Ci) to accompanying cold compound (mol), and was expressed in units of Ci/mol. There is substantial dilution of ^{11}C-labeled compound with cold compound. The amount of cold compound remains constant, while the radioactivity decays over time according to the half-life of the positron emitter. Therefore, the specific radioactivity for PET probe should be cited with respect to a particular time, such as the end of radionuclide production (EOB), EOS, or time of injection (TOI).

Positron emitting carbon-11 (^{11}C) was produced by a $^{14}N(p,a)^{11}C$ nuclear reaction using a cyclotron (HM-18; Sumitomo Heavy Industry, Tokyo, Japan) at Hamamatsu Photonics PET center and obtained as ^{11}C-CO_2. Labeled compounds were synthesized using a modified CUPID system (Sumitomo Heavy Industry, Tokyo, Japan). As isotopic labeling, ^{11}C-(+)3-MPB and its stereoisomer ^{11}C-(−)3-MPB were labeled by *N*-methylation of respective nor-compounds ((+)3-PB or (−)3-PB) with ^{11}C-methyl iodide (Fig. 2) [17, 19]. ^{11}C-CO_2 was converted to ^{11}C-methyl iodide by $LiAlH_4$ reduction followed by reaction with HI, and ^{11}C-methyl iodide was trapped in a reaction vial containing the free base (+) or (−)3-PB (0.5 mg) in DMF (200 μL) at −40 °C. When reaching the maximum radioactivity of ^{11}C-methyl iodide, the vial was hearted at 80 °C for 5 min. After cooling down, the reaction mixture was injected into an HPLC system with a C18 column (μBondapak-C18, Waters, Milford, MA) eluted with a mixture of acetonitrile/0.1 M AcONa/Acetic acid = 700/300/1 (v/v) under a flow rate of 7 ml/min. The fraction eluted at a retention time of ca. 10 min was transferred to an

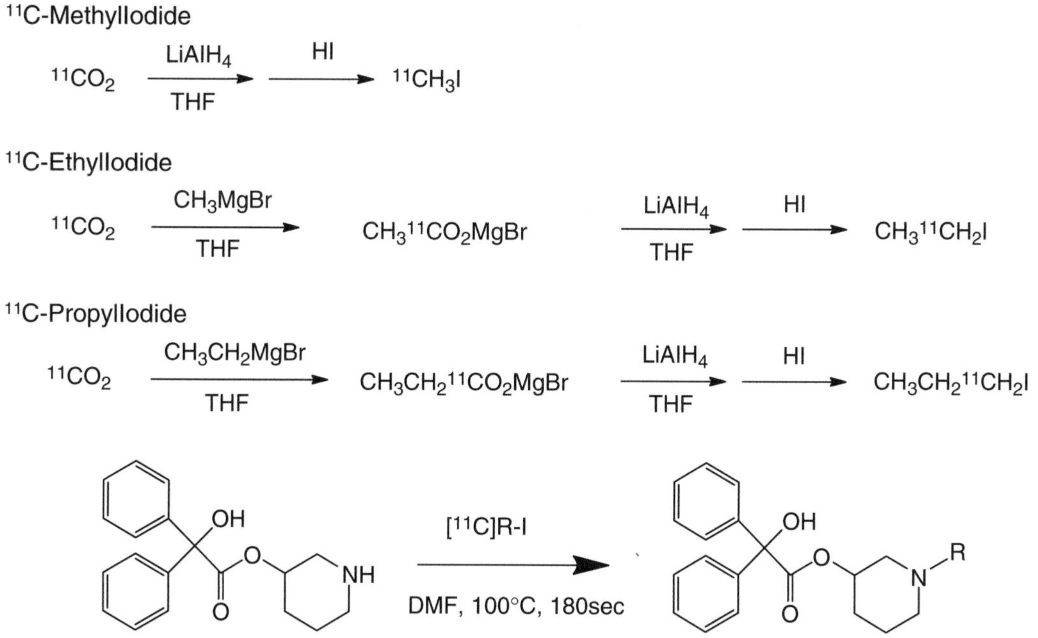

^{11}C-MethylIodide

$^{11}CO_2 \xrightarrow[THF]{LiAlH_4} \xrightarrow{HI} {}^{11}CH_3I$

^{11}C-EthylIodide

$^{11}CO_2 \xrightarrow[THF]{CH_3MgBr} CH_3{}^{11}CO_2MgBr \xrightarrow[THF]{LiAlH_4} \xrightarrow{HI} CH_3{}^{11}CH_2I$

^{11}C-PropylIodide

$^{11}CO_2 \xrightarrow[THF]{CH_3CH_2MgBr} CH_3CH_2{}^{11}CO_2MgBr \xrightarrow[THF]{LiAlH_4} \xrightarrow{HI} CH_3CH_2{}^{11}CH_2I$

OH, O, NH, O → [^{11}C]R-I, DMF, 100°C, 180sec → OH, O, N, R, O

Fig. 2 Radiolabeling of ^{11}C-(+)3-MPB, ^{11}C-(+)3-EPB, and ^{11}C-(+)3-PPB using (+)3-PB and ^{11}C-methyl iodide, ^{11}C-ethyl iodide, and ^{11}C-propyl iodide, respectively

evaporator for evaporation of elute solvent, dissolved in saline (10 ml), and filtered through a 0.22-μm pore size filter.

Chemical and radiochemical analysis of ^{11}C-(+)3-MPB and ^{11}C-(−)3-MPB was performed by HPLC in a system consisting of a column (Finepak SIL C18-S, 4.6 mm in diameter × 150 mm in length, Jasco, Tokyo, Japan), pump (CCPS, Tosoh, Tokyo, Japan), UV detector (UV-8020, Tosoh, Tokyo, Japan), and radio detector (RLC-700, Hitachi Aloka Medical, Inc., Tokyo, Japan) using CH_3CN/30 mM CH_3COONH_4/CH_3COOH (350/650/2) as a mobile phase at a flow rate of 1 ml/min. Analyses of the enantiomeric purity of ^{11}C-(+)-3-MPB and ^{11}C-(−)-3-MPB were performed by HPLC in a system consisting of a column (Chirobiotic V, 4.6 × 250 mm, Astec, NJ, USA), pump (DP8020, Tosoh, Tokyo, Japan), UV detector (UV-8020, Tosoh, Tokyo, Japan), and radio detector (TCS-713, Aloka, Tokyo, Japan) using CH_3OH/CH_3COOH/$(C_2H_5)_3N$ (1000/0.5/0.1) as a mobile phase at a flow rate of 1 ml/min. The retention times of ^{11}C-(+)3-MPB and ^{11}C-(−)3-MPB were 15.8 min and 13.2 min, respectively, both of which were identical to each corresponding standard compound.

The radioactive purity of each labeled compound used in this study was greater than 99 % and the specific radioactivity ranged from 61.7 to 92.4 GBq/μmol for ^{11}C-(+)3-MPB, and from 60.0 to 79.5 GBq/μmol for ^{11}C-(−)3-MPB at EOS. Enantiomeric purity of ^{11}C-(+)-3-MPB and ^{11}C-(−)-3-MPB was 100 %.

As for the previously reported reference, ^{11}C-4-MPB was prepared by the same method as used for ^{11}C-3-MPB labeling. The radioactive purity used in this study was greater than 99 % and the specific radioactivity ranged from 34.4 to 75.9 GBq/μmol for ^{11}C-4-MPB at EOS.

2.4 Radiosynthesis of ^{11}C-(+)3-EPB and ^{11}C-(+)3- PPB

In order to evaluate the effects of affinity to mAChR on kinetics as well as the sensitivity to changes in the synaptic endogenous ACh level, two *N*-alkyl substitution analogs of ^{11}C-(+)3-MPB, *N-11C-ethyl*-3-piperidyl benzilate (^{11}C-(+)3-EPB) and *N*-^{11}C-propyl-3-piperidyl benzilate (^{11}C-(+)3-PPB), were used [18]. Standard compounds, (+)3-EPB and (+)3-PPB, were synthesized by *N*-ethylation and *N*-propylation, respectively, of (+)3-PB. For the radiolabeling, C-11 was produced by $^{14}N(p,a)^{11}C$ nuclear reaction using a cyclotron (HM-18; Sumitomo Heavy Industry, Tokyo, Japan) and obtained as ^{11}C-CO_2. Labeled compounds were synthesized using a modified CUPID system (Sumitomo Heavy Industry, Tokyo, Japan). Instead of ^{11}C-methyl iodide, ^{11}C-ethyl iodide or ^{11}C-propyl iodide was prepared by reaction of Grignard reagent (Fig. 2) [18, 27]. In the synthesis of ^{11}C-ethyl iodide or ^{11}C-propyl iodide, 0.5 ml of 0.25 M methylmagnesium bromide in tetrahydrofuran (THF) or 0.5 ml of 0.25 M ethylmagnesium bromide in THF were used. After a reaction time for 5 min, 1 ml of lithium aluminum hydrate (1 M) in THF was added and the solvents were removed. Two milliliters of hydriodic acid (54 %) was then added, and the product was distilled off and transferred in a stream of nitrogen gas through a drying tower containing sodium hydroxide/phosphorus pentoxide to the reaction vessel. ^{11}C-(+)3-EPB or ^{11}C-(+)3-PPB was synthesized by *N*-ethylation or *N*-propylation of (+)3-PB with ^{11}C-ethyl iodide or ^{11}C-propyl iodide.

Chemical and radiochemical analysis of ^{11}C-(+)3-EPB and ^{11}C-(+)3-PPB was performed by HPLC in a system consisting of a column (Finepak SIL C18-S, 4.6 mm in diameter × 150 mm in length, Jasco, Tokyo, Japan), pump (CCPS, Tosoh, Tokyo, Japan), UV detector (UV-8020, Tosoh, Tokyo, Japan), and radio detector (RLC-700, Hitachi Aloka Medical, Inc., Tokyo, Japan) using CH_3CN/30 mM CH_3COONH_4/CH_3COOH (350/650/2) as a mobile phase at a flow rate of 1 ml/min. Analyses of the enantiomeric purity of ^{11}C-(+)-3-EPB and ^{11}C-(+)-3-PPB were performed by HPLC in a system consisting of a column (Chirobiotic V, 4.6 × 250 mm, Astec, NJ, USA), pump (DP8020, Tosoh, Tokyo, Japan), UV detector (UV-8020, Tosoh, Tokyo, Japan), and radio detector (TCS-713, Aloka, Tokyo, Japan) using CH_3OH/CH_3COOH/$(C_2H_5)_3N$ (1000/0.5/0.1) as a mobile phase at a flow rate of 1 ml/min. The retention times of ^{11}C-(+)3-EPB, ^{11}C-(−)3-EPB, ^{11}C-(+)3-PPB, and ^{11}C-(−)3-PPB were 15.3 min, 14.8 min, 12.7 min, and 13.7 min, respectively.

The radioactive purity of each labeled compound used in this study was greater than 99 %, and the specific radioactivity ranged from 34.1 (^{11}C-(+)3-PPB) to 66.3 GBq/μmol (^{11}C-(+)3-EPB) at EOS. The specific radioactivity levels of these two PET probes were comparable to that of ^{11}C-(+)3-MPB. Enantiomeric purity of ^{11}C-(+)-3-EPB and ^{11}C-(+)-3-PPB were 100 %.The solution of labeled compound was passed through a 0.22-μm pore size filter before intravenous administration to the subjects.

For the assessment of acetylcholinesterase (AChE) level in the living brain using PET, *N*-^{11}C-methyl-4-piperidyl acetate (^{11}C-MP4A) was labeled by *N*-methylation of its corresponding nor-compound, 4-piperidyl acetate (P4A) (ABX Advanced Biochemical Compounds, Radeberg, Germany), with ^{11}C-methyl iodide [28].

2.5 MRI Data Acquisition

Magnetic resonance images (MRI) of the monkeys were obtained with a 3.0 T MR imager (Signa Excite HDxt 3.0 T, GE Healthcare Japan, Tokyo, Japan) using a 3D-Spoiled Gradient Echo (SPGR) sequence (176 slices with a 256×256 image matrix, slice thickness/spacing of 1.4/0.7 mm, TE: 3.4–3.6 ms, TR: 7.7–8.0 ms, TI: 400 ms, and flip angle: 15°) under pentobarbital anesthesia.

2.6 PET Analysis

The PET studies in this chapter have been conducted with monkeys under conscious condition. Anesthetics have been used in non-human primate PET studies, because it is necessary to fix the animal during PET scanning. However, anesthetics have been reported to affect several neuronal functions, resulting in an alteration of neuronal activities in the central nervous system. Thus, the brain function as well as the pharmacological actions as measured by PET should be affected by anesthetics [29–32]. In order to avoid these anesthetic effects, we developed a PET system with transaxial resolution of 2.6 mm full width at half maximum (FWHM) and a center-to-center distance of 3.6 mm (SHR-7700, Hamamatsu Photonics, Hamamatsu, Japan) [33], and its gantry can be tilted up to 90° for brain imaging of monkey sitting on a monkey chair under conscious condition. In order to eliminate the stress caused by head motion restriction influence on measured parameters, as reported in the dopaminergic neuronal system [34], the monkeys were trained for more than 2 months and plasma cortisol was monitored.

After an overnight fast, a monkey (*Macaca mulatta*) was seated on a monkey chair under conscious condition and fixed with stereotactic coordinates aligned parallel to the orbitomeatal (OM) line. After a transmission scan for 30 min using a ^{68}Ge-^{68}Ga rotation rod source, an emission scan for 91 min was conducted after the injection of each labeled compound (100–120 MBq/kg body weight) through the venous cannula.

The PET data obtained were reconstructed by the filtered back projection (FBP) method with a Hanning filter of 4.5 mm FWHM. Volumes of interest (VOIs) in brain regions were drawn manually on the MRI referring regional information from BrainMaps.org, and VOIs of MRI were superimposed on the co-registered PET images to measure the time activity curves (TACs) of each PET probes for kinetic analysis.

For quantitative analysis of ^{11}C-(+)3-MPB, arterial blood samples were obtained every 8 s up to 64 s, followed by 90 and 150 s, and then 4, 6, 10, 20, 30, 45, 60, and 90 min after tracer injection, and the blood samples were centrifuged to separate plasma, weighed, and their radioactivity was measured. For metabolite analysis, methanol was added to some plasma samples (sample/methanol = 1/1) obtained at 16, 40, and 64 s, 6, 10, 30, and 45 min after the injection, followed by centrifugation. The obtained supernatants were developed using thin layer chromatography plates (AL SIL G/UV, Whatman, Kent, UK) with a mobile phase of ethyl acetate. The ratio of unmetabolized fraction was determined using a phosphor imaging plate (FLA-7000, Fuji Film, Tokyo, Japan). The input function of unmetabolized ^{11}C-(+)3-MPB was calculated using the data obtained by correction of the ratio of the unmetabolized fraction to total radioactivity, which was used as the arterial input function.

Logan [35] and Patlak [36] graphical analyses with metabolite-corrected plasma input were applied for quantitative measurements of ^{11}C-(+)3-MPB binding to mAChR in the living brain using PMOD software (PMOD Technologies Ltd., Zurich, Switzerland). The Logan graphical plot [35] directly gives a linear function of the free receptor concentration known as the distribution volume based on the following Eq. (1):

$$\int \mathrm{ROI}(t)\mathrm{d}t \,/\, \mathrm{ROI}(T) = \mathrm{DV}\int \mathrm{Cp}(t)\mathrm{d}t \,/\, \mathrm{ROI}(T) + C$$

where ROI(T) and Cp(T) represent tissue and metabolite-corrected arterial plasma radioactivity, respectively, at time T, DV is the slope and C is the intercept on the Y-axis. For the reversibly labeled compounds, this Logan plot becomes linear after some time with a slope (DV) that is equal to the steady-state distribution volume. The ratios of DV in each ROI (DV(ROI)) to K in the cerebellum (DV(CE)) were calculated to determine the distribution of mAChR in the living brain.

Patlak plot analysis [36] was used to measure the net accumulation of PET probe in the irreversible compartment based on the following Eq. (2):

$$\int \mathrm{ROI}(t)\mathrm{d}t \,/\, \mathrm{Cp}(T) = \mathrm{DV}\int \mathrm{Cp}(t)\mathrm{d}t \,/\, \mathrm{ROI}(T) + C$$

where ROI(*T*) and Cp(*T*) represent tissue and metabolite-corrected arterial plasma radioactivity, respectively, at time *T*, DV is the slope, and *C* is the intercept of the *Y*-axis. The slope (DV) was equal the parameter combination $K_1k_3/(k_2+k_3)$. The dissociation constant (k_4) is assumed to be negligible over the course of the scan. Patlak slope value is influenced by the blood–brain barrier (BBB), transport rate constants (K_1 and k_2), and thus does not provide a pure estimate of association constant (k_3).

2.7 Drug Perturbations on PET Probe Binding to mAChR

A PET probe is no-cold added (NCA) when no source of cold compound has been added deliberately during its production. In contrast, it is cold-added (CA) when a source of cold compound has been added deliberately during its production.

For the inhibition study of ^{11}C-(+)3-MPB binding to mAChR in the living brain, scopolamine, a specific mAChR antagonist, was administered at a dose of 50 μg/kg 30 min before PET probe injection under NCA condition.

In order to evaluate the sensitivity to changes in the synaptic endogenous ACh level induced by AChE inhibition, Aricept, an AChE inhibitor, was intravenously administered at doses of 50 and 250 μg/kg 30 min before ^{11}C-(+)3-MPB injection under NCA condition.

To determine the correlation between mAChR occupancy and cognitive impairment in monkeys, scopolamine was administered at the dose of 10 and 30 μg/kg, and then PET measurements with ^{11}C-(+)3-MPB injection under NCA condition and cognitive performance test were serially conducted 2, 6, 24 and 48 h after administration of scopolamine. The mAChR occupancy levels were determined from the degree of reduction (%) of the BP_{ND} by scopolamine as following Eq. (3):

$$\text{Occupancy}(\%) = \left(1 - \text{BP}_{\text{ND}}\text{post}(\text{ROI}) - \text{BP}_{\text{ND}}\text{pre}(\text{ROI})\right) \times 100$$

where BP_{ND}pre(ROI) and BP_{ND}post(ROI) are BP_{ND} pre- and post-scopolamine, respectively.

2.8 Effects of Aging on mAChR

Several clinical PET studies have attempted to determine quantitatively the age-related alterations of mAChR in the living brain using ^{11}C-benztropine [37], ^{11}C-4-MPB [38–40], and ^{11}C-TBZ [41]. These PET probes for mAChR, however, showed relatively low uptake to the brain and also slow dissociation rates from mAChR, which may limit estimation of the density of binding sites in vivo [15]. Therefore, we attempted to assess the age-related changes in mAChR with ^{11}C-(+)3-MPB, a reversible-type PET probe, in the living brains of young (ca. 6 years old) and aged (ca. 20 years old) monkeys in a conscious state [20]. The syntheses of ^{11}C-(+)3-MPB, PET data acquisitions, and the analysis of cortical TAC were conducted as described in Sections 2.3 and 2.5.

Saturation experiments were performed to examine the effects of aging on in vivo binding parameters (maximum binding capacity B_{MAX} and equilibrium dissociation constant K_D) of ^{11}C-(+)3-MPB as performed previously in the dopaminergic system [42, 43]. ^{11}C-(+)3-MPB was injected into monkeys under NCA and CA conditions and together with various amounts cold (+)3-MPB from 3 to 300 μg/kg. The total radioligand concentration of ^{11}C-(+)3-MPB in the cerebellum was used as an estimate of the free radioligand concentration (*F*) in each ROI. Specific binding (*B*) was defined as radioactivity in each ROI reduced by *F*. The curve for *B* was fitted to a set of three exponential functions to determine the time point at which *B* reached a peak. The values for *B* and *F* at these time points were used in the in vivo pseudo Scatchard plot analysis where the ratio of *B*/*F* was plotted against *B* [44]. The apparent in vivo B_{MAX} and K_D values were calculated using LIGAND software.

2.9 Cognitive Assessment of Monkeys

Although it had been well known that blockade of mAChR with scopolamine, a specific mAChR antagonist, resulted in transient cognitive impairment [45], there have been no study revealing the relationship between the occupancy level of central mAChRs and the degree of cognitive impairment induced by scopolamine in primates. Therefore, we attempted to evaluate the correlation between the mAChR occupancy level and cognitive impairment in conscious monkeys [22].

Cognitive impairment was determined by a titration version of delayed match to sample (T-DMS) task [46]. DMS task is one method for evaluating potential drug effects on cognitive functions. Typically, a sample visual stimulus is presented to the animal for a short period. Following a delay, the sample and another test stimulus are presented simultaneously. The subject is required to choose the sample visual stimulus in order to be rewarded. However, one limitation of the conventional DMS task is that the maximal delay is fixed for all subjects, although cognitive abilities differ among them. With T-DMS task test, under control condition, the monkeys' correct response levels depend on each animal's working memory capacity as determined by the maximal delay interval.

Sixty-four different visual stimuli comprising all combinations of eight distinct colors and eight distinct shapes were presented on a touch-sensitive screen placed in front of monkey. A sample stimulus appeared for 300 ms, disappeared, and then the sample stimulus and three other stimuli appeared after a delay. Monkeys were trained to touch the visual stimulus that matched the sample within 5 s, and if the answer was correct, water drops (0.2 ml) were given as a reward. If the trial was correct, the next trial was presented with a delay 1 s longer. The delay for the trial after an incorrect choice was decreased by 1 s. This process was repeated until the delay for a correct choice peaked, the delay interval over the last ten trials of each session was used as the cognitive index. Cognitive impairment was defined as following Eq. (4):

$$\text{Cognitive Impairment}\,(\%) = \left(1 - \text{Cognitive Index}_{\text{post}} - \text{Cognitive Index}_{\text{pre}}\right) \times 100$$

where Cognitive Index$_{pre}$ and Cognitive Index$_{post}$ are pre- and post-administration values for vehicle or scopolamine, respectively.

PET measurements and cognitive performance test were serially conducted 2, 6, 24, and 48 h after administration of scopolamine, and the degrees of impaired of cognitive function were plotted against mAChR occupancy degrees induced by scopolamine administration.

3 Experimental Results

3.1 *In Vitro* Assessment of (+)3-MPB Analogs

Synthesized 3-PB, 3-MPB, 3-EPB, and 3-PPB were analyzed to confirm their chemical structures as ^{1}H-NMR spectra (R-1200, Hitachi High-Tech Fielding Corporation, Tokyo, Japan). $^{1H\text{-}NMR}$ of 3-PB: 7.36 (m, 10H), 4.92 (m, 1H), 2.94 (m, 1H), 2.70 (m, 4H), 1.82 (m, 1H), 1.71 (m, 1H), 1.26 (m, 1H). 3-MPB: 7.36 (s, 10H), 5.02 (m, 1H), 2.69 (m, 1H), 2.46 (m, 1H), 2.24 (s, 3H), 2.17 (m, 1H), 1.94 (m, 1H), 1.82 (m, 1H), 1.71 (m, 1H), 1.58 (m, 1H), 1.41 (m, 1H), 1.26 (m, 1H); 3-EPB: 7.32 (s, 10H), 5.01 (m, 1H), 4.41 (s, 1H), 2.72 (m, 2H), 2.25 (m, 4H), 1.68 (m, 3H), 1.25 (s, 1H), 0.97 (t, 3H); 3-PPB: 7.36 (s, 10H), 5.01 (m, 1H), 4.37 (s, 1H), 2.74 (m, 2H), 2.27 (m, 4H), 1.60 (m, 6H), 0.92 (t, 3H).

As shown in Fig. 1, we designed methods to synthesize several analogs of (+)3-MPB, and evaluated their affinities to mAChR and lipophilicity. In competition assays with ^{3}H-QNB, the IC_{50} and Ki values of (+)3-MPB, (+)3-EPB, (+)3-PPB, and their inactive stereoisomers, (−)3-MPB, (−)3-EPB, and (−)3-PPB, for mAChR were determined using rat brain slices in vitro. Since the binding of ^{3}H-QNB showed a time-dependent increase, reaching equilibrium by about 30 min, an incubation time of 30 min was chosen for this competition assays. As a result, (+)3-MPB had the highest ($K_i = 1.69$ nM), (+)3-EPB had moderate (12.90 nM), and (+)3-PPB had the lowest affinity (47.53 nM) for mAChR in the neocortex of the rat brain (Fig. 1e), suggesting that the longer alkyl chain length caused the lower affinity [47]. The affinity of 4-MPB (0.51 nM) was 3.3-fold higher than that of (+)3-MPB. In contrast, (−)3MPB, (−)3-EPB, and (−)3-PPB showed remarkably lower affinity to the mAChR (1000 nM and more), which were two orders lower than those of their corresponding active (+) forms [18]. The log D at pH 7.4 of ^{11}C-(+)3-MPB, ^{11}C-(+)3-EPB, and ^{11}C-(+)3-PPB were 1.53, 1.68, and 2.37, respectively.

By using the five cloned human mACh receptors, the K_D values for (+)3-MPB exhibited no significant selectivity for any subtype, whereas the binding profile of K_i values for (+)3-MPB was $M_4 \geq M_1 > M_3 \geq M_2 > M_5$. A binding profile for 4-MPB was $M_4 > M_1 \geq M_3 \geq M_5 > M_2$. The largest difference in affinity was below ten times (between M_4 and M_5 for 3-MPB and between M_2 and M_4 for 4-MPB). These competition binding data of (+)3- and 4-MPB showed that neither of these substances has an affinity profile that makes them suitable for subtype-specific assays [26].

3.2 In Vivo Assessment of ^{11}C-(+)3-MPB Analogs

^{11}C-(+)3-MPB and its *N*-alkyl substitution analogs, (−)3-MPB, (+) and (−)3-EPB, (+) and (+)3-EPB, were labeled by using ^{11}C-methyl iodide, ^{11}C-ethyl iodide, or ^{11}C-propyl iodide for the alkylation of (+) or (−)3-PB as shown in Fig. 2. Each of these compounds was intravenously injected into monkey for scanning for 90 min, and then the acquired data were reconstructed to images for TAC determination as sequential images and ROI setting as summation images from 60 to 90 min after the injection.

In Fig. 3, typical PET images of ^{11}C-(+)3-MPB (B), ^{11}C-(−)3-MPB (C), ^{11}C-(+)3-EPB (D), ^{11}C-(+)3-PPB (E), and ^{11}C-4-MPB (A) as a reference are shown, which were obtained from the same monkey under conscious condition. The regional uptake pattern of ^{11}C-(+)3-MPB was high in the striatum; intermediate in the

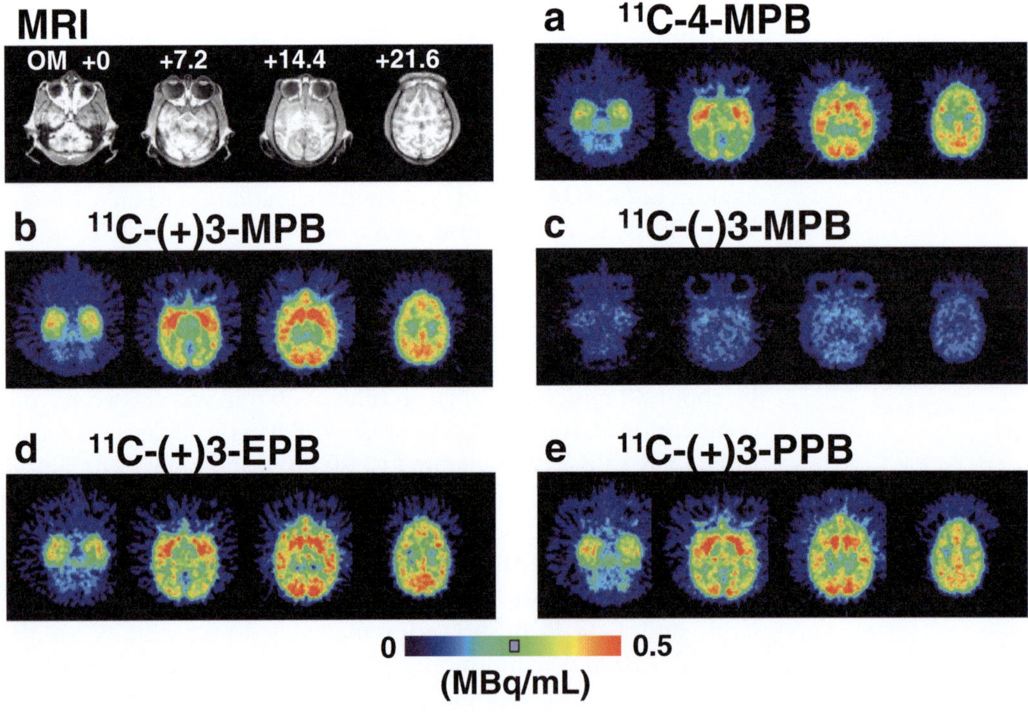

Fig. 3 MRI and PET images of ^{11}C-4-MPB (**a**), ^{11}C-(+)3-MPB (**b**), ^{11}C-(−)3-MPB (**c**), ^{11}C-(+)3-EPB (**d**), and ^{11}C-(+)3-PPB (**e**) in conscious monkey brain. PET data were collected in the conscious state with a high-resolution PET scanner (HAMAMATSU SHR-7700) with transaxial special resolution of 2.6 mm (FWHM). Each labeled compound (100–120 MBq/kg body weight) was injected through the venous cannula. PET scans were performed for 91 min, and each PET image was generated by summation of image data from 60 to 94 min post-injection

occipital, temporal, and frontal cortices, hippocampus, and thalamus; and low in the cerebellum (Fig. 3b). In contrast, the level of ^{11}C-(−)3-MPB was much lower than those of ^{11}C-(+)3-MPB showing a homogeneous distribution in all regions of the brain (Fig. 3c). The patterns of distribution of ^{11}C-(+)3-EPB (Fig. 3d), ^{11}C-(+)3-PPB (Fig. 3e), and ^{11}C-4-MPB (Fig. 3a) were almost identical to that of ^{11}C-(+)3-MPB. There results demonstrated that specificity of these PET probes except ^{11}C-(−)3-MPB remained constant even upon substitution of *N*-alkyl moiety.

The time activity curves of ^{11}C-(+)3-MPB in the frontal, temporal, and occipital cortices reached their peaks 40 min after injection, whereas the striatal and hippocampal regions reached peak values 60 min after injection. In contrast, the time–activity curves of ^{11}C-(−)3-MPB showed similar patterns in all regions of the brain [19]. The uptake of ^{11}C-4-MPB in all regions except the thalamus and cerebellum gradually increased over time during the scan until 91 min after injection [19].

As shown in Fig. 4a, Patlak plot graphical analysis demonstrated that ^{11}C-(+)3-MPB provided nonlinear curves showing slope = 0 in the late phase, suggesting that the dissociation rate constant (k_4) from mAChR was not negligible. Next, Logan plot graphical analysis was applied to investigate the in vivo binding of ^{11}C-(+)3-MPB (Fig. 4b). The ratio of Logan slopes (DV) of ^{11}C-(+)3-MPB in each region against the cerebellum was

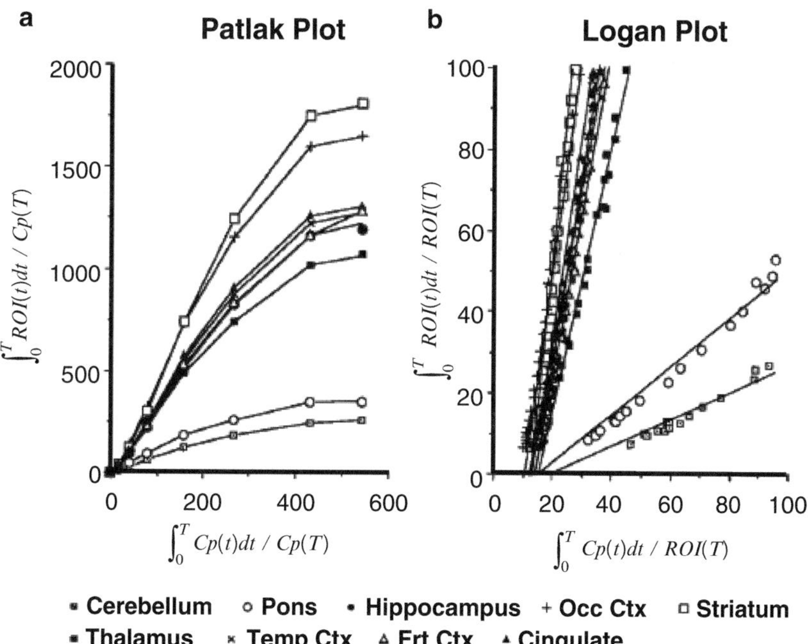

Fig. 4 Patlak plot graphical (**a**) and Logan plot graphical analysis (**b**) of ^{11}C-(+)3-MPB in conscious monkey brain. Regions of interest (ROIs) were identified according to an MRI scan of the each animal, and metabolite-corrected plasma input was applied for quantitative measurements of ^{11}C-(+)3-MPB binding to mAChR in the living brain

correlated with the data of mAChR density (B_{MAX}) as measured by the in vitro assay [48]. These results demonstrated that ^{11}C-(+)3-MPB was a more suitable PET probe for the quantification of mAChR in the living brain than conventional ^{11}C-labeled ligands such as ^{11}C-4-MPB and ^{11}C-scopolamine.

As described in Section 3.1, two *N*-alkyl substitution analogs of (+)3-MPB, (+)3-EPB, and (+)3-PPB revealed lower affinities to mAChR than 4-MPB and (+)3-MPB. In order to develop PET probes with moderate affinity to assess neurotransmitter release into the synaptic cleft [29, 43, 49], we attempted to synthesize ^{11}C-(+)3-EPB and ^{11}C-(+)3-PPB, and evaluated their specificity and kinetics in comparison with those of original ^{11}C-(+)3-MPB in conscious monkey brain.

The time–activity curves demonstrated that the peak times shifted to an earlier time with a higher clearance rate after injection of PET probes with longer ^{11}C-alkyl chains [18]. As shown in Fig. 5, kinetic analysis indicated that labeling with a longer ^{11}C-alkyl chain induced lower binding potential (Frontal cortex; 2.4, 2.0, and 1.4 for ^{11}C-(+)3-MPB, ^{11}C-(+)3-EPB, and ^{11}C-(+)3-PPB, respectively). The administration of Aricept, an AChE inhibitor, increased acetylcholine level in extracellular fluid of the frontal cortex (ca. 150 % and 175 % of baseline at 50 μg/kg and 250 μg/kg, respectively) [18]. The binding of ^{11}C-(+)3-PPB with the lowest affinity to mAChR was displaced by the endogenous ACh induced by Aricept, while ^{11}C-(+)3-MPB with the highest affinity was not significantly affected (Fig. 5). These results suggested that increasing ^{11}C-alkyl chain length did alter the kinetic properties of PET probes by reducing the affinity to mAChR, which might make it possible to assess the interaction between the endogenous neurotransmitter acetylcholine and ligand-receptor binding in vivo as measured by PET.

3.3 Effects of Aging on mAChR

In aged monkeys, the time–activity curves of ^{11}C-(+)3-MPB in regions rich in mAChR peaked at earlier time points with faster elimination rates than those in young monkeys, while curves in the cerebellum showed no significant difference between young and aged animals [20]. Significant age-related alterations of the in vivo binding of ^{11}C-(+)3-MPB were observed in the temporal and frontal cortices and the striatum (Fig. 6a). The Scatchard plot revealed a linear curve for ^{11}C-(+)3-MPB in all regions of young and aged monkeys (Fig. 6b). Aged animals showed the age-related reduction in the maximum number of binding sites (B_{MAX}) of mAChR, while there was no age-related alteration in the affinity ($1/K_D$) of mAChR for (+)3-MPB (Fig. 6b).

Previous studies demonstrated, when expressed relative to the cerebellum, which is assumed to be a nonspecific and tracer-free reference region, age-related reduction of mAChR in almost all cerebral regions in humans as measured by ^{11}C-benztropine [37] and ^{11}C-4-MPB [38–40]. There were trends toward reduced cerebral mAChR binding and toward elevated cerebellar binding with aging in humans,

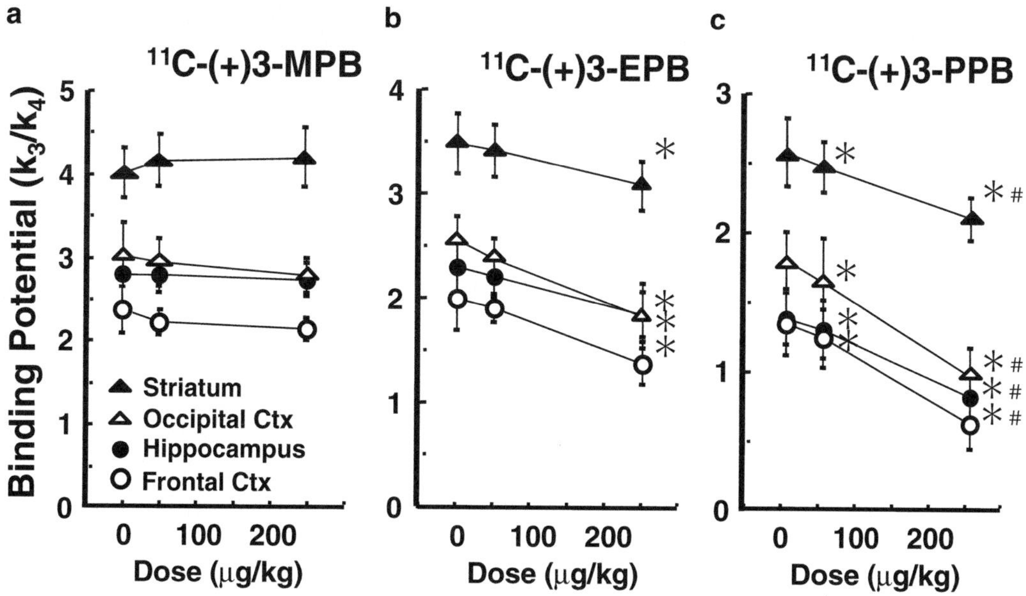

Fig. 5 Dose-dependent effects of Aricept on the binding of ^{11}C-(+)3-MPB (**a**), ^{11}C-(+)3-EPB (**b**), and ^{11}C-(+)3-PPB (**c**) to mAChR in the cortical regions of conscious monkey brain. PET scans were performed as shown in the legend of Fig. 3. Data are expressed as means ± SD for five animals per treatment condition. *$P<0.01$ vs. saline. #$P<0.01$ vs. 50 μg/kg Aricept

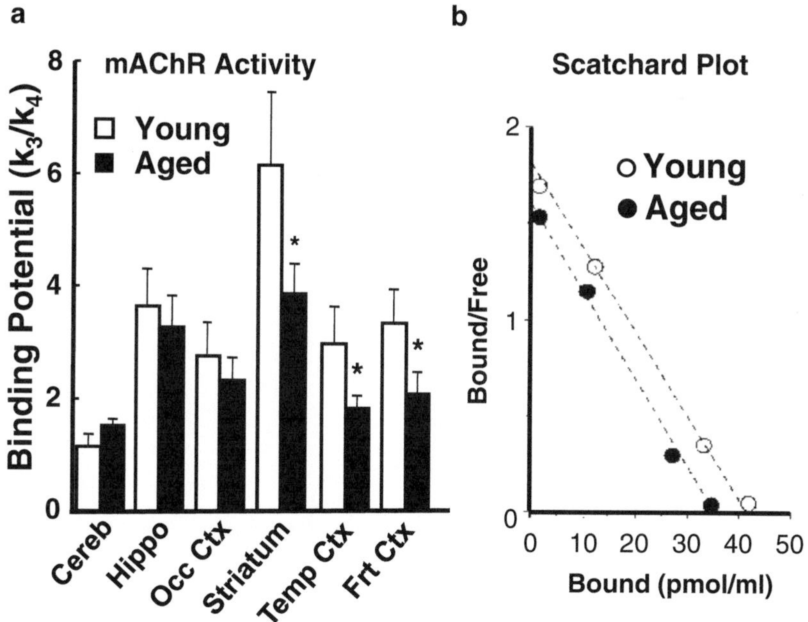

Fig. 6 Age-related changes in cerebral mAChR binding in vivo as measured with ^{11}C-(+)3-MPB. (**a**) PET scans in young (ca. 6 years old) and aged (ca. 20 years old) monkeys were performed as shown in the legend of Fig. 3. Data represent means ± SD for seven animals per group. *$P<0.05$ vs. young animals. (**b**) Scatchard plot analysis was conducted by injection of carrier-free ^{11}C-(+)3-MPB, and together with various amounts of cold (+)3-MPB to measure in vivo binding parameters (B_{MAX} and K_D)

as measured by a kinetic analysis method using ^{11}C-tropanyl benzilate using metabolite-corrected plasma TAC as input function, a gold standard kinetic analysis for reliable quantification of receptor binding in vivo [41]. This may suggest when used cerebellar TAC as input function, the opposing cortical and cerebellar changes of binding led to the apparent age-related loss of mAChR binding.

3.4 mAChR Occupancy and Cognition

With applying T-DMS task test, the memory assessment of young monkeys was serially conducted 2, 6, 24 and 48 h after administration of scopolamine at the doses of 10 and 30 μg/kg. Scopolamine impaired the memory performance in a dose-dependent manner 2 h after administration, followed by gradual recovery at 24 h and later after scopolamine administration (Fig. 7b). The time-dependent

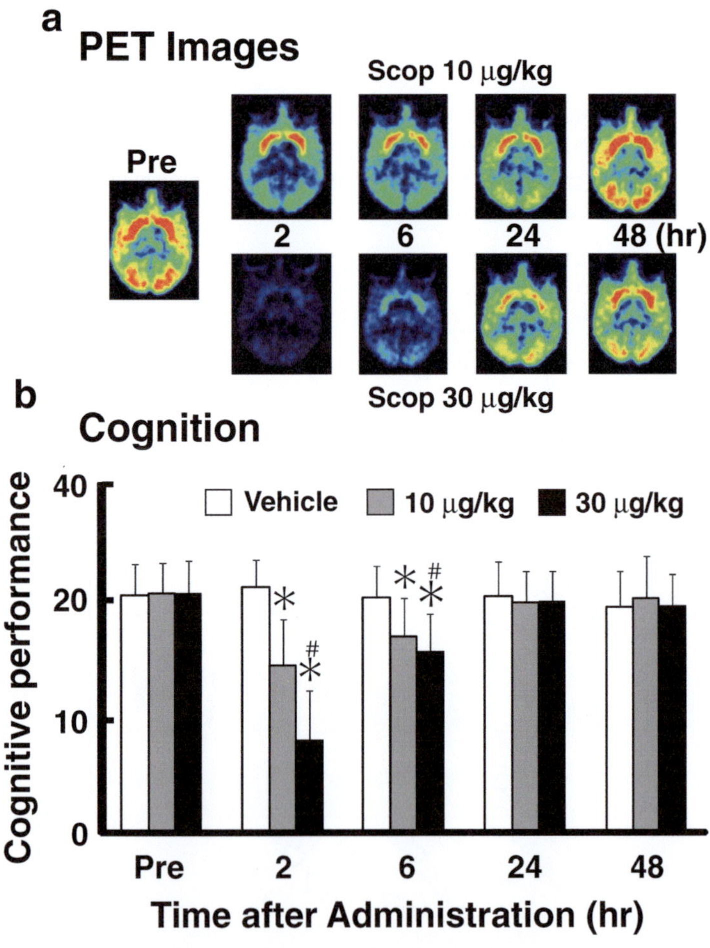

Fig. 7 Effects of scopolamine on ^{11}C-(+)3-MPB binding in conscious monkey brain (**a**) and on titration-delayed match to sample (T-DMS) task (**b**). Time courses of the mAChR occupancy (**a**) and cognitive index (**b**) were serially monitored 2, 6, 24, and 48 h after the administration of vehicle or scopolamine at doses of 10 and 30 μg/kg. **P*<0.01 vs. saline. #*P*<0.01 vs. 0.01 mg/kg scopolamine

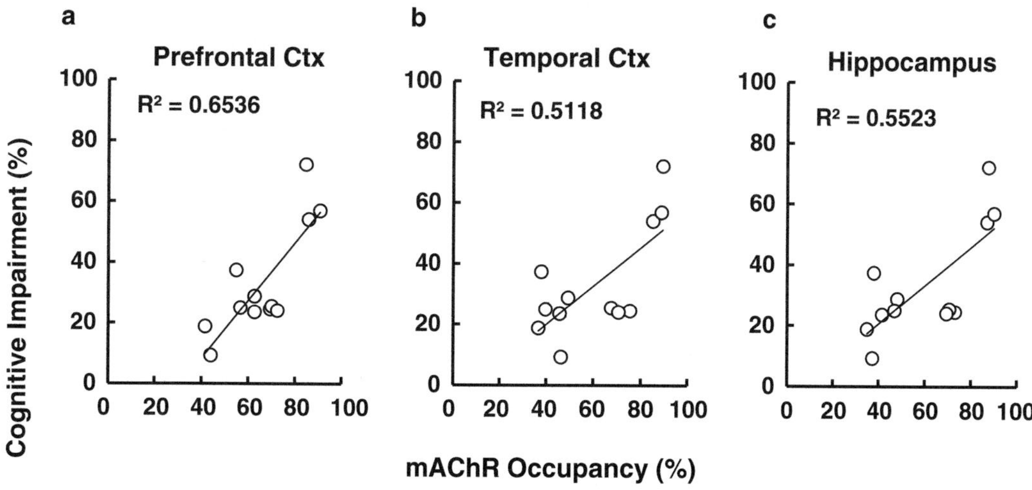

Fig. 8 Relationship between mAChR occupancy and cognitive impairment. Occupancy levels in prefrontal cortex (**a**), temporal cortex (**b**), and hippocampus (**c**) were assessed 2 and 6 h after scopolamine administration at the doses of 10 and 30 μg/kg

changes in occupancy of mAChR by scopolamine were simultaneously observed with serial PET measurements using ^{11}C-(+)3-MPB. Occupancy levels of mAChR peaked 2 h after scopolamine administration in the cortical regions innervated primarily by the basal forebrain, thalamus, and brainstem (Fig. 7a). As shown in Fig. 8, mAChR occupancy and the degree of cognitive impairment were significantly and positively correlated in the brain regions assessed. Cognitive impairment induced by scopolamine persisted for 6 h and was followed by complete recovery to normal levels 24 h later. It was very interesting that some scopolamine binding to mAChRs was still observed in most brain areas except the brainstem 24 h after scopolamine administration, suggesting the existence of a threshold (ca. 25 %) of mAChR occupancy to induce cognitive impairment.

4 Limitations

All these data described above indicate that ^{11}C-(+)3-MPB [17, 19] is superior to its previous alternatives, such as ^{11}C-scopolamine [8–10], ^{11}C-QNB [12], ^{11}C-dexetimide [11], ^{11}C-benztropine [13], ^{11}C-TRB [14, 15], and ^{11}C-4-MPB [16]. However, several limitations of ^{11}C-(+)3-MPB remain to be solved. One limitation may be poor subtype selectivity for mAChR, which consist of at least the M_1–M_5 subtypes. The in vitro radioligand binding experiments performed with five cloned human mAChR subtypes expressed in CHO-K1 cells demonstrated that (+)3-MPB have no apparent selectivity for M_1 and M_2 receptors (unpublished data), suggesting that ^{11}C-(+)3-MPB images the integration of both M_1

and M_2 receptors. To compensate the poor selectivity, a selective agonist-based PET probe for M_2 receptor, 3-(3-(3-^{18}F-flouropropyl) thio)-1,2,5-thiadiazol-4-yl)-1,2,5,6-tetrahydro-1-methylpyridine (^{18}F-FP-TZTP) was developed with K_i of 2.2 and 7.4 nM for M_2 and M_1, respectively [50]. By using mice with the knocked-out M_1, M_2, M_3, or M_4 receptors, it was confirmed that ^{18}F-FP-TZTP selectively bound to M_2 mAChR subtype [51].

The assessment of neurotransmitter release in the living brain should be very important from neurological and pharmacological points of view. In the last decade, interactions between endogenous neurotransmitters released into the synaptic cleft and the binding of labeled compounds have been examined by neuroimaging with PET/SPECT, especially in the dopaminergic system [29, 43, 49]. As demonstrated with ^{11}C-(+)3-EPB and ^{11}C-(+)3-PPB [18], the cortical binding of ^{18}F-FP-TZTP was also decreased by physostigmine, a AChE inhibitor, showing the competition between ^{18}F-FP-TZTP and increased synaptic endogenous ACh on mAChRs [52]. This PET probe seemed to be more favorable than previous ones; however uncertain properties have been revealed on ^{18}F-FP-TZTP as described later.

There is a loss of cholinergic neurons in the forebrain, reduced cholinergic activity in the hippocampus, and cortical loss of choline acetyltransferase in AD, and the severity of these cholinergic abnormalities is correlated well with the degree of dementia [3–7]. Noninvasive imaging of mAChRs in the brain could contribute to the diagnosis and monitoring of medications for treatment of these diseases. Of interest, an age-related "increased" binding of agonist-based ^{18}F-FP-TZTP was determined [53], which was completely the opposite result obtained with antagonist-based ^{11}C-benztropine [37], ^{11}C-4-MPB [38–40], ^{11}C-tropanyl benzilate [41], and ^{11}C-(+)3-MPB [20], showing "decreased" binding. The authors speculated that a lower concentration of ACh in the synapse was one possible explanation for the age-related increase in ^{18}F-FP-TZTP binding [53]. It remains controversial that opposite effects of aging on mAChR binding were observed between agonist-based and antagonist-based PET probes. We reported the similar opposite effects of stress induced by head motion restriction on dopamine D_2 receptor binding between agonist-based (^{11}C-MNPA) and antagonist-based PET probes (^{11}C-raclopride) [34]. Although it is well established that agonists can bind to a "high-affinity" state of the receptor, this phenomenon is conceptually unresolved because it has yet to be ascertained exactly what state of the receptor (conformational, coupled etc.) defines "high-affinity." These results suggest that when we attempt to develop novel PET probes, we should take into account the chemical properties of agonist or antagonist.

5 Conclusion

This chapter has introduced the procedure how to develop the novel PET probes for quantitative mAChR imaging with PET, and also showed how to utilize these PET probes in preclinical and clinical researches based on our own experiences. Thus, we developed novel PET probes for mAChR imaging, ^{11}C-(+)3-MPB and its *N*-alkyl substitution analogs with suitable kinetic properties for quantitative imaging of mAChRs in the living brain with PET. On the basis of these preclinical evaluations of ^{11}C-(+)3-MPB as a novel PET probe for mAChR imaging, we applied this PET probe to assess the changes in mAChR binding in the brain of chronic fatigue syndrome patients [54]. These results of our studies described in this chapter strongly suggested the usefulness of these PET probes for pathophysiological and neurological assessments of neurological/mental disorders.

References

1. Katzman R (1986) Alzheimer's disease. N Engl J Med 314:964–973
2. Selkoe DJ (1990) Deciphering Alzheimer's disease: the amyloid precursor protein yields new clues. Science 248:1058–1060
3. Höhmann C, Antuono P, Coyle JT (1998) Basal forebrain cholinergic neurons and Alzheimer's disease. In: Iversen LL, Iversen SD, Snyder SD (eds) Psychopharmacology of the aging nervous system. Plenum, New York, pp 69–106
4. Perry EK (1986) The cholinergic hypothesis—ten years on [Review]. Br Med Bull 42:63–69
5. Rinne JO, Laakso K, Lönnberg P et al (1985) Brain muscarinic receptors in senile dementia. Brain Res 336:19–25
6. Reinikainen KJ, Riekkinen PJ, Halonen T et al (1987) Decreased muscarinic receptor binding in cerebral cortex and hippocampus in Alzheimer's disease. Life Sci 41:453–461
7. Terry AV Jr, Buccafusco JJ (2003) The cholinergic hypothesis of age and Alzheimer's disease-related cognitive deficits: recent challenges and their implications for novel drug development. J Pharmacol Exp Ther 306:821–827
8. Vora MM, Finn RD, Boothe TE (1983) [*N*-methyl-^{11}C]Scopolamine: synthesis and distribution in rat brain. J Labelled Comp Radiopharm 20:1229–1234
9. Mulholland GK, Jewett DW, Toorongian SA (1988) Routine synthesis of *N*-[^{11}C-methyl] scopolamine by phosphate mediated reductive methylation with [^{11}C]formaldehyde. Appl Radiat Isot 39:373–379
10. Frey KA, Koeppe RA, Mulholland GK et al (1992) *In vivo* muscarinic cholinergic receptor imaging in human rain with [^{11}C]scopolamine and positron emission tomography. J Cereb Blood Flow Metab 12:147–1541
11. Dannals RF, Långström B, Ravert HT et al (1988) Synthesis of radiotracers for studying muscarinic cholinergic receptors in the living human brain using positron emission tomography: [^{11}C]dexetimide and [^{11}C]levetimide. Appl Radiat Isot 39:291–295
12. Prenant C, Barre L, Crouzel C (1989) Synthesis of n.c.a. [^{11}C]QNB. J Labelled Comp Radiopharm 26:199–201
13. Dewey SL, MacGregor RR, Brondie JD et al (1990) Mapping muscarinic receptors in human and baboon brain using [*N*^{11} C-methyl]benztropine. Synapse 5:213–223
14. Mulholland GK, Otto CA, Jewett DW et al (1992) Synthesis, rodent biodistribution, dosimetry, metabolism and monkey images of carbon-11-labeled (+)-2α-tropanyl benzilate: a central muscarinic receptor imaging agent. J Nucl Med 33:423–430
15. Koeppe RA, Frey KA, Mulholland GK et al (1994) [^{11}C]Tropanyl benzilate binding to muscarinic cholinergic receptors: methodology and kinetic modeling alterations. J Cereb Blood Flow Metab 14:85–99
16. Zubieta JK, Koeppe RA, Mulholland GK et al (1998) Quantification of muscarinic cholinergic receptors with [^{11}C]NMPB and positron emission tomography: method development and differentiation of tracer delivery from

receptor binding. J Cereb Blood Flow Metab 18:619–631

17. Takahashi K, Murakami M, Miura S et al (1999) Synthesis and autoradiographic localization of muscarinic cholinergic antagonist *N*-^{11}C-methyl-3-piperidyl benzilate as a potent radioligand for positron emission tomography. Appl Radiat Isot 50:521–525
18. Nishiyama S, Tsukada H, Sato K et al (2001) Evaluation of PET ligands (+)N-[^{11}C]ethyl-3-piperidyl benzilate and (+)*N*-[^{11}C]propyl-3-piperidyl benzilate for muscarinic cholinergic receptors: a PET study with microdialysis in comparison with (+)*N*-[^{11}C]methyl-3-piperidyl benzilate in the conscious monkey brain. Synapse 40:159–169
19. Tsukada H, Takahashi K, Miura S et al (2001) Evaluation of novel PET ligands (+)*N*-[^{11}C] methyl-3-piperidyl benzilate ([^{11}C](+)3-MPB) and its stereoisomer [^{11}C](−)3-MPB for muscarinic cholinergic receptors in the conscious monkey brain: a PET study in comparison with [^{11}C]4-MPB. Synapse 39:182–192
20. Tsukada H, Kakiuchi T, Nishiyama S et al (2001) Age differences in muscarinic cholinergic receptors assayed with (+)*N*-[^{11}C]methyl-3-piperidyl benzilate in the brains of conscious monkeys. Synapse 41:248–257
21. Tsukada H, Nishiyama S, Fukumoto D et al (2004) Effects of acute acetylcholinesterase inhibition on the cerebral cholinergic neuronal system and cognitive function: functional imaging of the conscious monkey brain using animal PET in combination with microdialysis. Synapse 52:1–10
22. Yamamoto S, Nishiyama S, Kawamata M et al (2011) Muscarinic receptor occupancy and cognitive impairment: A PET study with [^{11}C](+)3-MPB and scopolamine in conscious monkeys. Neuropsychopharmacology 36:1455–1465
23. Biel JH, Abood LG, Hoya WK et al (1961) Central stimulant. II, Cholinergic blocking agents. J Org Chem 26:4096–4103
24. Cheng YC, Prusoff WH (1973) Relationship between the inhibition constant (Ki) and the concentration of inhibitor which causes 50 percent inhibition (IC_{50}) of an enzymatic reaction. Biochem Pharmacol 22:3099–3108
25. Paxinos G, Watson C (1982) The rat brain in stereotaxic coordinates. Academic, New York
26. Sihver S (2000) Development of in vitro and ex vivo positron-emitting tracer techniques and their application to neurotrauma. Thesis in Faculty of Medicine, Uppsala University
27. Långström B, Antoni G, Gullberg P et al (1986) The synthesis of L-^{11}C-labeled ethyl, propyl, butyl and isobutyl iodides and examples of alkylation reactions. Appl Radiat Isot 37: 1141–1145
28. Irie T, Fukushi K, Namba H et al (1996) Brain acetylcholinesterase activity: validation of a PET tracer in a rat model of Alzheimer's disease. J Nucl Med 37:649–655
29. Tsukada H, Harada N, Nishiyama S et al (2000) Ketamine decreased striatal [^{11}C]raclopride binding with no alteration in static dopamine concentrations in the striatal extracellular fluid in the monkey brain: Multi-parametric PET studies combined with microdialysis analysis. Synapse 37:95–103
30. Tsukada H, Nishiyama S, Kakiuchi T et al (2001) Ketamine alters the availability of striatal dopamine transporter as measured by [^{11}C] β-CFT and [^{11}C]β-CIT-FE in the monkey brain. Synapse 42:273–280
31. Tsukada H, Miyasato K, Kakiuchi T et al (2002) Comparative effects of methamphetamine and nicotine on the striatal [^{11}C]raclopride binding in unanesthetized monkeys. Synapse 45:207–212
32. Ohba H, Harada N, Nishiyama S et al (2009) Ketamine/xylazine anesthesia alters [^{11}C] MNPA binding to dopamine D_2 receptors and response to methamphetamine challenge in monkey brain. Synapse 63:534–537
33. Watanabe M, Okada H, Shimizu K et al (1997) A high resolution animal PET scanner using compact PS-PMT detectors. IEEE Trans Nucl Sci 44:1277–1282
34. Tsukada H, Ohba H, Nishiyama S et al (2011) Differential effects of stress on [^{11}C]raclopride and [^{11}C]MNPA binding to striatal D_2/D_3 dopamine receptors: a PET study in conscious monkeys. Synapse 64:84–89
35. Logan J, Fowler J, Volkow N et al (1990) Graphical analysis of reversible radioligand binding from time-activity measurements applied to *N*-^{11}C-methyl-(−)-cocaine PET studies in human subjects. J Neurochem 10: 740–747
36. Patlak C, Blasberg RG, Fenstermacher JD (1983) Graphical evaluation of blood-to-brain transfer constants from multiple-time uptake data. J Cereb Blood Flow Metab 3:1–7
37. Dewey SL, Volkow ND, Logan J et al (1990) Age-related decrease in muscarinic cholinergic receptor binding in the human brain measured with positron emission tomography (PET). J Neurosci Res 27:569–575
38. Suhara T, Inoue O, Kobayashi K et al (1993) Age-related changes inhuman muscarinic acetylcholine receptors measured by positron emission tomography. Neurosci Lett 149: 225–228

39. Yoshida T, Kuwabara Y, Ichiya Y et al (1998) Cerebral muscarinic acetylcholinergic receptor measurement in Alzheimer's disease patient on 11C-methyl-4-piperidyl benzilate—comparison with cerebral blood flow and cerebral glucose metabolism. Ann Nucl Med 12:35–42
40. Zubieta JK, Koeppe RA, Frey KA et al (2001) Assessment of muscarinic receptor concentrations in aging and Alzheimer disease with ^{11}C-NMPB and PET. Synapse 39:275–287
41. Lee KS, Frey KA, Koeppe RA et al (1996) *In vivo* quantification of cerebral muscarinic receptors in normal human aging using positron emission tomography and [^{11}C]tropanyl benzilate. J Cereb Blood Flow Metab 16: 303–310
42. Tsukada H, Kreuter J, Maggos CE et al (1996) Effects of binge pattern cocaine administration on dopamine D_1 and D_2 receptors in the rat brain: an *in vivo* study using positron emission tomography. J Neurosci 16:7670–7677
43. Tsukada H, Harada N, Nishiyama S et al (2000) Cholinergic neuronal modulation alters dopamine D_2 receptor availability in vivo by regulating receptor affinity induced by facilitated synaptic dopamine turnover: PET studies with microdialysis in the conscious monkey brain. J Neurosci 20:7067–7073
44. Scatchard G (1949) The attractions of proteins for small molecules and ions. Ann N Y Acad Sci 51:660–672
45. Collerton D (1986) Cholinergic function and intellectual decline in Alzheimer's disease [Review]. Neuroscience 19:1–28
46. Hudzik TJ, Wenger GR (1993) Effects of drugs of abuse and cholinergic agents on delayed matching-to-sample responding in the squirrel monkey. J Pharmacol Exp Ther 265: 120–127
47. Tejani-Butt SM, Luthin GR, Wolfe BB et al (1990) *N*-substituted derivatives of 4-piperidinyl benzilate: affinities for brain muscarinic acetylcholine receptors. Life Sci 47:841–848
48. Snyder SH, Chang KJ, Kuhar MJ et al (1975) Biochemical identification of the mammalian muscarinic cholinergic receptor. Fed Proc 34:1919–1921
49. Laruelle M (2000) Imaging synaptic neurotransmission with in vivo binding competition techniques: a critical review. J Cereb Blood Flow Metab 20:423–451
50. Kiesewetter DO, Lee J, Lang L et al (1995) Preparation of ^{18}F-labeled muscarinic agonist with M_2 selectivity. J Med Chem 38:5–8
51. Jagoda EM, Kiesewetter DO, Shimoji K et al (2003) Regional brain uptake of the muscarinic ligand, [^{18}F]FP-TZTP, is greatly decreased in M_2 receptor knockout mice but not in M_1, M_3 and M_4 receptor knockout mice. Neuropharmacology 44:653–661
52. Carson RE, Kiesewetter DO, Jagoda E et al (1998) Muscarinic cholinergic receptor measurements with [^{18}F]FP-TZTP: control and competition studies. J Cereb Blood Flow Metab 18:1130–1142
53. Podruchny TA, Connolly C, Bokde A et al (2003) In vivo muscarinic 2 receptor imaging in cognitively normal young and older volunteers. Synapse 48:39–44
54. Yamamoto S, Ouchi Y, Nakatsuka D et al (2012) Reduction of [^{11}C](+)3-MPB binding in brain of chronic fatigue syndrome with serum autoantibody against muscarinic cholinergic receptor. PLoS One 7, e51515

Chapter 11

Detection of Non-neuronal Acetylcholine

Ignaz Karl Wessler and Charles James Kirkpatrick

Abstract

The biological role of acetylcholine and the cholinergic system has been revisited within the last 25 years. Acetylcholine and the pivotal components of the cholinergic system (high affinity choline uptake, choline acetyltransferase and its endproduct acetylcholine, muscarinic and nicotinic receptors, cholinesterases) are expressed by more or less all mammalian cells, i.e., cells not innervated by neurons at all. Moreover, acetylcholine and cholinergic binding sites have been described in plants. Acetylcholine is even detected in bacteria and algae and thus represents an extremely old signaling molecule on the evolutionary time scale. The following chapter summarizes the detection of acetylcholine beyond neurons with particular emphasis on the presence of acetylcholine in so-called primitive organisms. Finally, an overview is given about the detection in mammalian non-neuronal cells. The existence of the non-neuronal cholinergic system has identified an important new target to illuminate the pathophysiological background of acute and chronic inflammatory diseases as well as heart diseases and cancer.

Key words Non-neuronal acetylcholine, Non-neuronal cholinergic system, HPLC combined with bioreactors and electrochemical detection, Evolution, Bacteria, Plants, Unicellular organisms, Epithelial–mesothelial–endothelial and immune cells, Signaling via muscarinic and nicotinic receptors

1 Introduction

Even in our modern day and age textbooks as well as academic education are presenting acetylcholine as a neurotransmitter mediating the communication between neurons, interneurons and innervated effector cells such as muscle fibers and glandular cells. An actual upload (http://en.wikipedia.org/wiki/acetylcholine [1]) describes acetylcholine as follows: "*acetylcholine has functions both in the peripheral nervous system (PNS) and in the central nervous system (CNS) as a neuromodulator.*" An actual research in PubMed shows 150-fold more references in favor of the key words "acetylcholine and neurotransmitter" than given for the term "non-neuronal acetylcholine." Of course, our knowledge about acetylcholine and its biological functions has substantially increased in the last 130 years, when for the first time acetylcholine was extracted from the brain and called at first neurin and later on synthesized as acetylcholine [2, 3].

Jaromir Myslivecek and Jan Jakubik (eds.), *Muscarinic Receptor: From Structure to Animal Models*, Neuromethods, vol. 107, DOI 10.1007/978-1-4939-2858-3_11, © Springer Science+Business Media New York 2016

It is fascinating today to realize that all the biological systems/functions in which acetylcholine, released from central, parasympathetic, and peripheral intramural neurons as well as motoneurons, is involved by stimulating at least 11 different subunits of nicotinic receptors and five subtypes of muscarinic receptors. Thus, acetylcholine acts as neurotransmitter in the motoric and sensoric (i.e., thermal, pain, taste) system; acetylcholine is involved in complex integrative neuronal functions like memory, learning and sexual activity; acetylcholine, as a neurotransmitter of autonomic neurons, controls the cardiovascular, respiratory, gastrointestinal, and urogenital system. However, this summarizes only the role of acetylcholine within the nervous system, and the role of acetylcholine as a general signaling molecule beyond neurons has to be considered additionally.

To discriminate acetylcholine not synthesized by neuronal cells and not mediating nervous impulses but acting as autocrine/paracrine signaling molecule from acetylcholine acting as neurotransmitter, the nomenclature "non-neuronal acetylcholine" and "non-neuronal cholinergic system" has been introduced in 1998 and 1999 [4, 5]. One should consider that the existence of acetylcholine independent of neurons has been known for a long time before (for review see ref. [6]). Unfortunately, the scientific community has forgotten the first experiments by Ewins and Dale who investigated the effect of an extract of the ergot fungus (*Claviceps purpurea*) on the blood pressure in 1914 [7, 8]. Ergot grows on rye particularly during rainy periods in spring and can induce serious intoxications (called "St. Anthony's Fire") which were known during the Middle Ages and were based on the vasoactive effects of the ergot alkaloids. When Ewins and Dale investigated the hemodynamic effects of an extract of this fungus, they found a depressor effect [7, 8]. Later on they could attribute this depressor effect to acetylcholine [8]. In conclusion, the first experiments illuminating a biological role of acetylcholine, "the blood pressure lowering substance", the molecule was extracted from fungi, i.e., from non-neuronal organisms. Some years later (1921) Otto Loewi presented the first experimental evidence for the neurotransmitter role of acetylcholine, when he used a pair of isolated frog hearts [9]. The first heart with the nerves attached was stimulated and the second was used as detector to demonstrate the released substances from the first one. He postulated the so-called "Vagus-Stoff" or "parasympathin" acting as humoral transmission of nervous impulses [9] and 5 years later the vagus-substance was identified as acetylcholine [10]. Later on (1963) Whittaker stated that "acetylcholine occurs in non-nervous tissues and is so widely distributed in nature to suggest a non-nervous function of it" [11] and Koelle speculated about acetylcholine representing a phylogenetically very old molecule, which, in primitive organisms such as plants and unicellular organisms, might be involved in the regulation of transport processes [12]. Moreover, important contributions in the last century showed the synthesis of acetylcholine in bacteria, algae, yeast, fungi, protozoa, nematodes, sponges, and plants [4–6, 13–22] Thus, acetylcholine is as far as we know one of the oldest signaling molecules in the evolutionary process.

The present chapter is focused on the detection of non-neuronal acetylcholine. In the last years important review articles have been published to describe this topic in more detail [4–7, 23–35]. Moreover, in 2002, 2006, 2011, and 2014 international conferences on non-neuronal acetylcholine were held [36–38].

2 Detection Methods for Acetylcholine

For decades the most sensitive, but less specific, method for determination of acetylcholine was the bioassay, such as the leech longitudinal muscle, the guinea pig small intestine, the frog rectus abdominis muscle, and cat blood pressure. Using these detector systems together with specific antagonists the lower detection limit for acetylcholine amounted to about 0.2–5 ng (corresponding to about 1–20 μmol) [6]. In addition acetylcholine can be detected by gas chromatography combined with a preceding chemical transformation of the quaternary ammonium compound acetylcholine or by ion-pair extraction and using a nitrogen selective detector [39]. The detection limit for these methods is around 50–100 pmol acetylcholine.

Later on in the 1980 decade acetylcholine is detected by HPLC combined with bioreactors, i.e., an analytical column separates acetylcholine from choline and thereafter acetylcholine is converted by immobilized acetylcholinesterase to choline which reacts with immobilized choline-oxidase to H_2O_2 and betaine; H_2O_2 can then be detected either by luminescence or by electrochemical detection [40, 41]. The HPLC method was further optimized in the following years by using microbore columns with an internal diameter of 1 mm and attained a sensitivity of about 10–50 fmol/20 μl ([4], see also Fig. 1a). Finally, a radioimmunoassay for acetylcholine has also been established [42]. Very recently a highly sophisticated method has been developed which can visualize non-neuronal acetylcholine in the epithelial cell layer of the mouse small intestine [43]; matrix-assisted laser absorption/ionization (MALDI-TOF) imaging mass spectrometry has been optimized for the cellular detection and visualization of acetylcholine.

The increase in the detection limit was very important for progress in understanding the non-neuronal cholinergic system, because mammalian non-neuronal cells contain considerably less acetylcholine than neurons. The non-neuronal cells (see Table 3) do not concentrate acetylcholine in high quantities within small vesicles, where acetylcholine is highly concentrated and stored by

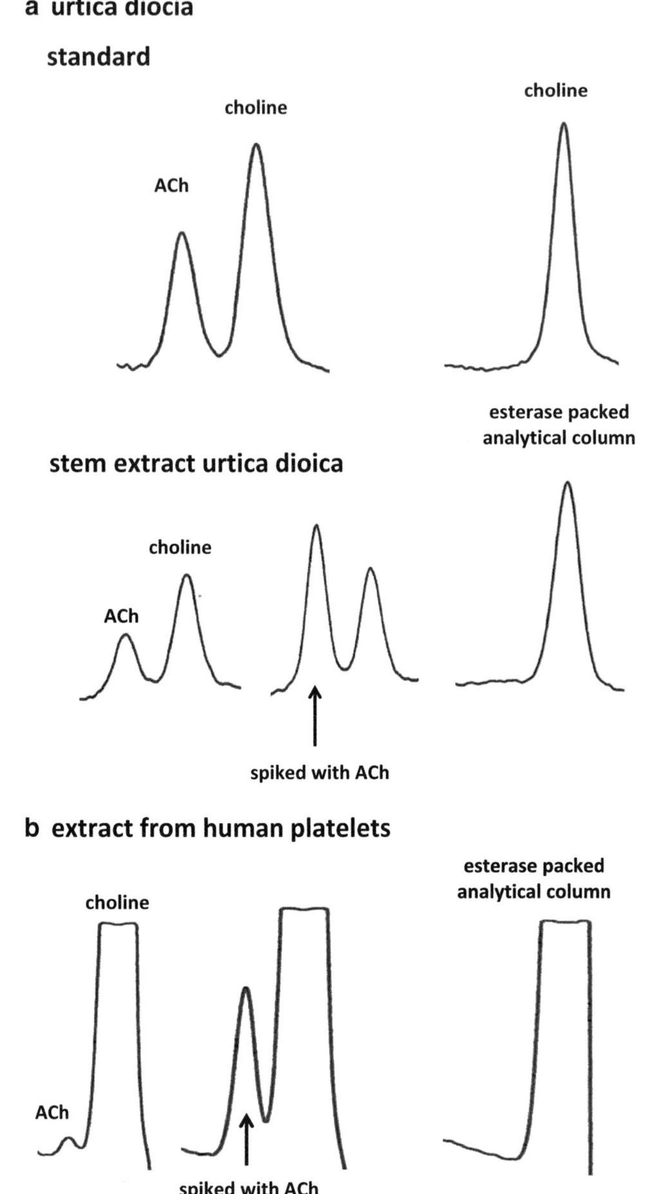

Fig. 1 Detection of non-neuronal ACh by HPLC with bioreactors and electrochemical detection. (**a**) 85 mg of a stem of *Urtica dioica* was placed in 1 ml of 15 vol.% formic acid in acetone and minced with scissors. After standing on ice (30 min) and centrifugation, the supernatant was evaporated to dryness by nitrogen. The dried sample was resuspended in 1 ml of a phosphate buffer, diluted by a factor of 200 and an aliquot (20 μl) was injected onto the HPLC-system (for details see ref. [44]). *First row*: chromatogram of a standard solution containing 1 pmol acetylcholine and choline/20 μl using the regular analytical column or, on the *right hand side*, an analytical column packed with acetylcholine-specific esterase, i.e., under this condition the first acetylcholine peak disappeared. *Second row*: chromatogram of the extract from *Urtica dioica*; the second chromatogram shows the same sample spiked with 0.6 pmol acetylcholine/20 μl; still only one peak appears at the retention time corresponding to acetylcholine; third chromatogram shows the same

neurons to generate a super threshold signal upon neuronal activity. In contrast, non-neuronal cells appear to release acetylcholine in small quantities more or less continuously to maintain cellular homeostasis by autocrine and paracrine signaling via muscarinic and nicotinic receptors which are abundantly expressed on more or less all cells. Thus, a very sensitive method is required to detect non-neuronal acetylcholine extracted either from human epithelial cells or human skin by dermal microdialysis [44, 45].

However, one has to consider that HPLC measurement combined with bioreactor and electrochemical detection does not represent a 100 % specific method, as other unknown compounds or other choline-esters can produce peaks with a retention rate similar to that of acetylcholine. Therefore, one has to prove the identity of the acetylcholine peak by spiking the sample with a low quantity of applied acetylcholine and by using an acetylcholinesterase-packed analytical column. Under this condition the acetylcholine peak must disappear and the choline peak should increase correspondingly. A typical example is shown in Fig. 1a using extracts of leaves of *Urtica dioica* and for reference a standard sample containing 1 pmol/20 μl of both acetylcholine and choline. Figure 1b demonstrates the presence of acetylcholine in human platelets.

For detection of non-neuronal acetylcholine in tissue or cells (freshly isolated or cultured) it is important to homogenize or lyse the cells/tissue in small volumes. For example, pieces of isolated airways or small intestine can be fixed in a Petri dish with the luminal surface facing upwards and a cotton-tipped applicator can gently rubbed for 5 s along the luminal surface. Using this approach the basal membrane of the surface airway epithelium is not penetrated, i.e., the underlying lamina propria remained intact [44]. Likewise, rubbing of the intestinal surface removed tips of villi only, the lamina muscularis mucosae with the underlying cholinergic submucosal plexus remained intact [44]. Corresponding samples can be taken from the lung surface (pulmonary pleura) or from the surface epithelium of oral and vaginal mucosa of volunteers [44]. After rubbing, the cotton-part of the applicator is placed in 1 ml ice-cold 15 % formic acid in acetone (v/v) for 30 min with intermittent vortexing. This medium mediates cell lysis and inactivates all enzymes immediately. Thereafter, the cotton is removed and the medium evaporated to dryness by a smooth nitrogen jet.

Likewise, tissue can be pulverized by means of liquid nitrogen and placed in 1 ml ice-cold 15 % formic acid in acetone (v/v). After standing on ice (30 min) with repeated vortexing, the samples

Fig. 1 (continued) sample using the esterase-packed analytical column; the first acetylcholine peak disappears. (**b**) 40 ml of a concentrate of human platelets were centrifuged and the pellet was analyzed as described above under a; the huge choline peak may be caused by activation of platelets by the separation procedure

can be centrifuged (10 min; 4000 rpm), and the supernatant is evaporated to dryness by a smooth gas jet of nitrogen. This will take about 30 min and within that time samples can again be placed on ice to prevent spontaneous hydrolysis of acetylcholine. The dried sample is resuspended in 300–1000 μl of the mobile phase of the HPLC system (70 mM phosphate buffer with 0.3 mM EDTA; pH 8.5 adjusted). The principle of the detection of acetylcholine by HPLC and the use of bioreactors and electrochemical detection is described above. The use of microbore columns is helpful because the flow-front of H_2O_2 originating from conversion of acetylcholine to choline and betaine is concentrated and then, can produce a measurable current at the working electrode. Moreover, the enzymatic reaction acts as an amplifier, because 1 mol acetylcholine produces 2 mol H_2O_2. It is difficult to use internal standard, therefore acetylcholine content has to be quantified by comparison with external acetylcholine standard which is measured before and after an individual sample.

In the following sections examples are given of the detection of non-neuronal acetylcholine, in which acetylcholine is either measured directly or the expression of one of the synthesizing enzymes is shown. Acetylcholine can be synthesized by choline acetyltransferase (ChAT) or carnitine acetyltransferase (CarAT). Both enzymes have been found to mediate the synthesis of non-neuronal acetylcholine, for example, in plants but also in vertebrates and invertebrates [6, 22, 46–48].

3 Detection of Acetylcholine in So-Called Primitive Organisms Generated Very Early on the Evolutionary Time Scale

Bacteria are regarded as one of the first forms of life on earth, arising about four billion years ago; also archaea represent prokaryotic microorganisms and are thought to have populated the earth about three billion years ago. Using a radioimmunoassay or HPLC combined with electrochemical detection acetylcholine has been detected in bacteria and archaea (see Table 1). An acetylcholine-synthesizing activity has been isolated from extracts of bacteria or archaea, but the properties differ from the mammalian ChAT enzyme. The function of acetylcholine in these microorganisms is unknown so far. However, bacteria show locomotion and it has been shown that motility of two photosynthetic bacteria (*Rhodospirillum rubrum*, *Thiospirillum jenense*) was stopped by 1 mM atropine, an antagonist of muscarinic receptors. Furthermore, physostigmine and other cholinesterase inhibitors also reduced motility [49]. It is probable that the system became desensitized in the presence of cholinesterase inhibitors. Table 1 gives an overview about the presence of acetylcholine in prokaryotic microorganisms and unicellular eukaryotic organisms as well as in fungi and lower

Table 1
Presence of non-neuronal acetylcholine early in evolution

		Amount	References
Bacteria	*Bacillus subtilis*, *Escherichia coli*, *Staphylococcus aureus*, *Lactobacillus plantarum*, *Lactobacillus bacillus*	0.39–55 pmol/10^{10} colony forming units (CFU)	[19, 22, 79]
Archaea	Hyperthermophiles *T. kodakaraensis* Euryarchaeota *S. tokodaii*; *P. calidifontis* Methanogens *M. thermautotrophicus* *M. barkeri* Halophiles *Halobacterium* sp., *H. volcanii*	0.05–1.2 pmol/g	[48]
Algae	Blue-green algae		[21]
Protozoa (Protista)	Paramecium, *Trypanosoma rhodesiense*		[14, 80]
Porifera (Sponge)	*Theonella swinhei*, *Xestospongia sapra*, *Halichondria japonica*, *Halichondria okudai*, *Halochondria permollis*	60–2200 pmol/g	[22]
Mussels	*Gill plates* (*Mytilis edulis*, *Anodonta*)	3.8 nmol/g	[16]
Fungi	Yeast *Saccharomyces cerevisiae* Shiitake mushroom *Lentinus eddoes* Matsutake mushroom *Tricholoma matsutake* *Agaricus bisporus*, *Cantharellus cibarius* Sordariomycetes *Claviceps* (*Ergot fungi*)	900–2800 pmol/g	[7, 22, 24]
"Lower plants"	Moss Spenophyta *Equisetum robustum*	2–124 nmol/g ~2 nmol/g	[50, 81]

plants, i.e., in biological systems (including the gill plates of mussels) not regulated by neurons. All these biological systems are created very early during evolution.

4 Detection of Acetylcholine in the Plant Kingdom

Table 2 gives an overview of the expression of acetylcholine in the plant kingdom, i.e., multiple examples for the expression of acetylcholine independent of any existing neuronal system. Obviously, acetylcholine is expressed in lower and higher plants. It seems inevitable that the list will be enlarged in the future. One has to consider that our knowledge about the biological function of acetylcholine is very scanty, and a systematic analysis of the expression and biological role of acetylcholine within the plant kingdom is lacking. Reports have indicated that the synthesis of acetylcholine in plants may be regulated by light [50]. One very interesting observation is that the acetylcholine content is very high in rapid growing plants like *bamboo*, *helianthus* and *Urtica dioica*. Moreover, it has been shown that at least in *Urtica dioica* acetylcholine is involved in the regulation of water homeostasis and photosynthesis [24]. Particularly, 1 μM atropine reduced the intracellular space, the cell vacuole, and cell size and mediated proliferation of the thylakoid membrane [24, 34]. In conclusion, also in the plant kingdom binding sites for acetylcholine exist, which can be blocked by atropine.

5 Detection of Acetylcholine or Positive Anti-ChAT Immunoreactivity in Mammalian Non-neuronal Cells

To demonstrate the existence of non-neuronal acetylcholine in mammalian cells without any doubt, a contamination of neuronal acetylcholine has to be excluded in the respective samples. Thus, it has to be shown conclusively that acetylcholine is synthesized by cells not innervated at all by neurons and that these cells/tissues cannot take up acetylcholine which may be released from possible adjacent neurons. The following findings demonstrate without any doubt that isolated cells or tissue which lacks any cholinergic innervation synthesize and release acetylcholine:

(a) Acetylcholine synthesis has been demonstrated in various cultured cells (keratinocytes, airway epithelial cells, and cardiomyocytes) and cell lines like leukemic T-cells (MOLT-3), embryonic stem cells (CGR8), colon (Caco-2; H508) or lung (H82) cancer cell lines [26, 51–60]. All these cultured cells are free of any neuronal input. Release of acetylcholine was demonstrated also from cultured bovine arterial endothelial cells [61].

Table 2
Presence of non-neuronal acetylcholine in the plant kingdom

Family	Genus, *Species*	Amount	References
Amaranthaceae	Spinacia *Spinacia oleracea*	~3–7 nmol/g	[21, 50]
Anthophyta	Arabidopsis *Arabidopsis thaliana* Eggplant *Solanum melongena* Bamboo shoot *Phyllostachys bambusoides* *Phyllostachys pubescens*	23.7 pmol/g 416 nmol/g 2.9 μmol/g 0.6–1.7 μmol/g	[22]
Apocynaceae	Amsonia *Amsonia angustifolia*		[82]
Araceae	Arum *Arum specificum*, *Arum maculatum*	~1.8 nmol/g	[24]
Asteraceae	Helianthus *Helianthus annuus* Senecio *Senecio vulgaris*	3–8 μmol/g ~6.5 nmol/g	[21, 50]
Brassicaceae	Capsella *Capsella bursa-pastoris* Sinapis *Sinapis alba*	~4.8 nmol/g	[24]
Bryophyta	Moss *Conocephalum conicum* *Polytrichum*, *Brachythecium*	0.03–8.0 nmol/g	[22, 24, 50, 83]
Coniferophyta	Cedar *Cryptomeria japonica* Hinoki *Chamaecyparis obtuse* Pine *Pinus thunbergii* Podocarp *Podocarpus macrophyllus*	120–343 pmol/g	[50]
Cucurbitales	Cucurbita *Cucurbita pepo*	3–10 nmol/g	[84]
Fabaceae	Phaseolus *Phaseolus vulgaris*, *Phaseolus aureus* Pisum *Pisum sativum*	~100 ng/g 1–8 nmol/g	[50, 85]
Moraceae	*Malayan jack-fruit Artocarpus integra*	564 μg/g	[84]
Plantaginaceae	Digitalis *Digitalis ferruginea*	1.6 mg/50 g pulverized leaves	[86]
Pterophyta	Fern *Pteridium*, *Gleichenia glauca*	0.07–1.6 nmol/g	[22]
Rosaceae	Fragaria *Fragaria vesca* Crataegus *Crataegus specificus*	~5.4 nmol/g	[24, 87]
Sphenophyta	Horsetail *Equisetum arvense* *Equisetum robustum*	38 pmol/g ~2.8 nmol/g	[22, 24]
Urticaceae	Urtica *Urtica dioica* *Girardinia heterophylla*	~0.5 μmol/g	[24, 88, 89] Fig. 1a

(b) Cultured epithelial cells isolated from the airways of monkeys release acetylcholine into the supernatant [62]. In organic-cation-transporter knockout mice, i.e., a condition limiting the release of non-neuronal acetylcholine, airway epithelial acetylcholine content was doubled, which indicates an in vivo release of acetylcholine from these cells [63, 64]. Likewise, the release of non-neuronal acetylcholine becomes evident by the inhibitory effect of nicotine receptor antagonists on the migration of cultured airway epithelial cells [65].

(c) The placenta of various species (human, monkey, cow, rabbit, rat, mouse), an organ free of cholinergic neurons, synthesizes, stores and releases acetylcholine [6, 15, 66–70].

(d) ChAT mRNA and ChAT protein have been demonstrated in most of these cells.

(e) In vivo release of acetylcholine from human skin has been demonstrated by dermal microdialysis. Botulinum toxin blocks neuronal acetylcholine release but does not inhibit acetylcholine release from the human skin [45].

In conclusion, convincing experimental data have been published since the third decade of the last century about the presence of non-neuronal acetylcholine in mammalian cells. Nevertheless, in the following decades the scientific community has focussed more or less exclusively on the role of neuronal acetylcholine in the brain and peripheral nervous system. Possibly, the brain and neurons may have drawn more attraction than apparently less specified non-neuronal cells, although the regulation and communication of these cells and their respective networks is already extremely complex. It is fascinating that epithelial or immune cells communicate by the same molecules and cholinergic receptors as do neurons in the brain. Both, the specific cholinesterase and the pseudocholinesterase, play an important role to clearly separate both systems (non-neuronal vs. neuronal) in vivo. This is operating because specific cholinesterase represents the enzyme with the highest turnover rate created by nature and because of the abundant presence of both enzymes in mammalian organisms thus limiting neuronal acetylcholine to act at hot spots only.

Table 3 gives an overview about the expression of non-neuronal acetylcholine in various mammalian cells. Accordingly, acetylcholine has been detected directly after extraction from these cells/ tissues or positive anti-choline acetyltransferase (ChAT) immunoreactivity has been found. However, in the case of using the method of immunohistochemistry alone some caution is required, because false positive staining has been found with antibodies directed against muscarinic receptors in corresponding knockout mice [71].

When non-neuronal acetylcholine is released and present in the extracellular space or plasma [26], it will diffuse in close proximity to its source but also to neighboring cells, because in principle the expression level of cholinesterase activity is lower in non-innervated than in innervated cells. For example release of acetylcholine from the isolated placenta can be measured without preceding inhibition of cholinesterase [6, 70]. Consequently, non-neuronal acetylcholine can mediate autocrine and paracrine effects by stimulating muscarinic and nicotinic receptors which are ubiquitously expressed in the majority of cells (for review see ref. [33]).

Table 3
Detection of ACh or positive anti-ChAT immunoreactivity independent of neuronal innervation in mammalian cells

Cell type	Location	References
Epithelial cells	*Airways* (basal, ciliated, secretory and brush cells of the airway mucosa) *Alimentary tract* (mucosa of oral cavity, esophagus, stomach (partially), jejunum, ileum, colon, sigmoid, gall bladder) *Skin* (keratinocytes, eccrine and sebaceous glands) *Urogenital tract* (urothelium, vaginal mucosa, tubuli of kidney, granulosa cells, spermatozoa, embryonic stem cells) *Eye* (cornea) *Placenta* (trophoblast) and *Amnion* *Glandular tissue* (female breast)	[4–6, 15, 33, 44, 51, 52, 90–101]
Endothelial cells	Skin, umbilical vein, pulmonary vessels	[4, 61, 102]
Immune cells	Leukocytes, bone marrow-derived dendritic cells, macrophages, skin mast cells, microglia	[6, 24, 54, 103–105]
Mesothelial cells	Pleura, pericardium, mesenteric root	[4, 44]
Various	Pinealocytes, astrocytes, embryonic stem cells, platelets	[4, 44, 55, 60, 106, 107] Fig. 1b
Mesenchymal cells	Adipocytes (skin), myoblasts and myotubes, smooth muscle fibers, fibroblasts (airways), tendon (tenocytes), cardiac muscle fibers, osteoblast-like cells	[4, 6, 33, 95, 108–114]

6 Conclusion

Research early and late in the last century has substantially broadened our understanding of the role of acetylcholine and the cholinergic system, i.e., choline uptake, synthesizing enzymes, muscarinic and nicotinic receptors, and the inactivating enzymes. This system has been created extremely early on the evolutionary time scale (about three to four billion years ago). Thus, it is not surprising to detect the cholinergic system in plants, unicellular organisms and in more or less all mammalian cells independently of neurons. Neurons have become specialized cells prepared for signaling on the milliseconds time scale. Therefore, neurons have taken an advantage of the already established cholinergic signaling system and have further specialized this system during evolution: storing in specialized organelles (vesicles), triggering vesicular release, establishing of hot spots for nicotinic receptors and acetylcholinesterase. In contrast, the non-neuronal cholinergic system does not mediate cellular communication on the millisecond time scale but establishes a variable cholinergic tone to co-regulate basic cell functions like proliferation, differentiation, cell–cell contact, secretion, and absorption. For example, acetylcholine via muscarinic receptors causes an increase of intracellular calcium within seconds but not within milliseconds [58]. It is important to learn more about the physiological and pathophysiological role of non-neuronal acetylcholine and the non-neuronal cholinergic system. Recent articles have described an important role of the non-neuronal cholinergic system in different diseases [30, 72–77]. Particularly, chronic airway diseases (COPD and bronchial asthma) have been identified to be treated with long acting muscarinic receptor antagonists (aclidinium, glycopyrronium, tiotropium) to induce therapeutic effects beyond bronchodilation. In animal models provoking COPD or asthma these receptor antagonists have been shown to reduce airway inflammation and airway remodeling by suppressing cellular effects of non-neuronal acetylcholine [78]. Moreover, intensive research is required to further illuminate the pathophysiological role of the non-neuronal cholinergic system in other inflammatory diseases, cancer, and heart diseases.

References

1. http://en.wikipedia.org/wiki/Acetylcholine. Accessed 18 Jan 2015
2. Liebreich O (1864) Über die chemische Beschaffenheit der Gehirnsubstanz. Ann Chem Pharm 54:29–44
3. Bayer A (1867) Synthese des Neurins. Ann Chem Pharm 140:306–313
4. Wessler I et al (1998) Non-neuronal acetylcholine, a locally acting molecule, widely distributed in biological systems: expression and function in humans. Pharmacol Ther 77(1):59–79
5. Wessler I et al (1999) The cholinergic 'pitfall': acetylcholine, a universal cell molecule in biological systems, including humans. Clin Exp Pharmacol Physiol 26(3):198–205
6. Sastry BV, Sadavongvivad C (1978) Cholinergic systems in non-nervous tissues. Pharmacol Rev 30(1):65–132

7. Dale HH (1914) The action of certain esters and ethers of choline and their relation to muscarine. J Pharmacol Exp Ther 6:147–190
8. Ewins AJ (1914) Acetylcholine, a new active principle of ergot. Biochem J 8:44–49
9. Loewi O (1921) Über humorale Übertragbarkeit der Herznervenwirkung. Plügers Arch 189:239–242
10. Loewi O, Navratil E (1926) Über humorale Übertragbarkeit der Herznervenwirkung. X. Über das Schicksal des Vagusstoff. Pflugers Arch Ges Physiol 214:678–688
11. Whittaker VP (1963) Identification of acetylcholine and related esters of biological origin. In: Eichler O, Farah A, Koelle GB (eds) Handbuch der experimentellen Pharmakologie, vol 15. Springer, Berlin, pp 1–39
12. Koelle G (1963) Cytological distributions and physiological functions of cholinesterases. In: Eichler O, Farah A, Koelle G (eds) Handbuch der experimentellen Pharmakologie, vol 15. Springer, Berlin, pp 187–298
13. Dale HH, Dudley HW (1929) The presence of histamine and acetylcholine in the spleen of the ox and the horse. J Physiol 68(2):97–123
14. Beyer G, Wense UT (1936) Über den Nachweis von Hormonen in einzelligen Tieren: Cholin und Acetylcholin in Paramecium. Plügers Arch Gesamte Physiol Menschen Tiere 237: 417–422
15. Comline RS (1946) Synthesis of acetylcholine by non-nervous tissue. J Physiol (Lond) 105: 6P–7P
16. Bülbring E, Burn JH, Shelley HJ (1953) Acetylcholine and ciliary movement in the gill plate of *Mytilus edulis*. Proc R Soc Lond B Biol Sci 141:445–466
17. Lentz TL (1966) Intramitochondrial glycogen granules in digestive cells of Hydra. J Cell Biol 29(1):162–167
18. Erzen I, Brzin M (1979) Cholinergic mechanisms in *Planaria torva*. Comp Biochem Physiol 64C:207–216
19. Stephenson M, Rowatt E (1947) The production of acetylcholine by a strain of Lactobacillus plantarum. J Gen Microbiol 1(3):279–298
20. Fischer H (1971) Vergleichende Pharmakologie von Überträgersubstanzen in tiersystematischer Darstellung. In: Eichler O, Farah A, Herken H, Welch A (eds) Handbuch der experimentellen Pharmakologie. Springer, Berlin
21. Smallman BN, Maneckjee A (1981) The synthesis of acetylcholine by plants. Biochem J 194(1):361–364
22. Horiuchi Y et al (2003) Evolutional study on acetylcholine expression. Life Sci 72(15): 1745–1756
23. Grando SA (1997) Biological functions of keratinocyte cholinergic receptors. J Investig Dermatol Symp Proc 2(1):41–48
24. Wessler I et al (2001) The biological role of non-neuronal acetylcholine in plants and humans. Jpn J Pharmacol 85(1):2–10
25. Wessler I et al (2003) The non-neuronal cholinergic system in humans: expression, function and pathophysiology. Life Sci 72(18–19): 2055–2061
26. Kawashima K, Fujii T (2000) Extraneuronal cholinergic system in lymphocytes. Pharmacol Ther 86(1):29–48
27. Kawashima K, Fujii T (2004) Expression of non-neuronal acetylcholine in lymphocytes and its contribution to the regulation of immune function. Front Biosci 9:2063–2085
28. Eglen RM (2006) Muscarinic receptor subtypes in neuronal and non-neuronal cholinergic function. Auton Autacoid Pharmacol 26(3):219–233
29. Grando SA et al (2006) Adrenergic and cholinergic control in the biology of epidermis: physiological and clinical significance. J Invest Dermatol 126(9):1948–1965
30. Grando SA et al (2007) Recent progress in understanding the non-neuronal cholinergic system in humans. Life Sci 80(24–25):2181–2185
31. Kummer W et al (2008) The epithelial cholinergic system of the airways. Histochem Cell Biol 130(2):219–234
32. Kurzen H et al (2007) The non-neuronal cholinergic system of human skin. Horm Metab Res 39(2):125–135
33. Wessler I, Kirkpatrick CJ (2008) Acetylcholine beyond neurons: the non-neuronal cholinergic system in humans. Br J Pharmacol 154(8): 1558–1571
34. Wessler I, Kirkpatrick CJ (2012) Activation of muscarinic receptors by non-neuronal acetylcholine. Handb Exp Pharmacol 208:469–491
35. Kawashima K, Fujii T (2008) Basic and clinical aspects of non-neuronal acetylcholine: overview of non-neuronal cholinergic systems and their biological significance. J Pharmacol Sci 106(2):167–173
36. Grando S, Kawashima K, Kirkpatrick C, Wessler I (eds) (2003) Proceedings of the 1st international symposium on non-neuronal acetylcholine. Life Sci, vol 72 (18–19), pp 2009–2182
37. Grando S, Kawashima K, Kirkpatrick C, Wessler I (eds) (2007) Proceedings of the 2nd international symposium on non-neuronal acetylcholine. Life Sci, vol 80 (24–25), pp 2181–2396
38. Grando S, Kawashima K, Kirkpatrick C, Meurs H, Wessler I (eds) (2012) Proceedings of the 3rd international symposium on non-neuronal

acetylcholine. Life Sci, vol 91 (21–22), pp 969–1137

39. Kilbinger H (1973) Gas chromatographic estimation of acetylcholine in the rabbit heart using a nitrogen selective detector. J Neurochem 21(2):421–429
40. Israel M, Lesbats B (1981) Chemiluminescent determination of acetylcholine, and continuous detection of its release from torpedo electric organ synapses and synaptosomes. Neurochem Int 3(1):81–90
41. Potter PE et al (1983) Acetylcholine and choline in neuronal tissue measured by HPLC with electrochemical detection. J Neurochem 41(1):188–194
42. Kawashima K et al (1980) Radioimmunoassay for acetylcholine in the rat brain. J Pharmacol Methods 3(2):115–123
43. Takahashi T et al (2014) Non-neuronal acetylcholine as an endogenous regulator of proliferation and differentiation of Lgr5-positive stem cells in mice. FEBS J 281(20):4672–4690
44. Klapproth H et al (1997) Non-neuronal acetylcholine, a signalling molecule synthesized by surface cells of rat and man. Naunyn Schmiedebergs Arch Pharmacol 355(4): 515–523
45. Schlereth T et al (2006) In vivo release of non-neuronal acetylcholine from the human skin as measured by dermal microdialysis: effect of botulinum toxin. Br J Pharmacol 147(2):183–187
46. White HL, Cavallito CJ (1970) Inhibition of bacterial and mammalian choline acetyltransferases by styrylpyridine analogues. J Neurochem 17(11):1579–1589
47. White HL, Wu JC (1973) Choline and carnitine acetyltransferases of heart. Biochemistry 12(5):841–846
48. Yamada T et al (2005) Expression of acetylcholine (ACh) and ACh-synthesizing activity in Archaea. Life Sci 77(16):1935–1944
49. Faust MA, Doetsch RN (1971) Effect of drugs that alter excitable membranes on the motility of Rhodospirillum rubrum and Thiospirillum jenense. Can J Microbiol 17(2): 191–196
50. Hartmann E, Kilbinger H (1974) Gas-liquid-chromatographic determination of light-dependent acetylcholine concentrations in moss callus. Biochem J 137(2):249–252
51. Grando SA et al (1993) Human keratinocytes synthesize, secrete, and degrade acetylcholine. J Invest Dermatol 101(1):32–36
52. Reinheimer T et al (1996) Acetylcholine in isolated airways of rat, guinea pig, and human: species differences in role of airway mucosa. Am J Physiol 270(5):L722–L728
53. Fujii T et al (1998) Induction of choline acetyltransferase mRNA in human mononuclear leukocytes stimulated by phytohemagglutinin, a T-cell activator. J Neuroimmunol 82(1): 101–107
54. Kawashima K, Fujii T (2003) The lymphocytic cholinergic system and its biological function. Life Sci 72(18–19):2101–2109
55. Paraoanu LE et al (2007) Expression and possible functions of the cholinergic system in a murine embryonic stem cell line. Life Sci 80(24–25):2375–2379
56. Cheng K et al (2008) Acetylcholine release by human colon cancer cells mediates autocrine stimulation of cell proliferation. Am J Physiol Gastrointest Liver Physiol 295(3): G591–G597
57. Song P et al (2003) Acetylcholine is synthesized by and acts as an autocrine growth factor for small cell lung carcinoma. Cancer Res 63(1):214–221
58. Song P et al (2007) M3 muscarinic receptor antagonists inhibit small cell lung carcinoma growth and mitogen-activated protein kinase phosphorylation induced by acetylcholine secretion. Cancer Res 67(8): 3936–3944
59. Kakinuma Y et al (2009) Cholinoceptive and cholinergic properties of cardiomyocytes involving an amplification mechanism for vagal efferent effects in sparsely innervated ventricular myocardium. FEBS J 276(18): 5111–5125
60. Wessler I et al (2012) Release of acetylcholine from murine embryonic stem cells: effect of nicotinic and muscarinic receptors and blockade of organic cation transporter. Life Sci 91(21–22):973–976
61. Kawashima K et al (1990) Synthesis and release of acetylcholine by cultured bovine arterial endothelial cells. Neurosci Lett 119(2):156–158
62. Proskocil BJ et al (2004) Acetylcholine is an autocrine or paracrine hormone synthesized and secreted by airway bronchial epithelial cells. Endocrinology 145(5):2498–2506
63. Kummer W et al (2006) Role of acetylcholine and muscarinic receptors in serotonin-induced bronchoconstriction in the mouse. J Mol Neurosci 30(1–2):67–68
64. Kummer W et al (2006) Role of acetylcholine and polyspecific cation transporters in serotonin-induced bronchoconstriction in the mouse. Respir Res 7:65. doi:10.1186/1465-9921-7-65
65. Tournier JM et al (2006) alpha3alpha5beta2-Nicotinic acetylcholine receptor contributes to the wound repair of the respiratory epithelium

by modulating intracellular calcium in migrating cells. Am J Pathol 168(1):55–68

66. Bischoff C et al (1932) Acetylcholin im Warmblüter. 4. Mitteilung. Hoppe Seyler's Z Physiol Chem 207:57–77
67. Haupstein P (1932) Acetylcholin in der menschlichen Placenta. Arch Gynaekol 152: 262–280
68. Chang HC, Gaddum JH (1933) Choline esters in tissue extracts. J Physiol 79(3): 255–285
69. Olubadewo JO, Rama Sastry BV (1978) Human placental cholinergic system: stimulation-secretion coupling for release of acetylcholine from isolated placental villus. J Pharmacol Exp Ther 204(2):433–445
70. Wessler I et al (2001) Release of non-neuronal acetylcholine from the human placenta: difference to neuronal acetylcholine. Naunyn Schmiedebergs Arch Pharmacol 364(3): 205–212
71. Jositsch G et al (2009) Suitability of muscarinic acetylcholine receptor antibodies for immunohistochemistry evaluated on tissue sections of receptor gene-deficient mice. Naunyn Schmiedebergs Arch Pharmacol 379(4):389–395
72. Grando SA et al (2012) The non-neuronal cholinergic system: basic science, therapeutic implications and new perspectives. Life Sci 91(21–22):969–972
73. Spindel ER (2012) Muscarinic receptor agonists and antagonists: effects on cancer. Handb Exp Pharmacol 208:451–468
74. Grando SA (2014) Connections of nicotine to cancer. Nat Rev Cancer 14(6):419–429
75. Beckmann J, Lips KS (2013) The non-neuronal cholinergic system in health and disease. Pharmacology 92(5–6):286–302
76. Roy A et al (2014) Cholinergic activity as a new target in diseases of the heart. Mol Med. doi:10.2119/molmed.2014.00125
77. Kistemaker LE, Gosens R (2014) Acetylcholine beyond bronchoconstriction: roles in inflammation and remodeling. Trends Pharmacol Sci. doi:10.1016/j.tips.2014.11.005
78. Kistemaker LE et al (2012) Regulation of airway inflammation and remodeling by muscarinic receptors: perspectives on anticholinergic therapy in asthma and COPD. Life Sci 91(21–22):1126–1133
79. Lembeck F, Schraven E (1964) The effect of hemicholinium No. 3 on acetylcholine formation in Lactobacillus arabinosus. Naunyn Schmiedebergs Arch Exp Pathol Pharmakol 247:93–99
80. Bülbring E, Lourie EM, Pardoe U (1949) Presence of acetylcholine in *Trypanosoma rhodesiense* and its absence from *Plasmodium gallinaceum*. Br J Pharmacol 4:290–294
81. Münch M et al (1996) Non-neuronal acetylcholine in plants, rodents and man. Naunyn Schmiedebergs Arch Pharmacol (Suppl) 354:R6
82. Tomczyk H (1964) Pharmacobotanical studies of *Amsonia angustifolia*. II. Phytochemical studies. Diss Pharm 16:297–310
83. Hartmann E (1971) Evidence of a neurohormone in moss callus and its regulation by the phytochrome. Planta 101(2):159–165
84. Lin RCY (1955) Presence of acetylcholine in the Malayan jack-fruit, *Artocarpus integra*. Br J Pharmacol Chemother 10(2):247–253
85. Jaffe MJ (1970) Evidence for the regulation of phytochrome-mediated processes in bean roots by the neurohumor, acetylcholine. Plant Physiol 46(6):768–777
86. Tulus MR et al (1961) Choline and acetylcholine in the leaves of Digitalis ferruginea L. Arch Pharm 294(66):11–17
87. Fiedler U et al (1953) Further constituents of the hawthorn: detection of choline and acetylcholine. Arzneimittelforschung 3(8): 436–437
88. Emmelin N, Feldberg W (1949) Distribution of acetylcholine and histamine in nettle plants. New Phytol 48:143–148
89. Saxena PR et al (1966) Identification of acetylcholine, histamine, and 5-hydroxytryptamine in Girardinia heterophylla (Decne.). Can J Physiol Pharmacol 44(4):621–627
90. Von Brücke H et al (1949) Azetylcholin- und Aneuringehalt der Hornhaut und seine Beziehungen zur Nervenversorgung. Ophthalmologica 117(1):19–35
91. Van Alphen GW (1957) Acetylcholine synthesis in corneal epithelium. AMA Arch Ophthalmol 58(3):449–451
92. Welsch F (1974) Choline acetyltransferase in aneural tissue: evidence for the presence of the enzyme in the placenta of the guinea pig and other species. Am J Obstet Gynecol 118(6):849–856
93. Bishop MR et al (1976) Occurrence of choline acetyltransferase and acetylcholine and other quaternary ammonium compounds in mammalian spermatozoa. Biochem Pharmacol 25(14):1617–1622
94. Bishop MR et al (1977) Identification of acetylcholine and propionylcholine in bull spermatozoa by integrated pyrolysis, gas chromatography and mass spectrometry. Biochim Biophys Acta 500(2):440–444
95. Wessler I et al (2003) Increased acetylcholine levels in skin biopsies of patients with atopic dermatitis. Life Sci 72(18-19):2169–2172

96. Nguyen VT et al (2000) Choline acetyltransferase, acetylcholinesterase, and nicotinic acetylcholine receptors of human gingival and esophageal epithelia. J Dent Res 79(4): 939–949
97. Fritz S et al (2001) Expression of muscarinic receptor types in the primate ovary and evidence for nonneuronal acetylcholine synthesis. J Clin Endocrinol Metab 86(1):349–354
98. Yoshida M et al (2004) Management of detrusor dysfunction in the elderly: changes in acetylcholine and adenosine triphosphate release during aging. Urology 63(3 Suppl 1): 17–23
99. Yoshida M et al (2006) Non-neuronal cholinergic system in human bladder urothelium. Urology 67(2):425–430
100. Hanna-Mitchell AT et al (2007) Non-neuronal acetylcholine and urinary bladder urothelium. Life Sci 80(24–25):2298–2302
101. Lips KS et al (2007) Acetylcholine and molecular components of its synthesis and release machinery in the urothelium. Eur Urol 51(4): 1042–1053
102. Kirkpatrick CJ et al (2001) The non-neuronal cholinergic system in the endothelium: evidence and possible pathobiological significance. Jpn J Pharmacol 85(1):24–28
103. Kawashima K et al (1993) Presence of acetylcholine in blood and its localization in circulating mononuclear leucocytes of humans. Biog Amines 9(4):251–258
104. Kawashima K et al (1998) Acetylcholine synthesis and muscarinic receptor subtype mRNA expression in T-lymphocytes. Life Sci 62(17–18):1701–1705
105. Fujii T et al (1996) Localization and synthesis of acetylcholine in human leukemic T cell lines. J Neurosci Res 44(1):66–72
106. Wessler I et al (1997) Day-night rhythm of acetylcholine in the rat pineal gland. Neurosci Lett 224(3):173–176
107. Wessler I et al (2013) Upregulated acetylcholine synthesis during early differentiation in the embryonic stem cell line CGR8. Neurosci Lett 547:32–36
108. Hamann M et al (1995) Synthesis and release of an acetylcholine-like compound by human myoblasts and myotubes. J Physiol 489 (Pt 3):791–803
109. Wessler I, Kirkpatrick CJ (2001) The non-neuronal cholinergic system: an emerging drug target in the airways. Pulm Pharmacol Ther 14(6):423–434
110. Danielson P et al (2007) Extensive expression of markers for acetylcholine synthesis and of M2 receptors in tenocytes in therapy-resistant chronic painful patellar tendon tendinosis—a pilot study. Life Sci 80(24–25): 2235–2238
111. Danielson P et al (2006) Immunohistochemical and histochemical findings favoring the occurrence of autocrine/paracrine as well as nerve-related cholinergic effects in chronic painful patellar tendon tendinosis. Microsc Res Tech 69(10):808–819
112. Rana OR et al (2010) Acetylcholine as an age-dependent non-neuronal source in the heart. Auton Neurosci 156(1–2):82–89
113. Rocha-Resende C et al (2012) Non-neuronal cholinergic machinery present in cardiomyocytes offsets hypertrophic signals. J Mol Cell Cardiol 53(2):206–216
114. En-Nosse M et al (2009) Expression of non-neuronal cholinergic system in osteoblast-like cells and its involvement in osteogenesis. Cell Tissue Res 338(2): 203–215

Chapter 12

Utilization of Superfused Cerebral Slices in Probing Muscarinic Receptor Autoregulation of Acetylcholine Release

Glenda Alquicer, Vladimír Doležal and Esam E. El-Fakahany

Abstract

Signal transmission from cholinergic nerves is mediated by two receptor families: ionotropic nicotinic and metabotropic (G-protein coupled) muscarinic receptors. The muscarinic receptor family comprises five receptor subtypes (M_1 through M_5), each encoded by its own gene. Individual subtypes play important specific roles in many physiological processes, ranging from vegetative to cognitive functions and memory. Cloning of individual muscarinic receptor subtypes has enabled targeted studies of ligand binding and activation characteristics of the recombinant protein and a rapid increase in the development of receptor selective drugs for various potential therapeutic uses. Results of testing new drugs obtained on recombinant proteins need to be verified using systems that better mimic a physiological environment. Here we provide a brief outline and examples of the utilization of superfused brain slices for probing drug effects on muscarinic autoreceptor-mediated regulation of acetylcholine release.

Key words Muscarinic receptors, Acetylcholine release, Autoregulation, Superfusion

1 Background

A hundred years ago acetylcholine (ACh) was identified and proposed as the first chemical neurotransmitter [1, 2]. The chemical transmission of nerve impulses by acetylcholine was unequivocally proven in superfused frog heart [3, 4]. The discovery of chemical neurotransmission by H.H. Dale and O. Loewi was awarded a Nobel Prize in medicine in 1936. Since the identification of ACh as a chemical neurotransmitter, multiple roles in the central as well as peripheral nervous systems have been revealed, and tremendous progress in the field of physiology and pharmacology of cholinergic neurotransmission has been achieved [5, 6].

Acetylcholine is synthesized and stored mainly in cholinergic nerve terminals from which it is liberated by nerve impulses [7–9]. Unlike other small neurotransmitters, quick hydrolysis by cholinesterases terminates ACh action in the absence of reuptake

Jaromir Myslivecek and Jan Jakubik (eds.), *Muscarinic Receptor: From Structure to Animal Models*, Neuromethods, vol. 107, DOI 10.1007/978-1-4939-2858-3_12, © Springer Science+Business Media New York 2016

of extracellular ACh back to the nerve terminal or surrounding glial cells. In addition to neuronal ACh some non-neuronal cells are also capable of synthesizing and releasing ACh [10–12]. Irrespective of its origin, ACh present in the extracellular fluid transmits signals via nicotinic and muscarinic receptors located in the plasma membranes of recipient cells [5]. Nicotinic receptors are a family of pentameric ligand-gated ionic channels that insure fast excitatory transmission. Skeletal muscle nicotinic receptors are composed of subunits α_1, β_1, γ, δ, or ε while neuronal nicotinic receptors are hetero or homopentamers composed of subunits α_2 through α_{11} and β_2 through β_4 [13–15]. Muscarinic receptors belong to the family of metabotropic receptors (GPCR, receptors that transduce extracellular signal to the cell interior via G-proteins) and comprise five subtypes denoted M_1 through M_5 [16]. Each muscarinic receptor subtype is a product of a single gene [17, 18]. Odd-numbered muscarinic receptors preferentially couple with $G_{q/11}$ G-proteins and activate phospholipase C signaling pathway. Even-numbered receptors preferentially utilize $G_{i/o}$ G-proteins to inhibit adenylyl cyclase and the cAPM signaling pathway [18]. In addition to the activation of these intracellular signaling pathways, odd-numbered muscarinic receptors regulate potassium conductance [19] and even-numbered muscarinic receptors directly regulate specific voltage-operated calcium channels [20, 21] by quick "membrane delimited" action through $G_{\beta\gamma}$ G-protein subunit dimers [22].

Muscarinic transmission subserves many diverse physiological roles both in the central and peripheral nervous systems ranging from vegetative to cognitive functions [6, 23]. With respect to synapse morphology, muscarinic receptors may be localized on presynaptic nerve terminals or postsynaptic membranes. Presynaptic muscarinic receptors may be further classified as heteroreceptors (located on nerve terminals releasing different transmitters than ACh) or autoreceptors (located on ACh-releasing terminals). Activation of presynaptic muscarinic receptors may increase or decrease release of different neurotransmitters. In addition to synaptic localization many cells (e.g., immune cells, endothelial cells, stem cells, cancer cells, keratinocytes) that do not form classical synapses express various subtypes of muscarinic receptors [10–12].

2 General Remarks

Acetylcholine is synthesized by the enzyme choline acetyltransferase in the cytoplasm of synaptic terminals from choline and acetylcoenzyme A [7–9]. Synthesis of ACh is entirely dependent on choline supply from the extracellular fluid while acetylcoenzyme A is formed in nerve terminal mitochondria. Choline is transported to the terminal from the extracellular fluid by a high-affinity choline transporter [24, 25]. Inhibition of the choline transporter results in inhibition of neuronal ACh synthesis. Acetylcoenzyme A originating from oxidative metabolism of glucose is supplied by nerve terminal mitochondria. Shortage of immediate pyruvate precursors (glucose and lactate) and oxygen strongly inhibits ACh synthesis and release [26–29]. Synthesized ACh is transported to synaptic vesicles by way of a vesicular acetylcholine transporter [30–32] and becomes available for vesicular release. Nerve stimulation-evoked release of neuronal ACh fully depends on the influx of extracellular calcium as a result of depolarization of nerve terminals and opening of specific voltage-operated calcium channels or artificially by calcium ionophores (Fig. 1).

Cerebral (tissue) slices ex vivo are useful for studies of presynaptic physiology and pharmacology of synthesis, storage, and release of ACh. The superfusion technique is particularly suitable for studies of presynaptic regulation of transmitter release. This approach is based on radioactive prelabeling of acetylcholine in the nerve terminal with its tritiated precursor choline followed by superfusion of slices under conditions that prevent re-uptake of choline and new synthesis of ACh. Measurement of the evoked release of preformed labeled ACh thus does not depend on changes in synthesis or storage of ACh but principally depends on characteristics of nerve terminal releasing machinery and its regulation by presynaptic receptors. Utilization of ex vivo brain slices represents a reasonable experimental preparation that allows good control of extracellular milieu and adequate degree of tissue integrity. However, synaptosomes [33, 34] and in vivo dialysis [35] are superior in these two aspects, respectively.

3 Preparation and Superfusion of Cerebral Slices

3.1 Stock Solutions

Stock solution A (tenfold concentrated) (mM): NaCl 1230, KCl 30, $CaCl_2$ 13, $MgSO_4$ 10 in redistilled water.

Stock solution B (tenfold concentrated) (mM): NaH_2PO_4 12, $NaHCO_3$ 250 in redistilled water.

Hemicholinium-3 (HC-3) 10 mM in redistilled water.

(Methyl ^{3}H)-choline chloride (^{3}H-Ch), SRA ~80 Ci/mmol, 1 mCi/ml, ~12.5 μM (e.g., ARC, USA).

Drugs to be tested 100- to 1000-fold concentrated in water or other solvent (e.g., DMSO, ethanol) as feasible.

3.2 Preparation of Krebs Buffer

Dilute 100 ml of stock solution B in approximately 750 ml of redistilled water and bubble with gas mixture of 95 % oxygen/5 % carbon dioxide for at least 20 min at room temperature. Add 100 ml of stock solution A, 2 g of glucose, adjust pH to 7.4 (using

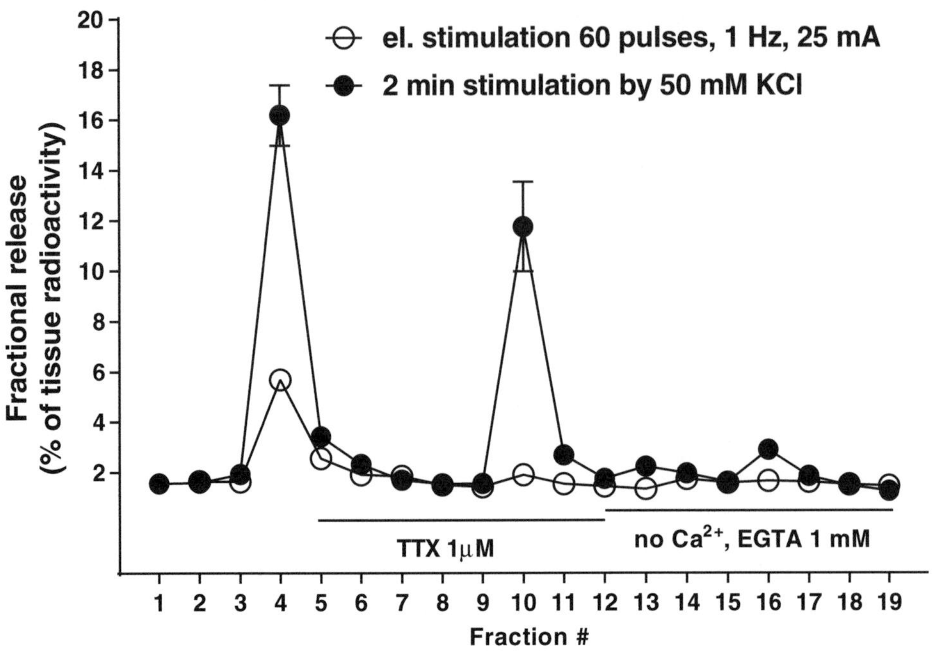

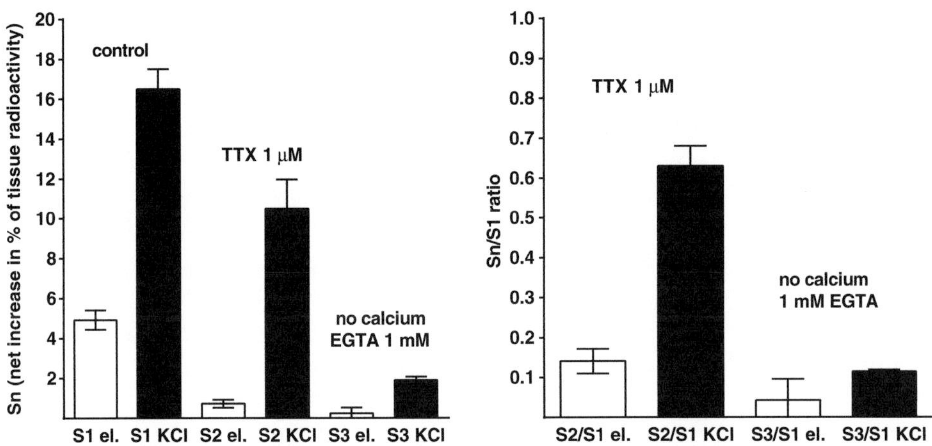

Fig. 1 Dependence of electrical stimulation-evoked ACh release on spreading of action potentials and extracellular calcium in rat striatal slices. *Upper panel*: Striatal slices were prelabeled with ^{3}H-choline, placed into superfusion apparatus, and superfused at a flow rate of 0.5 ml/min with Krebs buffer containing 10 μM hemicholinium-3 to prevent reuptake of choline. Collection of 4-min fractions (abscissa) started after 1 h washing. Slices were stimulated thrice, either electrically (2 ms monopolar pulses at 1 Hz for 1 min, 25 mA; *open symbols*) or by potassium depolarization (2 min using superfusion buffer with 50 mM KCl increased at the expense of NaCl to maintain osmolarity; *closed symbols*) at the beginning of the fourth (control stimulation), tenth (1 μM tetrodotoxin was present from sixth to twelfth fraction), and 16th fraction (calcium in media was replaced by 1 mM EGTA to chelate any remaining calcium from 13th fractions till the end of superfusion). Points are means ± SEM of three slices from a single experiment (SEM not shown in cases when they are smaller than the symbol). *Lower left panel*: The evoked release of labeled ACh was calculated as a difference between the release of radioactivity during stimulation and resting liberation of radioactivity. *Lower right panel*: The release of ACh during the second (S2) and third (S3) stimulation is normalized by the first stimulation (S1) and expressed as ratios S2/S1 and S3/S1. Prevention of action potential propagation by tetrodotoxin largely inhibits electrically evoked ACh release. Removal of extracellular calcium blocks electrically evoked and largely inhibits potassium depolarization-evoked release of ACh

sodium hydroxide or hydrochloric acid) if needed, and top up with redistilled water to 1000 ml. Keep for further use at room temperature under atmosphere 95 % oxygen–5 % carbon dioxide or continuous bubbling with 95 % oxygen–5 % carbon dioxide.

3.3 Labeling Medium

Add 10 μl of ^{3}H-Ch to 2 ml of Krebs buffer.

3.4 Superfusion Medim

Add HC-3 to Krebs solution at a final concentration of 10 μM. Drugs to be tested during the superfusion experiment are diluted from stock solutions in Krebs buffer containing HC-3.

3.5 Preparation of Cerebral Slices

Kill experimental animal (e.g., rat or mouse) by cervical dislocation and decapitation. Quickly remove the brain from the skull and dissect the selected brain region. Prepare slices from dissected tissue by McIlwain's tissue chopper set at a width of slices less than 0.4 mm. This is to ensure adequate substrate and oxygen saturation of all cells in the cerebral slices that solely depends on diffusion. Wash slices several times in surplus of fresh oxygenated Krebs buffer without HC-3 (at least three times in about 2 ml of fresh oxygenated buffer per 0.1 g of tissue) by gentle mixing and gravity sedimentation in order to remove tissue debris, extracellular fluid, and solutes released from damaged cells.

3.6 Labeling of Slices

Depending on the number of channels of the superfusion apparatus (e.g., Brandel, Gaithersburg, MD, USA), transfer 10–20 slices to 2 ml eppendorf test tube, add 1 ml of fresh oxygenated Krebs buffer (without HC-3) and ^{3}H-Ch to a final concentration around 100 nM (5 μl of ^{3}H-Ch stock solution per 1 ml of loading medium). Replace air atmosphere in test tube by 95 % oxygen–5 % carbon dioxide, close the tube, and incubate for typically 30 min at 37 °C. At the end of labeling wash slices at room temperature three to five times in 1 ml of fresh oxygenated Krebs buffer containing HC-3 to remove excess extracellular radioactivity.

3.7 Assembling Superfusion Chambers

Superfusion chambers are made of Teflon and after assembly incorporate platinum grid electrodes. Electrodes are physically separated from slices by Whatman GF/B or GF/C filter discs. First cover the lower electrode with filter disc, then place body of the chamber and randomly transfer 1–2 slices per chamber. Finally, cover the upper part of the chamber with filter disc and insert the upper electrode.

3.8 Superfusion

The whole superfusion apparatus including buffer vessels is placed in a temperature-controlled environmental cover set to 37 °C. Buffer flow is driven by peristaltic pump set to a flow rate of 0.5 ml/min. Superfusion chambers are oriented vertically with inflow from the bottom to minimize the risk of air bubble trapping inside the chamber that would impede electrical stimulation.

Electrical wiring of stimulation electrodes should be parallel so that occasional failures due to air bubble accumulation within chambers or incorrect assembly of a given chamber impacts only this chamber but not others.

Superfuse slices at a rate of 0.5 ml/min at 37 °C for 1 h with Krebs buffer containing 10 μM HC-3 that prevents further uptake of choline, to remove residues of extracellular and loosely bound tissue radioactivity, and to let slices recover after loading. Discard superfusate collected during this washing period.

Start collection of 4-min fractions to scintillation vials by means of a built-in fraction collector. The first two fractions before the first stimulation are sufficient to determine resting liberation of radioactivity (*see* Figs. 1, 2, and 3). Stimulation is then started with intervals between stimulations being dependent on the length of stimulation. Tested drugs should be present not earlier than baseline after preceding stimulation is reached and two fractions (8 min) before stimulation in the presence of tested drug to see a possible effect on resting liberation of radioactivity, and to assure sufficient equilibration of the drug within slices.

The following superfusion protocol was used for experiment shown in Fig. 2 (for details see text to Figure).

Apply the first electrical stimulation at the beginning of the third fraction collection.

Superfuse with carbachol from the beginning of the seventh fraction collection.

Apply the second stimulation at the beginning of the ninth fraction collection.

Superfuse with atropine from the beginning of the 13th fraction collection.

Apply the third stimulation at the beginning of the 15th fraction collection.

At the end of the superfusion stop the fraction collector and peristaltic pump. Remove filter discs with slices from superfusion chambers and let slices dissolve in scintillation vials in 0.5 ml of 1 M NaOH for 30 min at 50 °C. These samples serve to determine radioactivity remaining in slices after superfusion for calculation of fractional release of radioactivity in collected fractions (i.e., radioactivity in a given fraction divided by sum of radioactivity in this and all following fractions and slice at the end of superfusion).

3.9 Scintillation Counting

Add 3 ml of scintillation cocktail (hydrophilic scintillation cocktail with high water absorbing capacity, e.g., Rotiszint Eco Plus) to each of collected fractions and tissue lysates. Thoroughly mix to get homogenous solution, let stand for 2 h in the dark, then measure radioactive content in a liquid scintillation counter (e.g., Packard Tricarb, Perkin Elmer, USA) with counting each sample for 5 min.

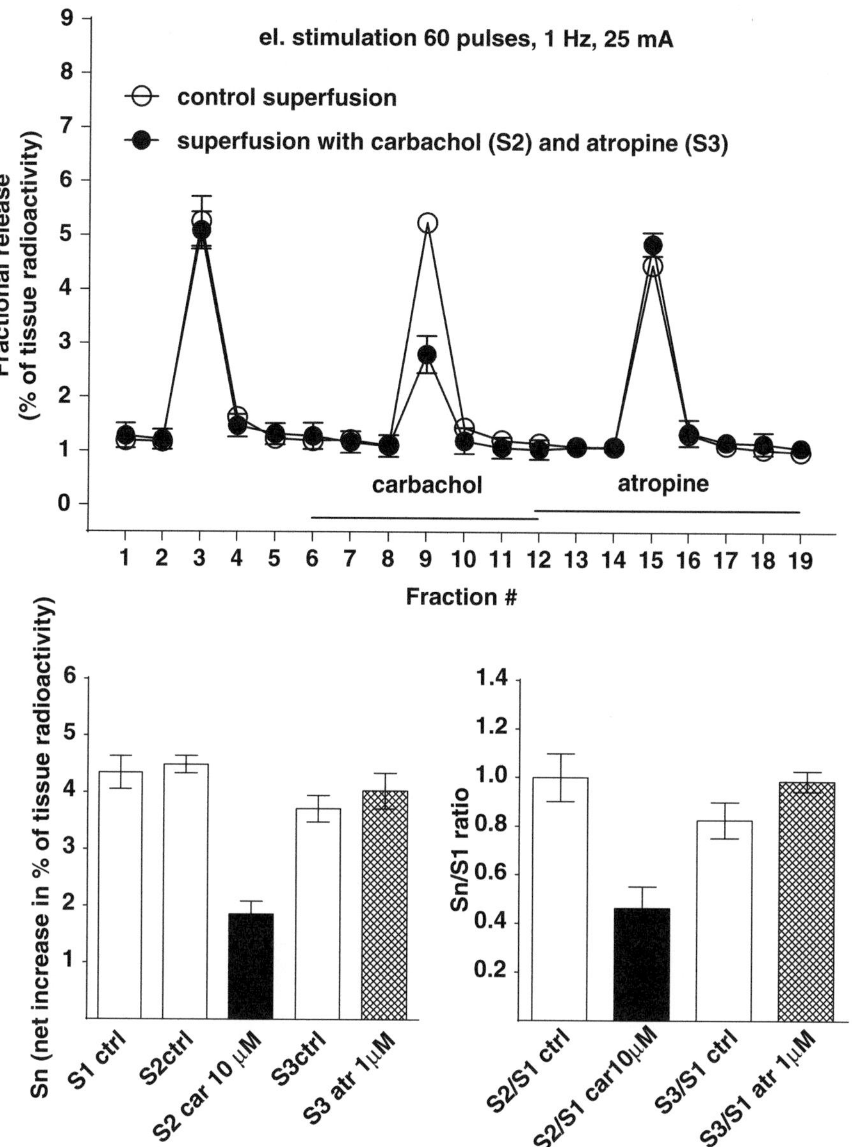

Fig. 2 Effects of carbachol and atropine on electrical stimulation-evoked ACh release from rat striatal slices. *Upper panel*: Striatal slices were prepared and superfused as described in Fig. 1. They were stimulated electrically thrice (2 ms monopolar pulses at 1 Hz for 1 min, 25 mA; *open symbols*) at the beginning of the third (always control stimulation), ninth (*open symbols*, control; *closed symbols*, 10 μM muscarinic agonist carbachol was present from the 6th to 11th fraction), and 15th (*open symbols*, control; *closed symbols*, 1 μM muscarinic antagonist atropine was present from the 12th fraction till the end of superfusion) fraction. Points are means ± SEM when bigger than symbol of three slices from a single experiment. *Lower left panel*: The evoked release of labeled ACh was calculated as the difference between the release of radioactivity during stimulation and resting liberation of radioactivity. *Lower right panel*: The release of ACh during the second (S2; control or in the presence of carbachol) and third (S3; control or in the presence of atropine) stimulation is normalized in relation to the first stimulation (S1) and expressed as ratios S2/S1 and S3/S1. Electrical stimulation evokes at least three successive reliable responses with a small decline of the third stimulation possibly due to autoinhibition. Carbachol inhibits electrically evoked release of ACh by about 55 % and washing with atropine fully reverses autoreceptor-mediated inhibition of the evoked release

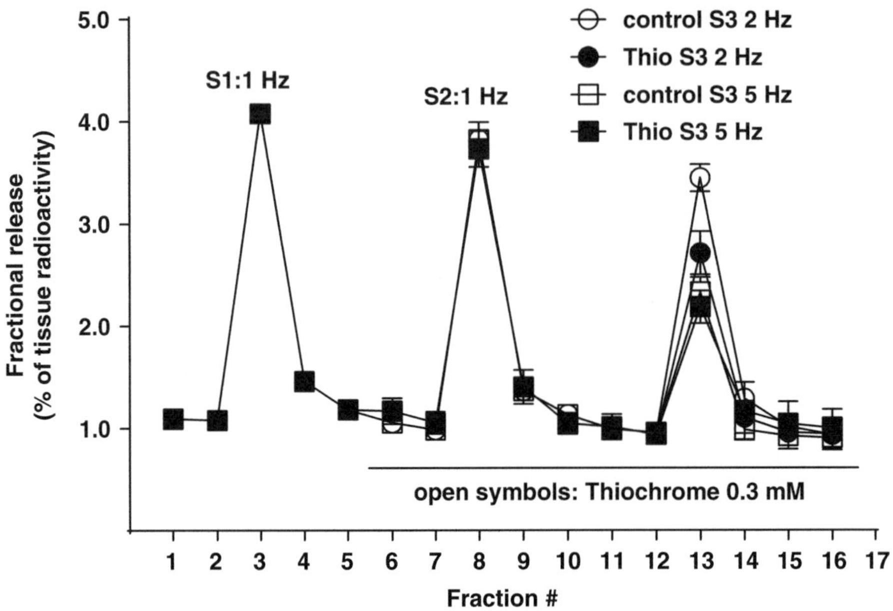

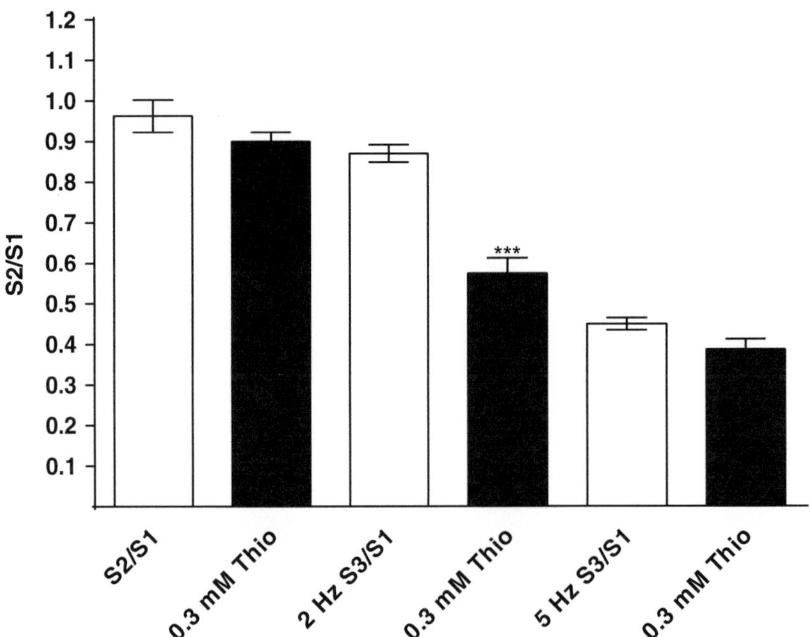

Fig. 3 Effects of thiochrome on electrical stimulation-evoked ACh release from rat striatal slices (modified from [44]). *Upper panel*: Striatal slices were prepared and superfused as described in Fig. 1. They were stimulated electrically thrice (forty 2 ms monopolar pulses, 35 mA delivered at frequency 1, 2, or 5 Hz as shown in Figure) at the beginning of the third (control stimulation, S1), eighth (S2), and 13th (S3) fraction. The allosteric modulator thiochrome (0.3 mM) was present in superfusion medium from the beginning of the sixth fraction till the end of superfusion in samples denoted by *closed symbols*. Points are means ± SEM when bigger than symbol of three to nine slices from three experiments. *Lower panel*: The release of ACh during the second (S2; *open columns* control medium, *black columns* medium with thiochrome) and third (S3; *open columns* control medium, *black columns* medium with thiochrome) stimulation is normalized relative to the first stimulation (S1, always in the absence of thiochrome) and expressed as the ratios S2/S1 and S3/S1. Thiochrome has no effect on resting liberation of radioactivity (see traces in *upper panel*). The release of ACh is not influenced by

3.10 Processing of Data

Unlike other small neurotransmitters, released ACh is quickly hydrolyzed to choline and acetate. Radioactivity found in the collected 4-min fractions thus consists of radioactive label in a rather small fraction of total ACh, labeled choline derived from hydrolysis of prelabeled ACh, and labeled choline and its metabolites derived from prelabeled tissue. For this reason the resting liberation of radioactivity is not a reliable measure of neuronal ACh release. However, the release of radioactivity evoked by electrical or chemical stimulation of nerve terminals over background liberation consists of labeled ACh and labeled choline derived from hydrolysis of released ACh. Stimulated release of ^{3}H is thus an adequate measure of [^{3}H]ACh release even in the absence of cholinesterase inhibitors [36, 37].

Typically, slices during collection period are exposed several times to electrical or chemical stimulations that are separated by resting periods for recovery before the next stimulation is delivered. The first stimulation always serves as control stimulation. Drugs to be tested are added to the superfusion fluid at a suitable time before the test stimulation. At the end of superfusion, slices are removed from superfusion chambers and dissolved. The radioactivity remaining in tissue lysates after superfusion and in each collected fraction is measured by liquid scintillation counting.

The release of radioactivity in individual fractions is expressed as fractional release, i.e., the proportion of released radioactivity from total radioactivity present in slices at the beginning of respective collection period. Afterwards stimulated release of ACh is calculated as the difference between the radioactivity collected during stimulation and that under resting conditions immediately before the stimulation and after the stimulated release returned back to resting liberation of radioactivity. Effects of tested drugs are further normalized to control (first) stimulation run in the absence of tested drug (ratio of stimulated release in the presence of drug/control stimulated release, Sn/S1) and compared to another measurement that includes only control stimulations (Figs. 2 and 3). A possible effect of drug solvent alone when different than water should be checked in a control experiment.

Fig. 3 (continued) thiochrome when evoked by 1 Hz stimulation that does not lead to autoinhibition of evoked ACh release, and 5 Hz stimulation that already induces robust autoinhibition. This demonstrates that thiochrome does not interact with the orthosteric binding site of muscarinic autoreceptors. Meanwhile, thiochrome significantly inhibits ACh release evoked by pulses delivered at 2 Hz indicating allosteric potentiation of the effects of released ACh

4 Typical Experiment

Superfusion experiments summarized in Figs. 2 and 3 demonstrate applicability of the method in the examination of muscarinic autoreceptors that mediate autoinhibition of electrically evoked ACh release from rat striatal slices. The autoinhibition is mediated by M_4 receptors [38, 39]. Electrical stimulation (60 pulses delivered at a frequency of 1 Hz) during superfusion in control conditions repeatedly evoke ACh release with a small decline during the third stimulation (Fig. 2). In parallel superfusion, application of the muscarinic agonist carbachol before the second stimulation results in a statistically significant decrease in evoked ACh release by about 55 %. The inhibition is abolished by atropine added before the third stimulation. Actually, compared to control superfusion, the release of ACh in the presence of atropine is slightly higher than in control conditions. Nonetheless, a small decline of the third control stimulation and a small increase of the third stimulation in the presence of atropine are not statistically significant, demonstrating that the increase in ACh concentration in the synaptic cleft evoked by this mild electrical stimulation is not big enough to activate presynaptic M_4 autoreceptors.

In addition to the orthosteric binding site for ACh, muscarinic receptors possess allosteric binding sites. Concurrent presence of orthosteric and allosteric ligands results in mutual interaction called cooperativity that can be positive (increase in binding affinity), negative (decrease in binding affinity), or neutral (no change in affinity). Allosteric modulators demonstrating positive cooperativity with ACh are particularly wanted for possible medical applications where there is deficiency in ACh release because they exhibit remarkable receptor subtype selectivity in influencing ACh affinity [40–45]. In contrast, orthosteric ligands exhibit lower selectivity due to the high degree of homology of the orthosteric binding site of muscarinic receptor subtypes [46]. An example of allosteric enhancement of endogenous ACh inhibitory effects on ACh release in striatal slices is given in Fig. 3 (*see* also [44]). The release of ACh was evoked by three consecutive stimulations by 40 pulses delivered at 1 Hz during the first and second stimulations, and by 2 or 5 Hz during the third stimulation. The allosteric ligand thiochrome was added to superfusion medium 8 min before the second stimulation and remained in medium till the end of superfusion. Thiochrome alone neither activates nor inhibits presynaptic receptors because it does not inhibit ACh release evoked by the second stimulation (unlike agonist carbachol; *see* Fig. 2) but inhibits ACh release evoked by 2 and 5 Hz stimulation during the third stimulation. The increase in stimulation frequency to 2 Hz during the third stimulation has no appreciable effect in control conditions while the release of ACh in the presence of thiochrome is significantly inhibited. Further increase in stimulation frequency to

5 Hz already leads to substantial autoinhibition of ACh release and thiochrome loses its potentiating effect observed at lower frequencies. This indicates that stimulation by 40 pulses delivered at 1 and 2 Hz does not induce sufficient accumulation of ACh in the synaptic cleft to induce autoinhibition before the next pulse arrives. However, the increase in residual concentration of ACh in the synaptic cleft remaining before arrival of the next pulse during 2 Hz stimulation that does not yet activate presynaptic receptors is high enough to induce autoinhibition in the presence of thiochrome that increases receptor affinity for ACh. Stimulation at 5 Hz in control medium demonstrates already substantial autoinhibition of ACh release (by about 50 % compared to control 2 Hz S3/S1 ratio; *see* Fig. 2, inhibition of ACh release by saturating concentration of carbachol is about 55 %) and the potentiating effect of thiochrome understandably disappears.

5 Practical Tips

- No drugs (e.g., general anesthetics) may be used before decapitation of the animal.
- Preparation of slices can be done at room temperature (around 20 °C) and should be as quick as possible (no longer than 5 min from killing animal and submersion of slices to oxygenated buffer). Alternatively, after removal from skull, the brain can be chilled in oxygenated buffer before dissection of selected regions.
- Loading of slices with labeled choline should start immediately after preparation and washing of slices. The content of cerebral tissue ACh after preparation of slices is much lower than in cerebral tissue in vivo but quickly increases.
- Cerebral tissues liberate high amounts of free choline. The specific radioactivity of used labeled choline should thus be high and its concentration low (at the beginning of loading around 100 nM) in order to get a fairly specific labeling of synthesized ACh that couples to the high-affinity choline transporter over other choline metabolite that are derived from choline transported by low affinity choline carriers [47, 48].

6 Conclusions

The superfusion technique of cerebral slices ex vivo after radioactive labeling of acetylcholine has proven a useful approach for studies of acetylcholine release. This approach is efficient tool for probing effects of both orthosteric and allosteric drugs on the muscarinic autoreceptor-mediated regulation of acetylcholine release.

Acknowledgments

This research was supported by projects AV0Z50110509, RVO: 67985823 and OP VK CZ.1.07/2.3.00/30.0025, and grant GACR 14-05696S.

References

1. Dale HH (1914) The action of certain esters and ethers of choline, and their relation to muscarine. J Pharmacol Exp Ther 6:147–190
2. Dale HH, Feldberg W, Vogt M (1936) Release of acetylcholine at voluntary motor nerve endings. J Physiol 86:353–380
3. Loewi O (1921) Über humorale Übertragbarkeit der Herznervenwirkung. I. Mitteilung. Pflügers Arch Ges Physiol 189:239–242
4. Loewi O, Navratil E (1926) Über humorale Übertragbarkeit der Herznervenwirkung. X. Mitteilung. Über das Schicksal des Vagusstoffes. Pflügers Arch Ges Physiol 214:678–688
5. Bennett MR (2000) The concept of transmitter receptors: 100 years on. Neuropharmacology 39:523–546
6. Fryer AD, Christopoulos A, Nathanson NM (eds) (2012) Muscarinic receptors. Handbook of experimental pharmacology, vol 208. Springer, New York, NY, pp 1–499
7. Tucek S (1978) Acetylcholine synthesis in neurons. Chapman and Hall Ltd., London
8. Tucek S (1985) Regulation of acetylcholine synthesis in the brain. J Neurochem 44: 11–24
9. Prado MA, Reis RA, Prado VF, de Mello MC, Gomez MV, de Mello FG (2002) Regulation of acetylcholine synthesis and storage. Neurochem Int 41:291–299
10. Eglen RM (2006) Muscarinic receptor subtypes in neuronal and non-neuronal cholinergic function. Auton Autacoid Pharmacol 26: 219–233
11. Kawashima K, Fujii T (2008) Basic and clinical aspects of non-neuronal acetylcholine: overview of non-neuronal cholinergic systems and their biological significance. J Pharmacol Sci 106:167–173
12. Wessler IK, Kirkpatrick CJ (2012) Activation of muscarinic receptors by non-neuronal acetylcholine. Handb Exp Pharmacol 208: 469–491
13. Dani JA, Bertrand D (2007) Nicotinic acetylcholine receptors and nicotinic cholinergic mechanisms of the central nervous system. Annu Rev Pharmacol Toxicol 47:699–729
14. Changeux JP (2012) The nicotinic acetylcholine receptor: the founding father of the pentameric ligand-gated ion channel superfamily. J Biol Chem 287:40207–40215
15. Alexander SP, Benson HE, Faccenda E et al (2013) The concise guide to PHARMACOLOGY 2013/14: ligand-gated ion channels. Br J Pharmacol 170:1582–1606
16. Alexander SP, Benson HE, Faccenda E et al (2013) The concise guide to PHARMACOLOGY 2013/14: G protein-coupled receptors. Br J Pharmacol 170:1459–1581
17. Buckley NJ, Bonner TI, Buckley CM, Brann MR (1989) Antagonist binding properties of five cloned muscarinic receptors expressed in CHO-K1 cells. Mol Pharmacol 35:469–476
18. Bonner TI (1989) The molecular basis of muscarinic receptor diversity. Trends Neurosci 12:148–151
19. Jones SV, Barker JL, Buckley NJ, Bonner TI, Collins RM, Brann MR (1988) Cloned muscarinic receptor subtypes expressed in A9 L cells differ in their coupling to electrical responses. Mol Pharmacol 34:421–426
20. Dolezal V, Tucek S, Hynie S (1989) Effects of pertussis toxin suggest a role for G-proteins in the inhibition of acetylcholine release from rat myenteric plexus by opioid and presynaptic muscarinic receptors. Eur J Neurosci 1:127–131
21. Dolezal V, Tucek S (1999) Calcium channels involved in the inhibition of acetylcholine release by presynaptic muscarinic receptors in rat striatum. Br J Pharmacol 127:1627–1632
22. Herlitze S, Garcia DE, Mackie K, Hille B, Scheuer T, Catterall WA (1996) Modulation of Ca2+ channels by G-protein beta gamma subunits. Nature 380:258–262
23. Caulfield MP (1993) Muscarinic receptors—characterization, coupling and function. Pharmacol Ther 58:319–379
24. Haga T, Noda H (1973) Choline uptake systems of rat brain synaptosomes. Biochim Biophys Acta 291:564–575
25. Yamamura HI, Snyder SH (1972) Choline: high-affinity uptake by rat brain synaptosomes. Science 178:626–628
26. Gibson GE, Blass JP (1976) Impaired synthesis of acetylcholine in brain accompanying mild hypoxia and hypoglycemia. J Neurochem 27: 37–42

27. Dolezal V, Tucek S (1981) Utilization of citrate, acetylcarnitine, acetate, pyruvate and glucose for the synthesis of acetylcholine in rat brain slices. J Neurochem 36:1323–1330
28. Ksiezak HJ, Gibson GE (1981) Oxygen dependence of glucose and acetylcholine metabolism in slices and synaptosomes from rat brain. J Neurochem 37:305–314
29. Tucek S, Dolezal V, Sullivan AC (1981) Inhibition of the synthesis of acetylcholine in rat brain slices by (−)-hydroxycitrate and citrate. J Neurochem 36:1331–1337
30. Erickson JD, Varoqui H, Schafer MK et al (1994) Functional identification of a vesicular acetylcholine transporter and its expression from a "cholinergic" gene locus. J Biol Chem 269:21929–21932
31. Roghani A, Feldman J, Kohan SA et al (1994) Molecular cloning of a putative vesicular transporter for acetylcholine. Proc Natl Acad Sci U S A 91:10620–10624
32. Berrard S, Varoqui H, Cervini R, Israel M, Mallet J, Diebler MF (1995) Coregulation of two embedded gene products, choline acetyltransferase and the vesicular acetylcholine transporter. J Neurochem 65:939–942
33. Raiteri M (2001) Presynaptic autoreceptors. J Neurochem 78:673–675
34. Nicholls DG (2010) Stochastic aspects of transmitter release and bioenergetic dysfunction in isolated nerve terminals. Biochem Soc Trans 38:457–459
35. Fadel JR (2011) Regulation of cortical acetylcholine release: insights from in vivo microdialysis studies. Behav Brain Res 221:527–536
36. Richardson IW, Szerb JC (1974) The release of labelled acetylcholine and choline from cerebral cortical slices stimulated electrically. Br J Pharmacol 52:499–507
37. Dolezal V, Jackisch R, Hertting G, Allgaier C (1992) Activation of dopamine D1 receptors does not affect D2 receptor-mediated inhibition of acetylcholine release in rabbit striatum. Naunyn Schmiedebergs Arch Pharmacol 345: 16–20
38. Dolezal V, Tucek S (1998) The effects of brucine and alcuronium on the inhibition of 3H-acetylcholine release from rat striatum by muscarinic receptor agonists. Br J Pharmacol 124:1213–1218
39. Zhang W, Basile AS, Gomeza J, Volpicelli LA, Levey AI, Wess J (2002) Characterization of central inhibitory muscarinic autoreceptors by the use of muscarinic acetylcholine receptor knock-out mice. J Neurosci 22: 1709–1717
40. Jakubik J, Bacakova L, El-Fakahany EE, Tucek S (1997) Positive cooperativity of acetylcholine and other agonists with allosteric ligands on muscarinic acetylcholine receptors. Mol Pharmacol 52:172–179
41. Lazareno S, Gharagozloo P, Kuonen D, Popham A, Birdsall NJ (1998) Subtype-selective positive cooperative interactions between brucine analogues and acetylcholine at muscarinic receptors: radioligand binding studies. Mol Pharmacol 53:573–589
42. Tucek S, Jakubik J, Dolezal V, el-Fakahany EE (1998) Positive effects of allosteric modulators on the binding properties and the function of muscarinic acetylcholine receptors. J Physiol Paris 92:241–243
43. Birdsall NJ, Farries T, Gharagozloo P, Kobayashi S, Lazareno S, Sugimoto M (1999) Subtype-selective positive cooperative interactions between brucine analogs and acetylcholine at muscarinic receptors: functional studies. Mol Pharmacol 55:778–786
44. Lazareno S, Dolezal V, Popham A, Birdsall NJ (2004) Thiochrome enhances acetylcholine affinity at muscarinic M4 receptors: receptor subtype selectivity via cooperativity rather than affinity. Mol Pharmacol 65:257–266
45. Jakubik J, El-Fakahany EE (2010) Allosteric modulation of muscarinic acetylcholine receptors. Pharmaceuticals 3:2838–2860
46. Hulme EC, Birdsall NJ, Buckley NJ (1990) Muscarinic receptor subtypes. Annu Rev Pharmacol Toxicol 30:633–673
47. Machova E, O'Regan S, Newcombe J et al (2009) Detection of choline transporter-like 1 protein CTL1 in neuroblastoma x glioma cells and in the CNS, and its role in choline uptake. J Neurochem 110:1297–1309
48. Machova E, Rudajev V, Smyckova H, Koivisto H, Tanila H, Dolezal V (2010) Functional cholinergic damage develops with amyloid accumulation in young adult APPswe/PS1dE9 transgenic mice. Neurobiol Dis 38:27–35

Chapter 13

Regulation of Heart Contractility by M_2 and M_3 Muscarinic Receptors: Functional Studies Using Muscarinic Receptor Knockout Mouse

Takio Kitazawa, Hiroki Teraoka, Nao Harada, Kenta Ochi, Tatsuro Nakamura, Koichi Asakawa, Shinya Kanegae, Noriko Yaosaka, Toshihiro Unno, Sei-ichi Komori, and Masahisa Yamada

Abstract

To investigate the functional roles of M_2 and M_3 muscarinic receptors in mouse atria, wild-type mice, muscarinic M_2 or M_3 single receptor knockout mice (M_2KO, M_3KO), and M_2 and M_3 muscarinic receptor double knockout mice (M_2/M_3KO) were used for pharmacological and molecular biological approaches. Effects of carbachol on spontaneous contraction (right atrium) or electrically evoked contraction (left atrium) were examined in the isolated atria of the respective mice. Presence of muscarinic receptor subtype mRNAs and proteins was determined by real time RT-PCR using specific primers and immunohistochemistry using specific anti-M_2 and anti-M_3 receptor antibodies. Quantitative real-time RT-PCR analysis showed that M_2 receptor mRNA was expressed dominantly in mouse atria but that the M_1, M_3, M_4, and M_5 receptor subtypes were also expressed at low levels. Carbachol decreased the frequency of spontaneous beating in right atria of mice through activation of the M_2 receptor subtype. In left atria of wild-type mice, carbachol decreased the amplitude of electrical field stimulation (EFS)-evoked contractions (M_2 receptors), but this inhibition was transient and was followed by a gradual increase in contraction amplitude (M_3 receptors). Cyclooxygenase-2 (COX-2) and prostaglandins in the endocardial endothelium were involved in the M_3 receptor-mediated positive inotropic actions. In conclusion, the present studies using isolated atria of muscarinic receptor knockout mice demonstrated that myocardial M_2 receptors mediate negative chronotropic/inotropic actions and that M_3 muscarinic receptors mediate positive chronotropic/inotropic actions in mouse atria. Physiologically, M_3 receptor-mediated excitatory cardiac effects might dampen the inhibitory effects of M_2 receptor activation on cardiac contractility.

Key words Atrial contraction, Muscarinic receptor knockout mouse, M_2 muscarinic receptor, M_3 muscarinic receptor, Inotropic action, Chronotropic action, Pertussis toxin, Endocardial endothelium, Cyclooxygenase-2, Cardiac intrinsic neuron

Jaromír Myslivecek and Jan Jakubik (eds.), *Muscarinic Receptor: From Structure to Animal Models*, Neuromethods, vol. 107, DOI 10.1007/978-1-4939-2858-3_13, © Springer Science+Business Media New York 2016

1 Introduction

Muscarinic receptor stimulation by acetylcholine plays an important role in parasympathetic control of cardiac functions such as heart rate (chronotropic action), conduction velocity (dromotropic action), and contractility (inotropic action). Muscarinic receptors are prototypic members of the superfamily of G protein-coupled receptors, and molecular cloning studies have demonstrated the existence of five distinct mammalian muscarinic receptor subtypes [1]. Based on their differential coupling to G-proteins and second messengers, the five receptors can be subdivided into two major functional classes. The M_1, M_3, and M_5 receptors preferentially couple to $G_{q/11}$ protein, increasing inositol-trisphosphate (IP_3) and diacylglycerol levels, while the M_2 and M_4 receptors are linked to $G_{i/o}$ protein decreasing level of cyclic AMP [1, 2]. It has been well documented that the heart predominantly expresses the M_2 receptor subtype [3, 4]. Following activation of cardiac M_2 receptors, the activated α subunit of $G_{i/o}$ proteins inhibits adenylyl cyclase activity, resulting in a decrease of cytoplasmic cyclic AMP, whereas the $\beta\gamma$ subunit of $G_{i/o}$ proteins directly activates the inwardly rectifying muscarinic K^+ channel to cause hyperpolarization of cardiac muscle. Hyperpolarization decreases the contractility of cardiac muscle [4, 5].

However, since many organs contain multiple subtypes of muscarinic receptors, it has been thought that the M_2 receptor is not the only muscarinic receptor subtype that is functional in the heart [6]. Pharmacological, biochemical, immunohistochemical and molecular biological studies using whole heart tissues or isolated cardiomyocytes indicate that the heart also expresses non-M_2 muscarinic receptors including the M_1 and M_3 receptor subtypes [6–16]. The M_3 receptor and associated proteins including connexin 43 [17] and β-catenin [18] have been reported to participate in ischemia and the induced infarction. A pervasive role of the M_3 receptor in cardiac diseases has been discussed recently [19]. Therefore, analysis of non-M_2 receptors in cardiac tissues is important both in the physiology and pathophysiology of the heart. However, expression of multiple muscarinic receptor subtypes (while M_2 subtype is dominant) in one organ and lack of selective potent muscarinic receptor agonists or antagonists for respective subtypes hamper functional analysis of non-M_2 muscarinic receptors in the heart.

Recently, mutant mice lacking M_1–M_5 muscarinic receptors (knockout mice) have become available as novel experimental tools to study the functional roles of individual muscarinic receptors in the heart and other organs [20, 21]. In spontaneously beating right atria from M_2 receptor knockout (M_2KO) mice, carbachol was devoid of any negative chronotropic activity [22, 23], indicating that the M_2

receptor mediates the negative chronotropic actions of muscarinic agonists in agreement with previous pharmacological evidence obtained using receptor subtype-preferring antagonists [1, 4].

We used muscarinic M_2 or M_3 single receptor knockout mice (M_2KO, M_3KO) and M_2 and M_3 muscarinic receptor double knockout mice (M_2/M_3KO) to investigate potential functional roles of the M_3 receptor subtype in regulation of atrial contractility. Isolated mouse atrium is a good animal model to examine function of muscarinic receptor in regulation of contractility. Each atrial preparation is isolated and suspended in an organ bath and spontaneous contraction and electrically evoked contraction are recorded. Muscarinic agonist-induced effects on the atrial contraction are observed and analyzed. Localization of M_3 receptors and downstream mechanisms of M_3 receptor activation have been also investigated using immunohistochemical methods and pharmacological characterization of the evoked muscarinic responses and reported in the following papers [23–25].

2 General Approach for Analysis of Muscarinic Receptor Subtypes in the Heart

Functional approaches (contraction, receptor binding and measurement of second messengers), immunological approaches (immunohistochemistry and Western blotting) and molecular biological approaches (measurements of receptor mRNAs) have been used to determine functional muscarinic receptor subtypes in heart (ventricles and atria). An outline of these approaches is given in this section.

2.1 Functional Approaches

Contraction study: Isolated cardiac muscle preparations (atrium and ventricle) are fixed in the organ bath and the mechanical changes in the preparations are detected by a force transducer. Our experimental procedures are shown in Protocol 1. Effects of some muscarinic receptor antagonists on muscarinic agonist-induced responses have been evaluated to determine muscarinic receptor subtypes in the heart. Receptor subtype-preferring muscarinic receptor antagonists, including M_1 antagonist pirenzepine [26], M_2 antagonists methoctramine [26, 27], AF-DX116 and 4-diphenylacetoxy-N-methylpiperidine methiodide (4-DAMP) [26, 28], M_3 antagonist hexahydro-sila-difenidol hydrochloride (p-F-HHSiD) and M_4 antagonist tropicamide [29], have been used in contractile functional studies. However, selectivity of these antagonists to respective receptor subtypes is largely dependent on the concentrations used, and these antagonists might to bind to all or some muscarinic receptor subtypes when affinity (pK_b values) to respective receptors are close. Receptor subtype-selective agonists are not available at present because structures of binding site of individual subtypes of muscarinic receptor are similar. Using these receptor antagonists, it has been proposed that M_1 or M_3 receptors mediate positive inotropic responses in isolated cardiac muscle strips from humans and mice [30, 31]. Nouchi et al. [32] analyzed the biphasic inotropic response in the developing chicken heart using antagonists and demonstrated M_4 receptor-mediated negative inotropic and M_1 receptor mediated-positive inotropic actions that are different from those in mammals.

Receptor binding study: Radioligand ([^{3}H]N-methylscopolamine and [^{3}H]quinuclidinyl benzilate) binding study using membrane preparations from atria or ventricles can reveal muscarinic receptor proteins as radioligand binding sites. Results of analysis of the saturation binding curve and competition curve by competitors (muscarinic receptor agonists or antagonists) indicate the characteristics of the radioligand binding sites (muscarinic receptors). Computer-aided analysis of binding competition curves shows how many binding sites are present in the membrane preparations. In the binding displacement studies using muscarinic receptor antagonists, Wang et al. [11], Pérez et al. [15], and Myslivecek et al. [16] showed that data from competition curves gave the best fit to a two-site binding model and suggested the presence of another muscarinic receptor subtype in addition to dominant M_2 receptor subtype. However, selectivity of muscarinic receptor agonists and antagonists is also a fundamentally important point in this kind of analysis.

Biochemical study: In general, the M_1, M_3, or M_5 receptors coupling with a $G_{q/11}$ protein activate phospholipase C to produce IP_3 and diacylglycerol, but the dominant M_2 receptor affects cyclic AMP contents in cardiac tissues. Therefore, muscarinic receptor agonist-induced increase of inositol phosphate formation in cardiac tissue suggests the expression of non-M_2 receptors. In isolated rat cardiomyocytes and slices of the human right atrium, stimulation of IP_3 formation via M_3 receptor subtype has been demonstrated [13, 14]. On the other hand, in guinea-pig isolated cardiomyocytes, it was suggested that the M_1 receptor caused accumulation of [^{3}H]inositol monophosphate [7]. In the rat cardiac homogenate, carbachol stimulated phospholipase C activity, which was inhibited by both M_1 and M_5 receptor antagonists but not by an M_3 receptor antagonist [16]. These biochemical approaches can indicate the presence of odd-numbered muscarinic receptors in the heart, but muscarinic receptor antagonists are generally used to identify receptor subtypes (M_1, M_3, and M_5 receptors). Therefore, as in contraction and binding studies, selectivity of muscarinic receptor antagonists is a basic problem for this kind of pharmacological analysis and suggests the limit of investigation.

2.2 Immunological Study

Immunohistochemical and Western blotting studies using respective muscarinic receptor antibodies might be useful for detecting the presence of muscarinic receptor proteins in cardiac tissues.

In immunostaining of muscarinic receptors, antibodies for M_1, M_2, M_3, and M_5 subtypes but not M_4 subtype showed positive staining in isolated human ventricular myocytes [11]. Abramochkin et al. [33] reported strong staining of M_3 receptors in the sinoatrial node of a murine heart and suggested possible regulation of heart rate by the M_3 receptor. Immunohistochemical analysis can show the localization of muscarinic receptors in the heart. M_5 receptor staining was largely restricted to intercalated discs, whereas other receptor subtypes were evenly distributed along the surface membranes of cardiac cells [11]. We have used the tissue section of mouse atrium for immunohistochemical study using antibodies for M_2 or M_3 receptors and have demonstrated that M_2 receptor immunoreactivity is restricted to the atrial muscles but that the M_3 receptor is expressed in both cardiac myocytes and the endocardial endothelium, Protocol 2 [24]. As selectivity of the antibody to G-protein coupled receptor, such as muscarinic receptors might be a problem in this kind of research [34], muscarinic receptor knockout mice are useful to check suitability of muscarinic receptor antibodies for immunohistochemical evaluation on tissue sections [35].

Immunoprecipitation study using each muscarinic receptor-selective antibody has shown the quantitative expression of muscarinic receptor subtypes in the rat heart [16]. The M_2 receptor was the dominant subtype (93–100 % of total muscarinic receptors) in both the atrium and ventricle, but a small population of M_3, M_4, and M_5 receptors (0.4–2.5 %) was also detected in the atria and ventricles [16].

2.3 Molecular-Biology Study

Investigation of the expression of individual muscarinic receptor subtype mRNAs might be crucial for evaluation and testing a hypothesis about their functional roles in the heart and other organs. Sharma et al. [8] showed the presence of M_1 and M_2 receptor mRNAs in the rat ventricle, and they observed M_1 receptor-mediated stimulatory responses in intracellular Ca^{2+} transient. Recently, the concentrations of mRNAs for muscarinic M_1–M_5 receptors in cardiac atria and ventricles were investigated and compared with each other using quantitative RT-PCR. Krejcí and Tucek [12] demonstrated that more than 90 % of total muscarinic receptor mRNAs in the rat heart were M_2 type and that the expression level of M_2 in the atria was two-times higher than that in the ventricle. However, expression of non-M_2 muscarinic receptors was found in the atria and ventricle. M_3 receptor mRNA expression level was 1–3 % and M_5 receptor mRNA expression level was 4–5 % of total mRNA for muscarinic receptors but M_1 or M_4 expression was less than 1 % (0.01–0.8 %). Pérez et al. [15] also provided molecular biological evidence for the presence of multiple subtypes of muscarinic receptor mRNAs in the human heart, though M_2 type was the dominant muscarinic receptor subtype. In mouse atria and ventricles, we carried out quantitative RT-PCR studies to examine the expression of M_1–M_5 muscarinic receptor mRNAs, Protocol 3. Total copies of muscarinic receptor transcripts in the atrium and ventricle were not significantly different. Similar to other animal species (rats and humans), M_2 receptor mRNA expression was dominant in both the ventricle and atrium (99.8 % of total muscarinic receptors), and M_1, M_3, M_4, and M_5 receptor mRNAs were also expressed at relatively low levels (0.02–0.06 %) [23]. In parallel with the quantitative data for expression levels of muscarinic receptor subtype mRNAs, functional (contraction) studies on the heart facilitate evaluation of the physiological and pathophysiological roles of individual muscarinic receptor subtypes in the heart.

3 Contraction Study Using Muscarinic Receptor Knockout (KO) Mice

Since expression levels of non-M_2 receptors in the heart are relatively low [23, 36], muscarinic receptor knockout (KO) mice are good animal models for determining the functional roles of muscarinic receptors. Stengel et al. [22, 37] has already shown that the negative chronotropic action of carbachol was abolished in M_2 receptor KO (M_2KO) mice but not in M_3 receptor KO (M_3KO) mice. Attenuation of carbachol-induced negative chronotropic actions in atria from M_4KO mice suggests possible involvement of the M_4 receptor in inhibition of heart rate in cooperation with the M_2 receptor [22]. We have also examined chronotropic and inotropic actions induced by the muscarinic receptor agonist carbachol (insensitive to acetylcholinesterase) using right and left atria isolated from M_2KO, M_3KO, and M_2/M_3 double KO mice, Protocol 1 [23, 24].

3.1 Chronotropic Actions

In spontaneously beating right atria of normal (wild type) mice, carbachol (10 nM–10 μM) caused concentration-dependent inhibition of spontaneous beating and finally abolished it (EC_{50} = 430 nM). The inhibition was competitively antagonized by AF-DX116 (pK_b = 7.54). Since pK_b for AF-DX116 was consistent with that for the M_2 muscarinic receptor [6], the negative chronotropic action was confirmed to be mediated by the M_2 receptor as previously described [4].

Negative chronotropic actions of carbachol were also examined in atria isolated from pertussis toxin-treated mice. Pertussis toxin is able to distinguish M_2/M_4 muscarinic receptors (pertussis toxin-sensitive G-protein pathway) and M_1/M_3/M_5 muscarinic receptors (pertussis toxin-insensitive G-protein pathway). Although spontaneous beating was almost the same in the atria from pertussis toxin-treated mice, the negative chronotropic action of carbachol was completely abolished, confirming the involvement of M_2 receptors (Fig. 1a). To obtain further evidence for the involvement

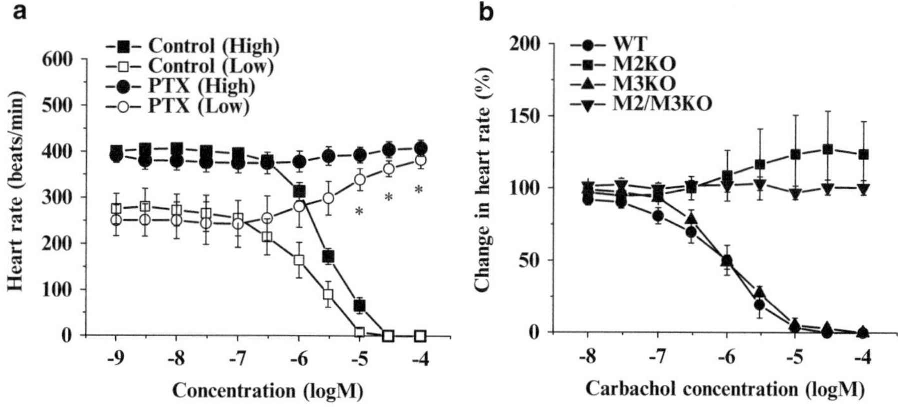

Fig. 1 Effects of carbachol on spontaneous contraction of isolated mouse right atria. (**a**) Carbachol-induced negative chronotropic actions and effects of pertussis toxin treatment on carbachol-induced actions. On the basis of average heart rate (beats/min), atrial preparations from wild-type and pertussis toxin-treated mice were divided into a high heart rate group (>350 beats/min) and a low heart rate group (<350 beats/min) (see the text). In these two groups, the effect of increasing concentration of carbachol was examined and concentration–response curves were compared. Ordinate: spontaneous contraction (beats/min). *$p<0.05$; significant increase of heart rate compared with that in the absence of carbachol. (**b**) Comparison of chronotropic actions of carbachol in the right atria from wild-type (WT) and muscarinic receptor KO mice (M_2KO, M_3KO, and M_2/M_3KO). Heart rates (beats/min) of isolated atria were 378 ± 22 (wild-type, $n=5$), 406 ± 20 (M_2KO, $n=5$), 420 ± 32 (M_3KO, $n=5$) and 385 ± 35 (M_2/M_3KO, $n=5$). Ordinate: heart rate change expressed as percentage of heart rate just before application of carbachol. Values are means ± S.E.M. of at least 4 experiments. These figures are from Kitazawa et al. [23] and Harada et al. [24]

of M_2 receptors in the negative chronotropic actions, effects of carbachol on spontaneous beating of right atria were compared among M_2KO, M_3KO, and M_2/M_3KO mice (Fig. 1b). The spontaneous beating rates were almost the same for wild-type mice and muscarinic receptor KO mice. Carbachol concentration-dependently decreased spontaneous beating of atria from wild-type and M_3KO mice and finally abolished spontaneous beating at concentrations of 10–30 μM. On the other hand, in atria of M_2KO mice, the negative chronotropic actions of carbachol were abolished and heart rate tended to increase at high concentrations of carbachol (30–100 μM). The carbachol-induced negative chronotropic actions were also not observed in the right atria from M_2/M_3KO mice (Fig. 1b).

Effects of carbachol on respective spontaneously beating atria are shown in Fig. 2. In atria from wild-type and M_3KO mice, carbachol only decreased the spontaneous contraction regardless of high or low heart rate preparations. However, in M_2KO mice, carbachol did not induce any chronotropic actions in the high heart rate atrium (400 beats/min), but when the heart rate was low (under 300 beats/min), carbachol caused an increase in heart rate (positive chronotropic actions). Since carbachol did not cause any chronotropic actions in either high or low heart rate atria from

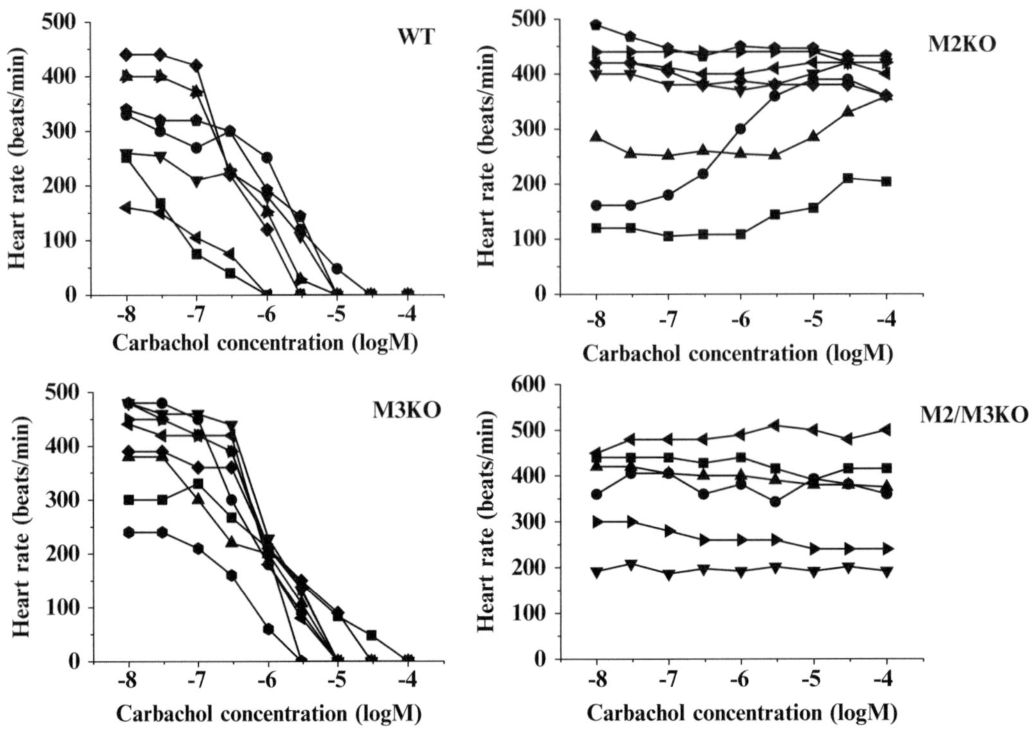

Fig. 2 Effects of carbachol on spontaneous contraction of right atria from wild-type (WT) and muscarinic receptor KO mice (M_2KO, M_3KO, and M_2/M_3KO). Each symbol indicates the data of an atrium from each mouse. Ordinate: spontaneous contraction (beats/min). In the atria from M_2KO mice, the effects of carbachol were different depending on heart rate

M_2/M_3KO mice, it is thought that the positive chronotropic actions were due to activation of the M_3 receptor. When the heart rate was high, positive chronotropic action by M_3 receptor activation was masked, but it was revealed when the heart rate was low. Similar heart rate-dependent chronotropic actions of carbachol were also observed in the atria from pertussis toxin-treated mice (Fig. 1a). Average spontaneous beating rate in the atria of wild-type, pertussis toxin-treated, M_2KO, M_3KO, and M_2/M_3KO mice was 348 ± 12 beats/min (n = 59). Therefore, we divided the atrial preparations of wild-type and pertussis toxin-treated mice into a high heart rate group (>350 beats/min) and a low heart rate group (<350 beats/min). Within 14 atrial preparations from 14 pertussis toxin-treated mice, carbachol failed to cause any chronotropic actions in the high heart rate group (404 ± 14 beats/min, n = 8) but caused concentration-dependent positive chronotropic actions in the low heart rate group (256 ± 34 beats/min, n = 6). Until now, positive chronotropic actions by M_3 receptor activation have not been demonstrated in atria of M_2KO mice probably due to high heart rate (396 beats/min) [22], but the present results indicated that activation of the M_3 receptor could cause a positive chronotropic action under certain experimental conditions. Since the M_2 receptor is involved in negative chronotropic actions and is the dominant receptor subtype expressed in the heart [11, 23], after reduction of M_2 receptor function (by pertussis toxin treatment or knockout of M_2R) it is possible to observe M_3 receptor-mediated actions. In addition, the results suggest that a low heart rate condition is required for detecting M_3 receptor-mediated positive chronotropic actions. Highly positive immunostaining of M_3 receptors has been demonstrated in pacemaker cells of a murine atrium [33]. This histological observation supports M_3 receptor-mediated positive chronotropic actions of a muscarinic receptor agonist in M_2KO mice.

3.2 Inotropic Actions

Electrical field stimulation (EFS, 1 Hz) was applied to the isolated left atrium and the inotropic action of carbachol was measured as a change in amplitude of EFS-induced contraction. As previously described [31, 38], carbachol induced negative inotropic actions (decrease in EFS-induced contraction) followed by slowly developed positive inotropic actions (increase in EFS-induced contraction) in the mouse left atrium (Fig. 3). Both inotropic actions increased depending on the carbachol concentrations, and atropine abolished both actions, indicating the involvement of muscarinic receptors in both actions. Positive and negative inotropic actions induced by carbachol were characterized using muscarinic receptor KO mice. As shown in Fig. 3, the first negative inotropic actions by carbachol were abolished in the atria of M_2KO and M_2/M_3KO mice. Slowly developed positive inotropic actions of carbachol were observed in M_2KO mice but not in M_2/M_3KO mice. In the atria of M_3KO mice, carbachol caused only negative inotropic actions and a slowly developed positive-inotropic actions were abolished. In the atria from pertussis toxin-treated wild-type mice, the negative inotropic action induced by carbachol was abolished and only a positive inotropic action remained. In addition, a low concentration of 4-DAMP (M_3 receptor preferential antagonist) decreased the positive inotropic actions without affecting the negative inotropic actions (Fig. 3). Taken together, the results of the functional study using muscarinic receptor KO mice indicated that the M_2 receptor mediates the first negative inotropic action and that the M_3 receptor mediates the second positive inotropic action in the mouse atrium. Using M_2KO and M_3KO mice, the concentration–response relationships of M_2-mediated inhibition and M_3-mediated potentiation of contraction were compared (Fig. 4). The EC_{50} value of negative inotropic action by the M_2 receptor in M_3KO mice was 1.5 μM and that of positive inotropic action by the M_3 receptor in M_2KO mice was 3.7 μM. Although the difference in the EC_{50} values might be due to the difference in M_2 and M_3 receptor expression levels, a low concentration of acetylcholine decreases cardiac contractility by the M_2 receptor, and if

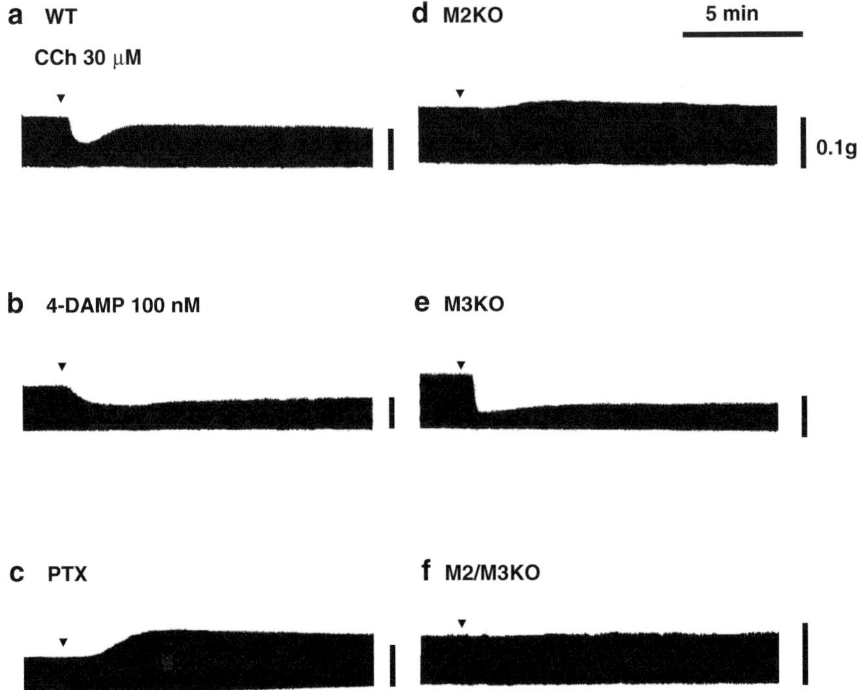

Fig. 3 Effects of carbachol on EFS-induced contraction of isolated left atria from wild-type, M_2KO, M_3KO, and M_2/M_3KO mice. Carbachol (CCh, 30 μM) induced biphasic inotropic actions in atria of wild-type mice (**a**). Slowly developed positive inotropic actions were decreased by 4-DAMP (100 nM) (**b**). In atria from pertussis toxin-treated mice, the action of carbachol was reversed and only positive inotropic action was observed (**c**). In atria from M_2KO mice, negative inotropic action was abolished and only positive inotropic action was evoked (**d**), but in atria from M_3KO mice, slowly developed positive inotropic action was abolished and the negative inotropic action remained (**e**). Carbachol did not cause any inotropic actions in atria from M_2/M_3KO mice (**f**). The figure is from Harada et al. [24]

the acetylcholine concentration increases, the M_3 receptor is possibly stimulated to decelerate the excessive inhibition by M_2 receptor activation.

3.3 Distribution of M_2 and M_3 Muscarinic Receptors in the Mouse Atrium

M_3 receptor-mediated positive inotropic action in the mouse atrium has been shown to be decreased by treatment with indomethacin [23, 38] and with prostanoid receptor antagonists (EP and FP receptors) [39]. Prostaglandins are effective for inducing positive inotropic actions in the mouse atrium [38, 39]. Therefore, it is thought that M_3 receptor activation stimulates the synthesis of prostaglandins through activation of cyclooxygenase (COX) and that released prostaglandins cause positive inotropic actions. COXs are divided into COX-1 (constitutive type) and COX-2 (inducible type), and indomethacin non-selectively inhibits both COX isozymes. To clarify functional roles of COX isozymes in the carbachol-induced positive inotropic actions, the effects of selective COX-1

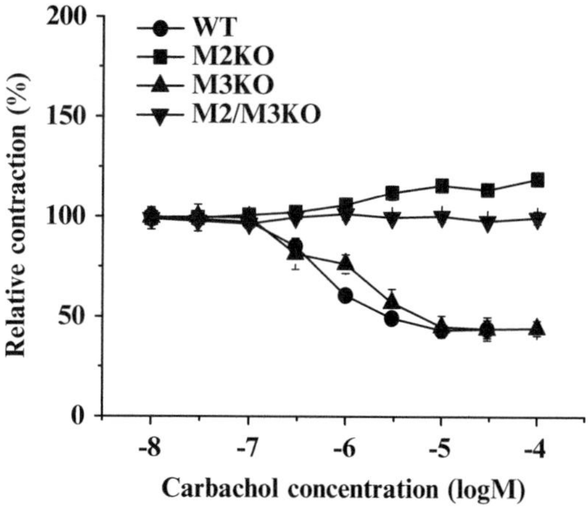

Fig. 4 Comparison of inotropic actions induced by carbachol in left atria isolated from wild-type, M_2KO, M_3KO, and M_2/M_3KO mice. After establishing reproducible EFS-induced atrial contractions, carbachol (10 nM–100 μM) was applied non-cumulatively at 30-min intervals and concentration–response relationships were constructed. Values are means and SEM of 5–6 experiments. The figure is from Kitazawa et al. [23]

and COX-2 inhibitors on carbachol-induced inotropic actions were examined. Carbachol-induced negative and positive inotropic actions were not affected by SC560 and FR122047 (COX-1 inhibitors). However, NS398 (a COX-2 inhibitor) decreased the positive inotropic action induced by carbachol and the responses changed to a sustained negative inotropic action without affecting the first negative inotropic actions (Fig. 5). Hara et al. [39] have already reported the same results. Therefore, COX-2, but not COX-1 is involved in the positive inotropic actions induced by carbachol in the mouse atrium. Our results also indicated that nitric oxide was not involved in the carbachol-induced action because treatment with L-nitroarginine methyl ester (L-NAME) failed to change the EFS-induced contraction (Fig. 5). In addition, the nitric oxide donor nitroprusside did not cause any inotropic actions. Nitric oxide might not be an important modulator of contractility in the mouse atrium.

The positive inotropic action induced by carbachol was also abolished by mechanical removal or chemical destruction of the endocardial endothelium [24, 38, 39], suggesting that the M_3 receptor to COX-2 pathway is located in endocardial endothelial cells.

For detecting M_2R and M_3R immunoreactivities in the mouse heart, immunohistochemical study was carried out, Protocol 2. In an immunohistochemical study using the mouse atrium, M_2 receptor immunoreactivity was shown to be localized only in the myocardial cells, whereas M_3 receptor immunoreactivity was distributed in both endocardial endothelium and myocardium of the atrium (Fig. 6, Harada et al. [24]). Distribution of M_3 receptor in

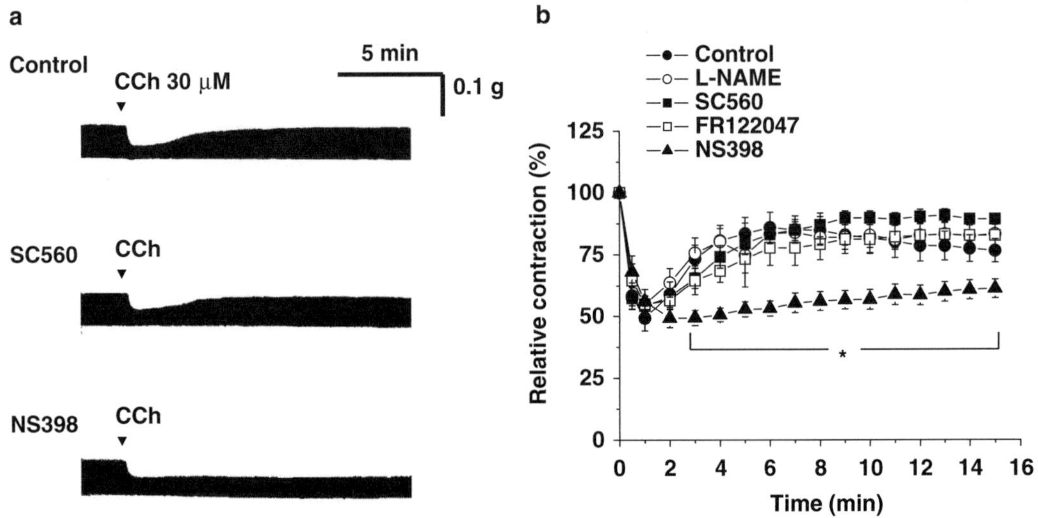

Fig. 5 Effects of L-NAME and COX inhibitors on carbachol-induced inotropic actions in the isolated left atrium of a mouse. (**a**) Typical effects of SC560 (1 μM) and NS398 (1 μM) on carbachol-induced inotropic action (30 μM). (**b**) Comparison of the time courses of carbachol-induced inotropic actions in the absence (Control, *filled circle*) and presence of L-NAME (*open circle*, 100 μM), SC560 (*filled square*, 1 μM), FR122047 (*open square*, 1 μM), and NS398 (*filled triangle*, 1 μM). Ordinate: relative amplitude of EFS-induced contraction expressed as percentage of the amplitude just before application of carbachol. Abscissa: incubation time with carbachol (min). Values are means ± S.E.M. of at least 5 experiments. *$p < 0.05$, significantly different from the values of the control at the corresponding time. The figure is from Harada et al. [24]

the ventricle was the same with that in the atrium and both endocardial endothelium and myocardium were stained by M_3 receptor antibody. Atria from M_2KO and M_2/M_3KO mice lacked M_2 immunoreactivity in the myocardium, but atria from M_3KO mice showed positive immunostaining of the M_2 receptor. On the other hand, M_3R immunoreactivity was observed in the myocardium and endocardial endothelium of wild-type and M_2KO mice [24]. These results for muscarinic receptor KO mice guaranteed the specificity of both the anti-M_2 receptor antibody and anti-M_3 receptor antibody in the mouse atrium.

Localization of COX-1 and COX-2 in the mouse atrium was also examined immunohistochemically. Briefly, the left atria were isolated from wild-type and rinsed with fresh Krebs solution. Each atrium was fixed with 4 % paraformaldehyde-Tris buffer (pH = 7.4) and finally embedded in paraffin. All tissue blocks were cut into 6 μm section using microtome. The sections were immersed in 0.5 % hydrogen peroxide in methanol at room temperature for 10 min and were then incubated with 10 % normal goat serum (Histofine SAB Kit, Nichirei, Tokyo, Japan) to block nonspecific binding at room temperature for 20 min, followed by incubation with respective antibodies for COX (mouse anti-COX-1 antibody 1:50 and mouse anti-COX-2 antibody 1:300, Cayman Chemical,

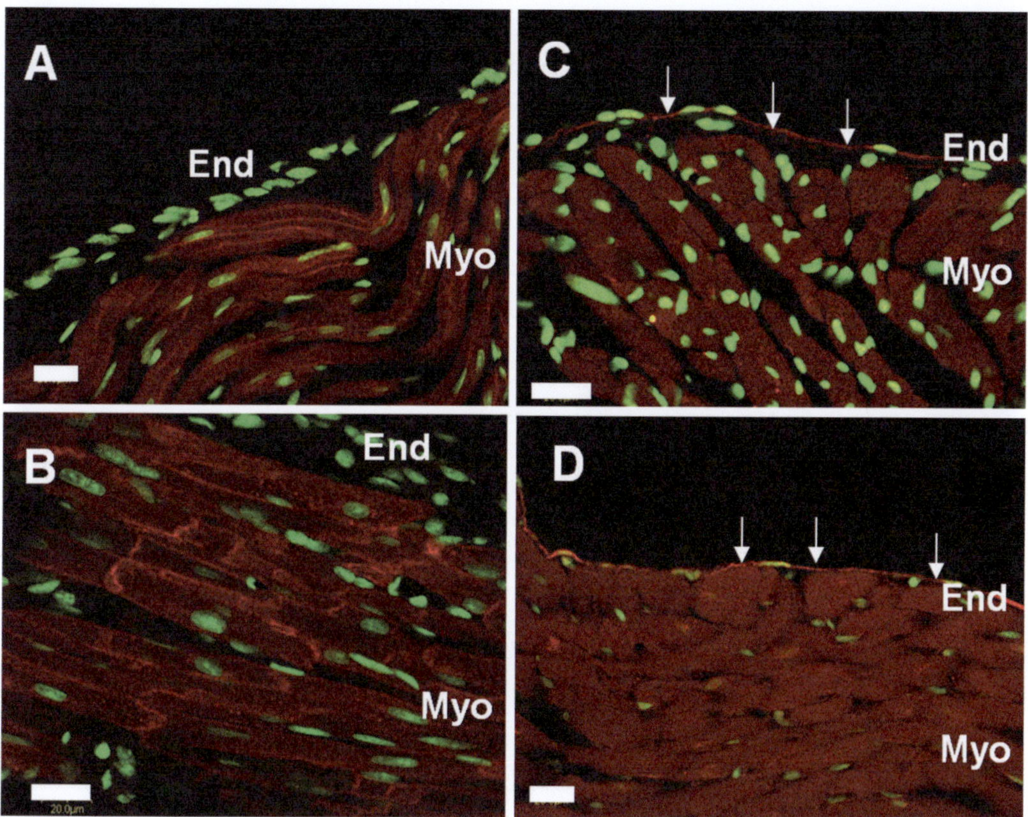

Fig. 6 Localization of M_2R and M_3R immunoreactivities in the atrium and ventricle of wild-type mice. Immunofluorescence of muscarinic receptors was detected by confocal microscopy. Tissue sections from the atrium (**a**, **c**) and ventricle (**b**, **d**) were treated with an anti-M_2R antibody (**a**, **b**) or anti-M_3R antibody (**c**, **d**) and stained with Cy3-labeled anti-IgG (*red*). The nucleus was stained using SOYTOX Green Nucleic Acid Stain. M_3R immunoreactivity was detected in the endocardial endothelium of both the atrium and ventricle (*arrows*). Myo, myocardium; End, endocardial endothelium. Scale bars = 20 μm. The figure is from Harada et al. [24]

USA) at 4 °C, overnight. Subsequently, the sections were washed with phosphate buffered saline and incubated with second antibodies (anti-rabbit IgG 1:300, Cayman Chemical, USA) for 2 h at room temperature. After washing out, the sections were incubated with avidin-biotinylated peroxidase complex for 30 min and washed out. Finally, the sections were incubated with hydrogen peroxide conjugated 3-3′-diaminobenzidine tetrahydrochloride (DAB) (Histofine Simple Stain DAB solution, Nichirei, Japan) and counterstained with hematoxylin and eosin. COX-1 immunoreactivity was observed in both the myocardium and endocardial endothelium of wild-type mice. COX-2 immunoreactivity was also localized in the atrial myocardium and endocardial endothelium. Distributions of the M_2 receptor, M_3 receptor, COX-1 and COX-2 in the mouse atrium are shown in Table 1. Since the

Table 1 Summary of distribution of M2 receptor, M3 receptor, COX-1 and COX-2 immunoreactivity in the endocardial endothelium and myocardium of the mouse atrium

	Wild-type		M2KO		M3KO		M2/M3KO	
	Endothelium	Myocardium	Endothelium	Myocardium	Endothelium	Myocardium	Endothelium	Myocardium
M2 receptor	–	+	–	–	–	+	–	–
M3 receptor	+	+	+	+	–	–	–	–
COX-1	+	+	ND	ND	ND	ND	ND	ND
COX-2	+	+	ND	ND	ND	ND	ND	ND

+ presence of immunoreactivity, – absence of immunoreactivity, *ND* not determined
The table was referred the results of Harada et al. (2012) [24].

endocardial endothelium and COX-2 are necessary for M_3 receptor-mediated positive inotropic actions, it is thought that the M_3 receptor-COX-2-prostaglandin pathway in endocardial endothelial cells participates in the carbachol-induced positive inotropic actions in the mouse atrium.

Immunohistochemical evidence for localization of muscarinic receptors and COXs raises two questions to be solved. First, the M_3 receptor-COX-2-prostaglandin pathway is also present in the myocardium, but the functional roles of this pathway were not explained by the results of present experiment. Second, both COX-1 and COX-2 are present in the endocardial endothelium. Why is only COX-2 functional and what is the physiological role of COX-1 (in both the myocardium and endothelium) in the mouse atrium? Further studies are needed to answer these questions.

3.4 Regulation of COX-2 mRNA Expression in the Isolated Mouse Atrium

The involvement of COX-2 in M_3 receptor-mediated inotropic actions was described in the previous section. Generally, COX-2 is thought to be an inducible isozyme, but its expression in fresh mouse atria has already been demonstrated by Western blotting issues [39, 40] such as the kidney, brain and heart, COX-2 may be expressed constitutively to produce endogenous prostaglandins [40]. In the rat heart, expression of COX-2, but not that of COX-1, increased in an age-dependent manner, suggesting that COX-1 expression and COX-2 expression in cardiac tissues are regulated by different mechanisms [41]. Recently, Hara et al. [39] suggested that COX-2 protein in the mouse atrium increased in an incubation time-dependent manner during continuous electrical stimulation, and we also confirmed incubation time-dependent increase in M_3 receptor-COX-2-mediated positive inotropic actions [24]. Therefore, we examined the relationships between expression levels of COX-1 and COX-2 mRNAs and incubation time. For quantitative analysis of COX-1 and COX-2 mRNAs, real-time RT-PCR was conducted using SYBR Green, Protocol 3. Primer pair sequences were 5′-ACTATCCGTGCCAGAACCAG-3′ (forward) and 5′-ATTCCCAGAGCCAGTATCCA-3′ (reverse) for COX-1 (192 bp), and 5′-TGCGACATACTCAAGCAGGA-3′ (forward) and 5′-CAATGCGGTTCTGATACTGG-3′ (reverse) for COX-2 (196 bp), respectively. As shown in Fig. 7, COX-2 mRNA expression level, but not COX-1 mRNA expression level, significantly increased depending on incubation time in the mouse atrial preparations. The increases in COX-2 mRNA expression levels both in stimulated and non-stimulated left atrial preparations were almost the same, suggesting that continuous contraction of the atrium is not an inducer of COX-2 mRNA expression but that incubation in Krebs solution itself might stimulate expression of COX-2 mRNA. Incubation time-dependent increase in COX-2 mRNA was also observed in other isolated preparations such as the aorta, gastric strips and colonic strips. As a bacterial endotoxin, lipopolysaccha-

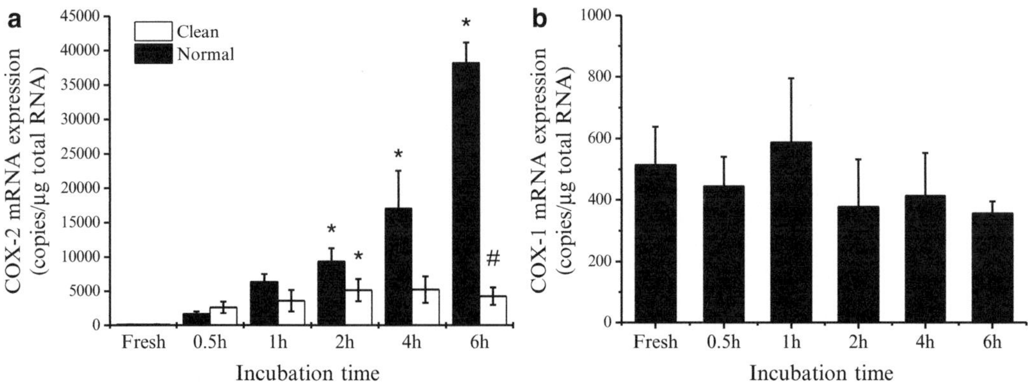

Fig. 7 Incubation time-dependent increase of COX-2 mRNA expression in the left atria of mice. The left atrium was isolated from a wild-type mouse and the preparation was electrically stimulated. After 0.5, 1, 2, 4, and 6 h of incubation in warmed Krebs solution, COX-2 mRNA in the atrium was extracted and measured by quantitative real-time RT-PCR using specific primers (**a**, *black column*). COX-1 mRNA expression in the atrium was also measured using real-time RT-PCR (**b**, *black column*). When sterile water and sterilized apparatus (scissors, forceps, organ bath) were used in the preparation of the atrium and organ bath study, incubation time-dependent increase in COX-2 mRNA expression decreased markedly (*open column*). $^*p<0.05$, significantly different from the values of fresh atrium. $^{\#}p<0.05$, significantly different from the corresponding normal values. Values are means ± S.E.M. of at least 4 experiments

ride is known to stimulate COX-2 mRNA expression [42], it is hypothesized that growth of bacteria and increase in their toxins in the medium during incubation might stimulate the expression of COX-2 mRNA and that the synthesized prostaglandins by COX-2 also enhance the expression of COX-2 mRNA. To investigate this hypothesis, we changed the experimental conditions and used sterile water and sterilized apparatus (scissors, forceps and organ bath) for making preparations and doing the organ bath experiment. As shown in Fig. 7a, COX-2 mRNA expression level was decreased by sterilization. However, a slight increase in COX-2 mRNA expression remained probably due to some contamination of bacteria in the incubation medium. To determine the involvement of endogenous prostaglandins in COX-2 mRNA expression, the effects of COX inhibitors and bath-applied prostaglandin on the expression level of COX-2 mRNA were examined. Either indomethacin or COX-2 inhibitor (NS398) tended to decrease the COX-2 mRNA expression. However, COX-1 inhibitor (FR122047) did not affect the COX-2 mRNA expression level (Fig. 8a). In addition, application of prostaglandin E_2 (1 μM) and lipopolysaccharide (1 μg/ml) significantly increased COX-2 mRNA expression (Fig. 8b). Although further experiments are needed, these results support our hypothesis and indicated that some contamination of bacteria in the incubation medium used in vitro is likely to affect functional actions mediated by the M_3 receptor and COX-2 in the mouse atrium.

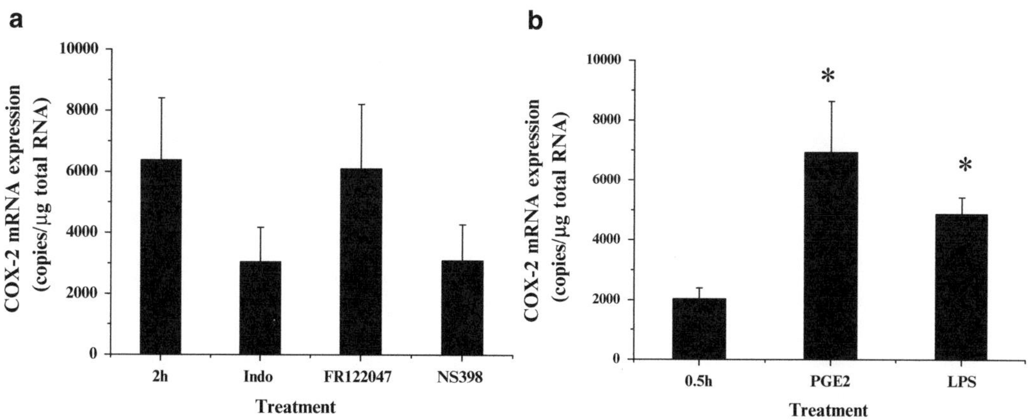

Fig. 8 Effects of COX inhibitors, prostaglandins and lipopolysaccharide on COX-2 mRNA expression in the mouse left atrium. (**a**) Increase in COX-2 mRNA expression after 2 h of incubation in Krebs solution was decreased by indomethacin (Indo, 1 μM, nonselective COX inhibitor) and NS398 (1 μM, COX-2 inhibitor), but FR122047 (1 μM, COX-1 inhibitor) did not change COX-2 mRNA expression. The inhibition by indomethacin and NS398 did not reach significance in the present experimental conditions. (**b**) COX-2 mRNA expression increased significantly by incubation with prostaglandin E_2 (1 μM) and lipopolysaccharide (LPS, 1 μg/ml) for 30 min. *$p < 0.05$, significantly different from the values of the control (0.5 h). Values are means ± S.E.M. of at least 4 experiments

3.5 Cardiac Intrinsic Neurons Regulating Mouse Atrial Contractility

Intrinsic cardiac ganglia were initially regarded as simple relay stations in parasympathetic preganglionic cholinergic neurons to postganglionic cholinergic neurons innervating the myocardium. However, cardiac ganglia have been demonstrated to have a complex neurochemical phenotype different from the classical postganglionic cholinergic neurons. Morphological study has shown that some cholinergic nerve cell bodies contain noradrenergic neurons markers, such as tyrosine hydroxylase, dopamine-β-hydroxylase, and noradrenaline transporters. The presence of such hybrid (cholinergic/adrenergic) intrinsic cardiac neurons has been reported in several animal species [43–46]. Although regulation of cardiac contractility through these kinds of intrinsic neurons is interesting the functional role of these neurons has not been clarified in detail. Isolated atrium is a good experimental model to investigate the function of intrinsic nerves because the atrium contains postsynaptic ganglionic nerves. The atrial preparations are also suspended in the organ bath and their contractility is measured and recorded. We used 1,1-dimethyl-4-phenylpiperazinium (DMPP) to stimulate nicotinic receptor on intrinsic nerves [25].

Chronotopic actions: In wild-type mice, DMPP caused a biphasic response consisting of an initial transient decrease in heart rate and amplitude (from 5 to 20 s after application, first phase) followed by increase in heart rate and amplitude (from 20 to 60 s after application, second phase) in a concentration-dependent manner

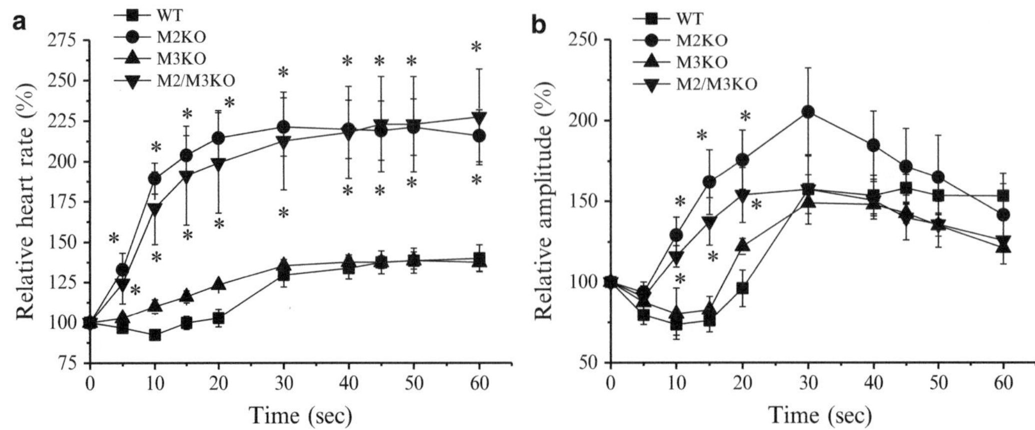

Fig. 9 Comparison of DMPP-induced actions on heart rate and amplitude of spontaneous contraction in the right atria from wild-type and muscarinic receptor knockout mice. Each graph shows the time course of DMPP-induced (100 μM) chronotropic action (**a**) and inotropic action (**b**) observed in wild-type (WT, *filled square*), M_2KO (*filled circle*), M_3KO (*filled triangle*) and M_2/M_3KO mice (*filled inverse triangle*). Ordinate: relative heart rate and amplitude of spontaneous contraction (absence of DMPP = 100 %). *$p < 0.05$, significantly different from the values of wild-type mice (*filled square*). Values are means ± S.E.M. of at least 5 experiments. The figures are from Ochi et al. [25]

(1–100 μM). In the first phase, inhibition of amplitude (20–25 % inhibition) was more marked than that in heart rate (10 % inhibition) (Fig. 9). The first phase inhibition was abolished by atropine, hexamethonium and pertussis toxin treatment, and the following phase increase in contractility was decreased by hexamethonium and atenolol (β_1-adrenoceptor antagonist). In the atria from reserpine-treated mice, increase of spontaneous contraction by DMPP decreased markedly. These results suggested that nicotinic receptor activation caused both cholinergic nerve excitation to decrease contractility and adrenergic nerve excitation to increase contractility of the mouse atrium.

In the atria from M_2KO mice, the DMPP-induced inhibition of heart rate and amplitude was abolished and changed to enhancement of the EFS-induced contraction. The first inhibitory phase was also abolished in the atria from M_2/M_3KO mice and changed to the excitatory responses (Fig. 9). Due to abolition of the first inhibitory phase, the second increase phase of contractility in M_2KO and M_2/M_3KO mice was significantly greater than that in wild-type mice. In the atria from M_3KO mice, time courses of changes in heart rate and amplitude were comparable with those of wild-type mice (Fig. 9). The results from muscarinic receptor KO mice confirmed that the initial inhibition of atrial contractility is due to activation of intrinsic cholinergic nerves followed by acetylcholine release and activation of M_2 receptors.

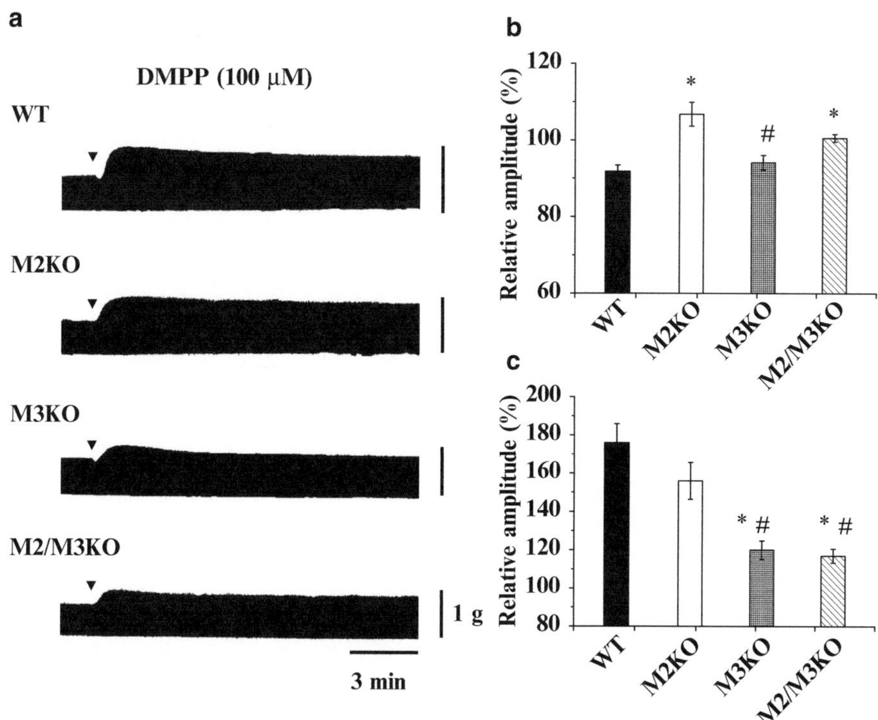

Fig. 10 Effects of DMPP on EFS-induced contraction of the left atrium. (**a**) Typical inotropic action of DMPP (100 μM) in left atria from wild-type (WT), M_2KO, M_3KO, and M_2/M_3KO mice. Comparison of negative (**b**) and positive inotropic actions (**c**) induced by DMPP in wild-type and muscarinic receptor KO mice. Transient negative inotropic actions were evaluated at 15 s after application, and positive inotropic actions were evaluated at the maximum amplitude of contraction. Each column indicates relative amplitude of EFS-induced atrial contraction. $^*p<0.05$, significantly different from the values of wild-type mice (*black column*). $^\#p<0.05$, significantly different from the values of M_2KO mice (*open column*). Values are means ± S.E.M. of more than 4 experiments. The figures are from Ochi et al. [25]

Inotropic actions: Effects of DMPP on EFS-induced contraction of the left atria were also examined to characterize inotropic responses evoked by activation of intrinsic nerves. As shown in Fig. 10, DMPP (100 μM) caused transient small negative inotropic actions followed by long-lasting positive inotropic actions. The negative inotropic action was short-lived and was changed to positive inotropic action after 30 s of application, and the positive inotropic action peaked at 3–6 min after application. Negative inotropic actions of DMPP were decreased by treatment with atropine or hexamethonium but tended to be potentiated by atenolol. In the atria from pertussis toxin-treated mice, the DMPP-induced negative inotropic action was abolished. The positive inotropic actions were also characterized using some autonomic drugs. Both atropine and hexamethonium decreased the onset and amplitude of positive inotropic actions. Atenolol and reserpine treatments were also effective in decreasing positive inotropic actions.

Pharmacological characterization indicated that the negative transient inotropic action is due to activation of cholinergic nerves and M_2 receptors but that both adrenergic and cholinergic muscarinic mechanisms are involved in the slowly developed positive inotropic actions.

The biphasic inotropic actions induced by DMPP were compared in the atria from muscarinic receptor KO mice. The initial inhibition of EFS-induced contraction was abolished in the atria from M_2KO and M_2/M_3KO mice (Fig. 10b). On the other hand, positive inotropic actions were significantly decreased in the atria from M_3KO and M_2/M_3KO mice but not in the atria from M_2KO mice (Fig. 10c). Therefore, the results indicated that the M_2 receptor is involved in the initial negative inotropic actions and that the M_3 receptor is involved in the second positive inotropic actions in the mouse atrium.

The functional study for characterization of the cardiac intrinsic nerves using a nicotinic receptor stimulant indicated that both cholinergic and adrenergic intrinsic nerves are present in the mouse atrium and participate in the biphasic inotropic actions consisting of inhibition of atrial contraction (acetylcholine-M_2 receptor pathway) and potentiation of atrial contraction (acetylcholine-M_3 receptor pathway and noradrenaline-β_1 receptor pathway). Extrinsic neurons innervating cardiac ganglia of the mouse atrium exhibit both excitatory and inhibitory actions through activation of cardiac intrinsic neurons releasing acetylcholine or noradrenaline, and these neurons regulate cardiac contraction.

4 Conclusions

Parasympathetic control of visceral organs is accomplished by acetylcholine acting via muscarinic receptors. Dysfunction of neural control is related to the specific pathophysiology. Therefore, it is necessary to examine the function of muscarinic receptors in each organ. There are five subtypes of muscarinic receptor. Their distributions are different and multiple subtypes of muscarinic receptors are present in each organ. In addition, lack of selectivity of muscarinic receptor agonists and antagonists makes pharmacological analysis of receptor subtypes difficult. However, the combination of classical pharmacological analysis and useful muscarinic receptor knockout mouse can overcome the difficulty for clarification of functional roles of individual muscarinic receptor subtypes. In the isolated mouse atrium, our data demonstrated that although the expression level is low, M_3 muscarinic receptors expressed on the endocardial endothelium cause positive inotropic actions in response to both muscarinic receptor agonists and endogenous acetylcholine to antagonize the excessive inhibition by M_2 muscarinic receptors (Fig. 11). Therefore, muscarinic receptor knockout mice

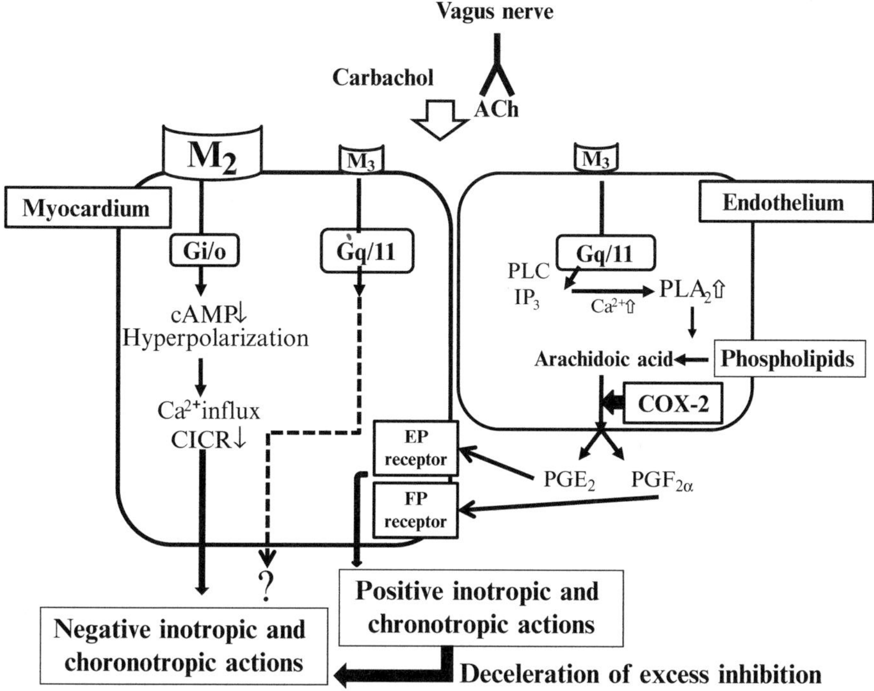

Fig. 11 Parasympathetic control of mouse atrial contractility by M_2 and M_3 muscarinic receptors. In the mouse atrium, M_2 receptors are the dominant muscarinic receptor subtype and are located on the myocardium. Although the expression level of M_3 receptors is low, M_3 receptors are located in both the myocardium and endocardial endothelium. Acetylcholine released from postganglionic cholinergic nerves first acts on the dominant M_2 receptors and induces negative inotropic/chronotropic actions through activation of $G_{i/o}$ protein (hyperpolarization, decrease in Ca^{2+} influx and Ca^{2+}-induced Ca^{2+} release, CICR). However, a higher concentration of acetylcholine can act on the small population of M_3 receptors. Endothelial M_3 receptor activation increases Ca^{2+} concentration and activates phospholipase A_2 to produce arachidonic acids. Arachidonic acids are converted to prostaglandins (PGE_2 and $PGF_{2\alpha}$) through COX-2 in the endothelium. Synthesized prostaglandins act in a paracrine fashion and are effective for producing positive inotropic/chronotropic actions to decelerate the M_2-mediated inhibitory actions. However the function of M_3 receptors on the myocardium is unknown at present

such as M_2KO mice open up a new research filed for physiological and pathophysiological functions of non-M_2 receptors in the heart.

5 Protocols

5.1 Protocol 1: In Vitro Contraction Study for Analysis of Functional Muscarinic Receptor Subtypes in the Mouse Heart

1. Mice are killed by cervical dislocation and beating hearts are isolated from animals and immersed in warmed bubbling Krebs solution.
2. Left and right atria are dissected together from ventricles and their lumen is rinsed well by Krebs solution to remove blood. Then right atrium and left atrium are separated and are conducted to the contraction study.

3. The atrial muscle preparations are fixed in the organ bath (20 ml) containing gassed (95 $\%O_2+5$ $\%CO_2$) and warmed (37 °C) Krebs solution (NaCl, 118 mM; KCl, 4.75 mM; $MgSO_4$, 1.2 mM; KH_2PO_4, 1.2 mM; $CaCl_2$, 2.5 mM; $NaHCO_3$, 25 mM; and glucose, 11.5 mM).
4. Mechanical changes in the preparations are detected by a force transducer (TB-612 T, Nihon Kohden, Japan). Right atrial preparations show spontaneous contractions. Left atrial preparations are electrically stimulated (1 Hz, 2 ms in duration, 1.5 × threshold voltage) using a pair of platinum electrodes connected to Stimulator (SEN-3301, Nihon Kohden).
5. Effects of muscarinic agonist (carbachol) on the spontaneous contraction (chronotropic action) and the electrical stimulation evoked contraction (inotropic action) are compared among wild-type, drug-treated, and muscarinic receptor knockout mice.
6. Effects of muscarinic receptor antagonists and drugs on the carbachol-induced actions are also examined to clarify the mechanisms of muscarinic action in the heart.

5.2 Protocol 2: Immunohistochemical Study for Muscarinic Receptor Subtypes in the Mouse Heart

1. Mice are deeply anesthetized with phenobarbital (i.p., 25 μg/kg) and perfused with physiological saline via the aorta, followed with 4 % formaldehyde plus 0.2 % picric acid in 0.1 M phosphate buffer, pH 7.4.
2. After perfusion, the heart is removed and immersed in the same fixative solution for an additional 6 h at 4 °C.
3. The formaldehyde-fixed tissues are dipped in 30 % sucrose solution overnight at 4 °C, embedded in OCT compound, and quickly frozen in liquid nitrogen.
4. Frozen sections of 12 μm in thickness are mounted on poly-l-lysine-coated glass slides, incubated with a guinea-pig anti-mouse M_2R antibody (Frontier Institute Co. Ltd, Japan, 1:200) or a rabbit anti-mouse M_3R antibody (Acris antibodies GmbH, Germany, 1:400) overnight, followed by incubation with Cy3-labeled anti-rabbit IgG or anti-guinea-pig IgG (Jackson Immuno Research, West Grove, PA, USA, 1:200) for 2 h.
5. Nucleus is counterstained using SOYTOX Green Nucleic Acid Stain (Invitrogen, Carlsbad, CA, USA, 1:30,000, incubation time: 10 min).
6. The stained sections are mounted with glycerin-PBS and observed under a confocal laser scanning microscope (Fluoview; Olympus, Tokyo, Japan).

5.3 Protocol 3: Molecular-Biology Study for Expression of Muscarinic Receptor mRNAs in the Mouse Heart

1. Isolated mouse ventricles and atria are cut into small pieces and were immediately immersed in RNA Later (Takara, Japan) for 12 h at 4 °C and then stored at −30 °C until use.
2. Total RNA of cardiac tissues is extracted by the acid guanidine–phenol–chloroform method (TRIzol reagent, Invitrogen, Carlsbad, CA, USA).
3. For quantitative analysis of muscarinic receptor mRNA expression levels, real-time RT-PCR is conducted using SYBR Green. Sequences of primer pairs used are 5′-TCAGGACTCC TCTGGCTTC-3′ (forward) and 5′-CCGGGTTTCACTCT CTGTCT-3′ (reverse) for the M_1 receptor (Gene Bank, NM 007698), 5′-CCGGTGTCTCCCAGTCTAGT-3′ (forward) and 5′-CAGACGTGGAGTCATTGGAG-3′ (reverse) for the M_2 receptor (Gene Bank, NM 203491), 5′-ACCAAGCTACCC TCCTCAGA-3′ (forward) and 5′-GACAGTTGTCACGGT CATCC-3′ (reverse) for the M_3 receptor (Gene Bank, NM 033269), 5′-ATGGTGTTCATTGCGACAGT-3′ (forward) and 5′-GACTGTCTGCAACTGCCTGT-3′ (reverse) for the M_4 receptor (Gene Bank, NM 007699), and 5′-CGATCATGA TGCCAGCCCTCT-3′ (forward) and 5′-GACTGTCTGCAA CTGCCTGT-3′ (reverse) for the M_5 receptor (Gene Bank, NM 205783).
4. PCR amplification is carried out in a total volume of 20 μl containing Platinum SYBR Green qPCR SuperMix-UDG (Invitrogen). The assay is performed using an Opticon Chromo 4 real-time PCR detection system (Bio-Rad), and the PCR cycling program consisted of 2 min at 50 °C, 2 min at 95 °C, 50 cycles of 15 s at 95 °C, and 1 min at 61.4 °C. Plate reading is carried out at 61.4 °C. Melting curve analysis following PCR amplification confirms the specificity of the primers by detection of a single PCR product of the expected size on 2 % agarose gels.
5. PCR reactions are carried out in duplicate in 96-well plates. Standard curves are obtained using PCR fragments that had been isolated using a PCR purification kit, resuspended in 10 mM Tris-EDTA buffer (pH = 8.0), and quantified with a spectrophotometer to calculate cDNA concentration. Sample concentrations calculated from the standard curves are converted into copies of receptor cDNA per 1 μg RNA.

References

1. Caulfield MP, Birdsall NJ (1998) International union of pharmacology. VXII. classification of muscarinic acetylcholine receptors. Pharmacol Rev 50:279–290
2. Lanzafame AA, Christopoulos A, Mitchelson F (2003) Cellular signaling mechanisms for muscarinic acetylcholine receptors. Receptors Channels 9:241–260

3. Brodde OE, Michel MC (1999) Adrenergic and muscarinic receptors in the human heart. Pharmacol Rev 51:651–690
4. Dhein S, van Koppen CJ, Brodde OE (2001) Muscarinic receptors in the mammalian heart. Pharmacol Res 44:161–182
5. Yamada M, Inanobe A, Kurachi Y (1998) G protein regulation of potassium ion channels. Pharmacol Rev 50:723–760
6. Wang Z, Shi H, Wang H (2004) Functional M3 muscarinic acetylcholine receptors in mammalian hearts. Br J Pharmacol 142:395–408
7. Gallo MP, Alloatti G, Eva C, Oberto A, Levi RC (1993) M1 muscarinic receptors increase calcium current and phosphoinositide turnover in guinea-pig ventricular cardiocytes. J Physiol 471:41–60
8. Sharma VK, Colecraft HM, Wang DX, Levey AI, Grigorenko EV, Yeh HH, Shue SS (1996) Molecular and functional identification of m1 muscarinic acetylcholine receptors in rat ventricular myocytos. Circ Res 79:86–93
9. Hardouin S, Bourgeois F, Toraasson M, Oubenaissa A, Elalouf JM, Fellmann D, Dakhli T, Swynghedauw B, Moalic JM (1998) Beta-adrenergic and muscarinic receptor mRNA accumulation in the sinoatrial node area of adult and senescent rat hearts. Mech Ageing Dev 100:277–297
10. Hellgren I, Mustafa A, Riazi M, Suliman I, Sylvén C, Adem A (2000) Muscarinic M3 receptor subtype gene expression in the human heart. Cell Mol Life Sci 57:175–180
11. Wang H, Han H, Zhang L, Shi H, Schram G, Nattel S, Wang Z (2001) Expression of multiple subtypes of muscarinic receptors and cellular distribution in the human heart. Mol Pharmacol 59:1029–1036
12. Krejcí A, Tucek S (2002) Quantitation of mrnas for M(1) to M(5) subtypes of muscarinic receptors in rat heart and brain cortex. Mol Pharmacol 61:1267–1272
13. Pönicke K, Heinroth-Hoffmann I, Brodde O (2003) Demonstration of functional M3-muscarinic receptors in ventricular cardiomyocytes of adult rats. Br J Pharmacol 138:156–160
14. Willmy-Matthes P, Leineweber K, Wangemann T, Silber R, Brodde O (2003) Existence of functional M3-muscarinic receptors in the human heart. Naunyn Schmiedebergs Arch Pharmacol 368:316–319
15. Pérez CCN, Tobar IDB, Jiménez E, Castañeda D, Rivero MB, Concepción JL, Chiurillo MA, Bonfante-Cabarcas R (2006) Kinetic and molecular evidences that human cardiac muscle express non-M2 muscarinic receptor subtypes that are able to interact themselves. Pharmacol Res 54:345–355
16. Myslivecek J, Klein M, Novakova M, Ricny J (2008) The detection of the non-M2 muscarinic receptor subtype in the rat heart atria and ventricles. Naunyn Schmiedebergs Arch Pharmacol 378:103–116
17. Yue P, Zhang Y, Du Z, Xiao J, Pan Z, Wang N, Yu H, Ma W, Qin H, Wang W, Lin D, Yang B (2006) Ischemia impairs the association between connexin 43 and M3 subtype of acetylcholine muscarinic receptor (M3-mAChR) in ventricular myocytes. Cell Physiol Biochem 17:129–136
18. Wang Y, Hang P, Sun L, Zhang Y, Zhao J, Pan Z, Ji H, Wang L, Bi H, Du Z (2009) M3 muscarinic acetylcholine receptor is associated with beta-catenin in ventricular myocytes during myocardial infarction in the rat. Clin Exp Pharmacol Physiol 36:995–1001
19. Hang P, Zhao J, Qi J, Wang Y, Wu J, Du Z (2013) Novel insights into the pervasive role of M(3) muscarinic receptor in cardiac diseases. Curr Drug Targets 14:372–377
20. Wess J (2004) Muscarinic acetylcholine receptor knockout mice: novel phenotypes and clinical implications. Annu Rev Pharmacol Toxicol 44:423–450
21. Wess J, Eglen RM, Gautam D (2007) Muscarinic acetylcholine receptors: mutant mice provide new insights for drug development. Nat Rev Drug Discov 6:721–733
22. Stengel PW, Gomeza J, Wess J, Cohen ML (2000) M(2) and M(4) receptor knockout mice: muscarinic receptor function in cardiac and smooth muscle in vitro. J Pharmacol Exp Ther 292:877–885
23. Kitazawa T, Asakawa K, Nakamura T, Teraoka H, Unno T, Komori S, Yamada M, Wess J (2009) M3 muscarinic receptors mediate positive inotropic responses in mouse atria: a study with muscarinic receptor knockout mice. J Pharmacol Exp Ther 330:487–493
24. Harada N, Ochi K, Yaosaka N, Teraoka H, Hiraga T, Iwanaga T, Unno T, Komori S, Yamada M, Kitazawa T (2012) Immunohistochemical and functional studies for M_3 muscarinic receptors and cyclo-oxygenase-2 expressed in the mouse atrium. Auton Autacoid Pharmacol 32:41–52
25. Ochi K, Teraoka H, Unno T, Komori S, Yamada M, Kitazawa T (2013) A ganglionic stimulant, 1,1-dimethyl-4-phenylpiperazinium, caused both cholinergic and adrenergic responses in the isolated mouse atrium. Eur J Pharmacol 704:7–14

26. van Zwieten PA, Doods HN (1995) Muscarinic receptors and drugs in cardiovascular medicine. Cardiovasc Drugs Ther 9:159–167
27. Michel AD, Whiting RL (1988) Methoctramine reveals heterogeneity of M2 muscarinic receptors in longitudinal ileal smooth muscle membranes. Eur J Pharmacol 145:305–311
28. Araujo DM, Lapchak PA, Quirion R (1991) Heterogeneous binding of [3H]4-DAMP to muscarinic cholinergic sites in the rat brain: evidence from membrane binding and autoradiographic studies. Synapse 9:165–176
29. Lazareno S, Birdsall NJ (1993) Pharmacological characterization of acetylcholine-stimulated [35S]-GTP gamma S binding mediated by human muscarinic M1-M4 receptors: antagonist studies. Br J Pharmacol 109:1120–1127
30. Du XY, Schoemaker RG, Bos E, Saxena PR (1995) Characterization of the positive and negative inotropic effects of acetylcholine in the human myocardium. Eur J Pharmacol 284:119–127
31. Nishimaru K, Tanaka Y, Tanaka H, Shigenobu K (2000) Positive and negative inotropic effects of muscarinic receptor stimulation in mouse left atria. Life Sci 66:607–615
32. Nouchi H, Kaeriyama S, Muramatsu A, Sato M, Hirose K, Shimizu N, Tanaka H, Shigenobu K (2007) Muscarinic receptor subtypes mediating positive and negative inotropy in the developing chick ventricle. J Pharmacol Sci 103:75–82
33. Abramochkin DV, Tapilina SV, Sukhova GS, Nikolsky EE, Nurullin LF (2012) Functional M3 cholinoreceptors are present in pacemaker and working myocardium of murine heart. Pflugers Arch 463:523–529
34. Michel MC, Wieland T, Tsujimoto G (2009) How reliable are G-protein-coupled receptor antibodies? Naunyn Schmiedebergs Arch Pharmacol 379:385–388
35. Jositsch G, Papadakis T, Haberberger RV, Wolff M, Wess J, Kummer W (2009) Suitability of muscarinic acetylcholine receptor antibodies for immunohistochemistry evaluated on tissue sections of receptor gene-deficient mice. Naunyn Schmiedebergs Arch Pharmacol 379:389–395
36. Ito Y, Oyunzul L, Seki M, Fujino Oki T, Matsui M, Yamada S (2009) Quantitative analysis of the loss of muscarinic receptors in various peripheral tissues in M1-M5 receptor single knockout mice. Br J Pharmacol 156:1147–1153
37. Stengel PW, Yamada M, Wess J, Cohen ML (2002) M(3)-receptor knockout mice: muscarinic receptor function in atria, stomach fundus, urinary bladder, and trachea. Am J Physiol Regul Integr Comp Physiol 282:R1443–R1449
38. Tanaka H, Nishimaru K, Kobayashi M, Matsuda T, Tanaka Y, Shigenobu K (2001) Acetylcholine-induced positive inotropy mediated by prostaglandin released from endocardial endothelium in mouse left atrium. Naunyn Schmiedebergs Arch Pharmacol 363:577–582
39. Hara Y, Ike A, Tanida R, Okada M, Yamawaki H (2009) Involvement of cyclooxygenase-2 in carbachol-induced positive inotropic response in mouse isolated left atrium. J Pharmacol Exp Ther 331:808–815
40. Testa M, Rocca B, Spath L, Ranelletti FO, Petrucci G, Ciabattoni G, Naro F, Schiaffino S, Volpe M, Reggiani C (2007) Expression and activity of cyclooxygenase isoforms in skeletal muscles and myocardium of humans and rodents. J Appl Physiol 103:1412–1418
41. Kim JW, Baek BS, Kim YK, Herlihy JT, Ikeno Y, Yu BP, Chung HY (2001) Gene expression of cyclooxygenase in the aging heart. J Gerontol A Biol Sci Med Sci 56:B350–B355
42. Eliopoulos AG, Dumitru CD, Wang C, Cho J, Tsichlis PN (2002) Induction of COX-2 by LPS in macrophages is regulated by Tpl2-dependent CREB activation signals. EMBO J 21:4831–4840
43. Bałuk P, Gabella G (1990) Some parasympathetic neurons in the guinea-pig heart express aspects of the catecholaminergic phenotype in vivo. Cell Tissue Res 261:275–285
44. Hoard JL, Hoover DB, Mabe AM, Blakely RD, Feng N, Paolocci N (2008) Cholinergic neurons of mouse intrinsic cardiac ganglia contain noradrenergic enzymes, norepinephrine transporters, and the neurotrophin receptors tropomyosin-related kinase A and p75. Neuroscience 156:129–142
45. Hoover DB, Isaacs ER, Jacques F, Hoard JL, Pagé P, Armour JA (2009) Localization of multiple neurotransmitters in surgically derived specimens of human atrial ganglia. Neuroscience 164:1170–1179
46. Weihe E, Schütz B, Hartschuh W, Anlauf M, Schäfer MK, Eiden LE (2005) Coexpression of cholinergic and noradrenergic phenotypes in human and nonhuman autonomic nervous system. J Comp Neurol 492:370–379

Chapter 14

Muscarinic Receptor Gene Transfections and In Vivo Dopamine Electrochemistry: Muscarinic Receptor Control of Dopamine-Dependent Reward and Locomotion

Stephan Steidl, David Ian Wasserman, Charles D. Blaha, and John Yeomans

Abstract

Cholinergic neurons in laterodorsal (LDT) and pedunculopontine (PPT) tegmental nuclei respond to novel, arousing stimuli, then directly activate dopamine neurons, and increase dopamine outputs as measured by either in vivo microdialysis or by electrochemistry (described here). These mesopontine cholinergic neurons also directly activate superior colliculus and thalamic systems important for attention to novel stimuli, and for reward-seeking behaviors. M_5 muscarinic receptors that activate dopamine neurons and reward-seeking behaviors have been studied using pharmacology, knockout mice, oligonucleotide knockdown, and with electrochemistry and Herpes simplex viral gene transfections (HSV-M_5) protocols described here. Protocols for using HSV-M_5 genes, and designed M_4D and M_3D muscarinic receptor genes in behaving mice and for dopamine electrochemistry are presented, along with consequences for drug and gene therapy.

Key words Chronoamperometry, Viral transfections, DREADDs, Gene therapy, HSV, AAV, Reward, Locomotion, M_5 muscarinic receptors, Dopamine

1 Background and Historical Overview

Dopamine neural output is important for human diseases, such as Parkinson's, schizophrenia, and drug abuse. Drugs that influence dopamine output have powerful effects on behaviors in animals (such as locomotion, reward seeking, and motor control) and similarly powerful effects on confidence, mood and motor control in humans important for the therapeutic value of pharmaceuticals. We have studied how dopamine neural outputs and reward-seeking behaviors in rodents are influenced by cholinergic and GABAergic neurons of the midbrain and pons, and then how M_2, M_3, M_4, and especially M_5 muscarinic receptors and gene manipulations influence dopamine efflux and behaviors.

Jaromir Myslivecek and Jan Jakubik (eds.), *Muscarinic Receptor: From Structure to Animal Models*, Neuromethods, vol. 107, DOI 10.1007/978-1-4939-2858-3_14,

1.1 Localization of Muscarinic M_5 mRNA and Receptors in the Rodent Brain

The m5 receptor gene was discovered by Bonner's group [1], shortly after the identification of m1–m4 genes. Complete maps of m1–m5 mRNA expression are now available on the Allen Mouse Brain Atlas. Regional localization of M_1–M_5 proteins in brain stem has been reviewed [2].

Immunoprecipitation studies show that M_5 receptors account for only about 1 % of all muscarinic receptors in the rat brain. Low receptor numbers make it difficult to localize M_5 receptors by immunohistochemistry [3, 4]. Expression of m5 mRNA is localized to ventral tegmental area (VTA) and substantia nigra, zona compacta (SNc). Following 6-hydroxydopamine dopamine lesions m5 mRNA is no longer detected in midbrain [5, 6]. This suggests that excitatory M_5 receptors are important for activation of dopamine neurons, and inspired the creation of M_5 knockout (M_5-KO) mice [7, 8].

1.2 Muscarinic Receptors in VTA Activate Dopamine Rewards

Cholinergic input to dopamine reward systems was first studied using VTA microinjections in rats bar-pressing for rewarding hypothalamic brain stimulation. Microinjections of acetylcholine, nicotine or carbachol into the VTA were found to increase sensitivity to rewarding brain stimulation [9–11], as well as to increase mesolimbic dopamine efflux [12]. Microinjections of muscarinic antagonists into the VTA were found to reduce sensitivity to rewarding brain stimulation more than nicotinic antagonists [13]. Similarly, selective knockdown of M_5 muscarinic receptors with antisense oligonucleotides in the VTA reduced sensitivity to rewarding brain stimulation [14].

Food- and opioid-motivated rewards were also strongly reduced by these same muscarinic antagonists in the VTA [15, 16]. Acetylcholine levels in VTA are increased during eating, drinking and brain-stimulation rewards as measured by in vivo microdialysis [17].

1.3 Ch5 and Ch6 Cholinergic Neurons Activate Dopamine Neurons

Cholinergic neurons (Ch1–8) are defined by the presence of the synthesizing enzyme choline acetyltransferase (ChAT) [18]. The only direct projections of cholinergic neurons to the VTA and SNc are from pedunculopontine (PPT; Ch5) and laterodorsal tegmental nuclei (LDT; Ch6) [19, 20]. The direct projections to VTA dopamine neurons from these nuclei are especially from LDT and caudal PPT cholinergic and glutamate neurons [21, 22]. The Ch5/Ch6 cholinergic projections to VTA and SNc are less dense than to thalamus or superior colliculus intermediate layers, however.

These anatomical connections suggest that cholinergic activation of dopamine systems is part of a larger cholinergic arousal system [2]. So, novel and rewarding stimuli activate fast saccadic eye movements and approach turns via the superior colliculus, facilitated by nicotinic input from Ch5 [23, 24]. At the same time, Ch5/6 cholinergic activation of thalamus facilitates cortical arousal and attention to these stimuli [25] via fast nicotinic and slower M_1-type muscarinic receptors [26, 27]. Ch5 and Ch6 neurons also provide weaker anatomical projections to Ch1–4 basal forebrain nuclei supporting cortical arousal [28–30]. Finally, SNc and VTA dopamine activation provides sustained motor arousal and motivated reward-seeking behaviors.

Excitotoxic lesions of PPT and LDT neurons reduce behavioral arousal and reward sensitivity in several ways. For example, bilateral lesions of caudal PPT block acquisition of brain-stimulation reward [31] and morphine conditioned place preference [32]. Cholinergic output can be inhibited pharmacologically in PPT or LDT by local infusion of the cholinergic agonist carbachol, which inhibits Ch5 and Ch6 neurons via M_2 and M_4 inhibitory receptors [33, 34]. Carbachol in the PPT similarly inhibits dopamine output [35], brain-stimulation reward or locomotion [36, 37]. By contrast, muscarinic antagonists in the PPT or LDT strongly facilitate dopamine outputs, rewarding brain stimulation, or locomotion, apparently via M_2 receptors [38]. M_4 receptors in VTA also appear to inhibit ACh release from VTA cholinergic terminals [39].

1.4 Mesopontine Activation of Dopamine Output via M_5 Muscarinic Receptors in Tegmentum

In vivo electrochemistry allows temporal resolution of the many different neural pathways and receptors that influence dopamine transmission (e.g., ref. 38). Following LDT or PPT electrical stimulation (see Protocol and Fig. 1) dopamine levels change in three phases in rats: (1) Dopamine increases for 2–4 min due to both VTA nicotinic and AMPA glutamate receptors, (2) Dopamine decreases for 5–10 min due to LDT/PPT M_2-like receptors, followed by (3) sustained increases in dopamine (from 10 to 60 min post stimulation), due to M_5 muscarinic receptors in VTA or SNc [38, 40]. The third phase is completely removed by muscarinic receptor antagonists infused into the VTA or SNc, or by m5 gene knockout (Fig. 1a; [38, 40–42]). Therefore, M_5 receptors on dopamine neurons have very long-lasting and powerful activating effects on dopamine outputs to the nucleus accumbens and striatum.

Although excitatory M_3 receptor mRNA is found near orexin/hypocretin neurons in the lateral hypothalamus, carbachol inhibits, and the M_3 antagonist 4-DAMP excites, orexin neurons [43]. Therefore M_3 receptors may excite GABA interneurons in the hypothalamus as they do in midbrain near dopamine neurons [44].

2 Electrochemical Methods to Study Muscarinic Modulation of Brain Dopamine Signaling

Oxidization of an electroactive compound such as dopamine results in measurable current flow (Fig. 1b bottom). The methods outlined below describe an experimental setup that uses chronoamperometry

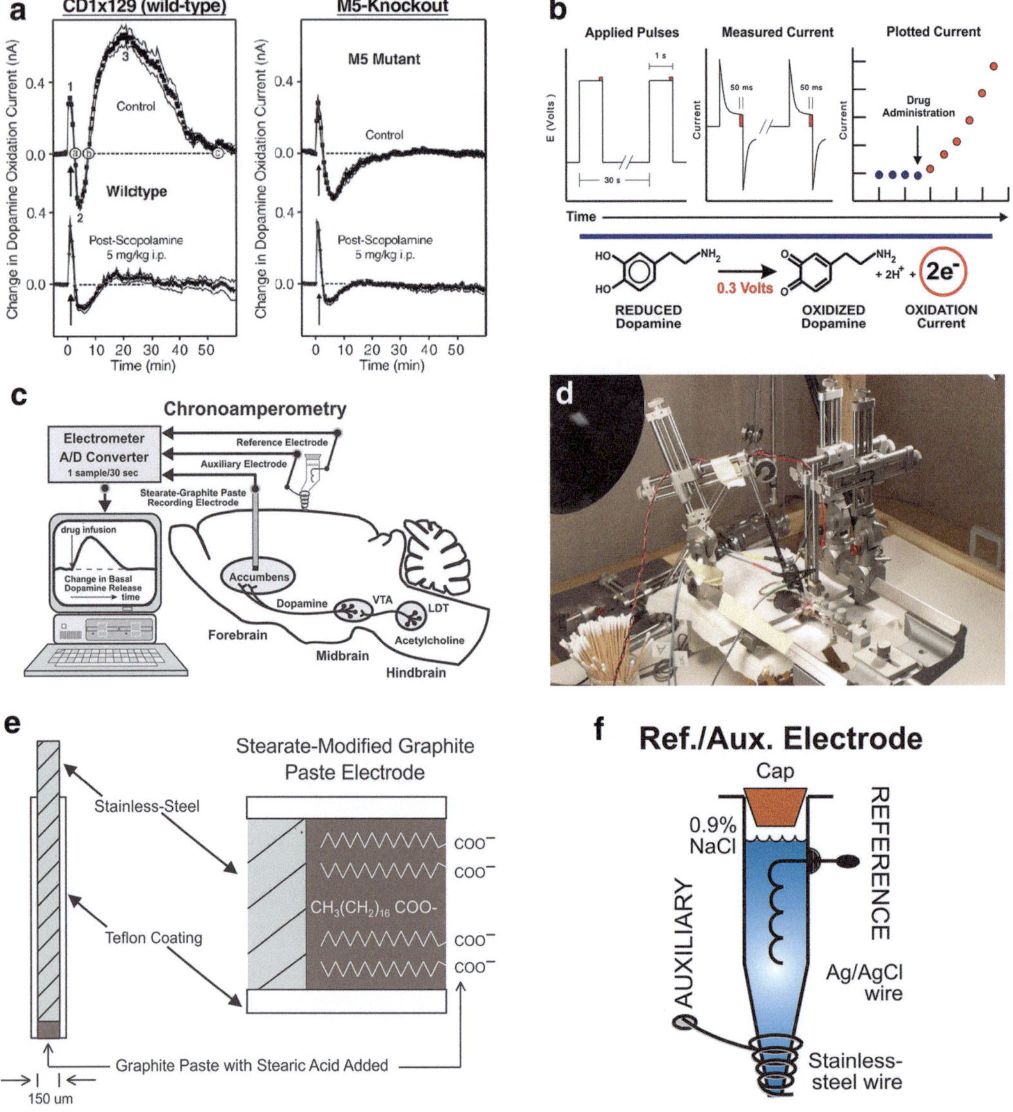

Fig. 1 (**a**) Nucleus Accumbens (NAc) Dopamine Efflux Measured by Chronoamperometry in Mice after Electrical Stimulation of the LDT (*arrow*). *Left Panel Top*: In wild-type mice LDT stimulation induces a rapid increase in NAc dopamine (1) followed by a decrease in dopamine efflux from 3 to 8 min (2). Dopamine efflux is again increased from 10 to 60 min following LDT stimulation (3). *Left Panel Bottom*: Systemic pretreatment with the muscarinic receptor antagonist scopolamine in wild-type mice selectively blocks the third phase of increased dopamine efflux. *Right Panel*: In M_5 knockout mice the third phase of increased dopamine efflux following LDT electrical stimulation is completely absent and is not further reduced by systemic scopolamine pretreatment (Figure adapted from [41]). (**b**) In chronoamperometry the applied potential to the working electrode is near-instantaneously stepped up from a resting potential to a value above the dopamine oxidation potential (−0.15 V to 0.25 V) for 1 s duration at a fixed interval. To measure dopamine with carbon paste electrodes, the potential is instantly stepped from a resting value of −0.15 V to 0.25 V for 1 s, every 30–60 s (*left*). As the potential is stepped up, the recorded current increases dramatically, and this is due to charging current of the electrode and oxidation of the electroactive species. The charging current decays as the potential is held, and the faradaic (i.e., dopamine oxidation) current (*bottom*) is measured and integrated over the final 50 ms of the 1000 ms pulse (*middle*; [45]). Oxidation current is then plotted across time. Dopamine oxidation current is proportional to in vivo dopamine concentration (*right*). (**c**) Chronoamperometry in combination with carbon paste electrodes to measure in vivo dopamine: the stearate-modified graphite paste recording electrode is implanted in the

(one of several available electrochemical techniques) in combination with stearate-modified graphite paste electrodes to measure dopamine in vivo. Three electrodes are needed: the working electrode is implanted into the brain area of interest (see section 2.2.2) and a combined reference and auxiliary electrode is placed into contact with the cortex (see section 2.2.3). A potentiostat, or electrometer, provides a circuit between the three electrodes and allows for the application of a voltage to the recording electrode via the auxiliary electrode, as well as the maintenance of a potential difference between the recording and reference electrodes (Fig. 1c; [45]). When a positive voltage of sufficient amplitude is applied to the working electrode it causes the oxidation of electroactive compounds (e.g., dopamine) at its surface. Oxidation results in the transfer of electrons producing a measurable current (Fig. 1b bottom). The current flow, termed the faradaic current, is proportional to the concentration of the neurochemical of interest [45, 46]. In chronoamperometry the applied potential is near-instantaneously stepped from a resting value of −0.15 V to 0.25 V for 1 s, every 30–60 s. The faradaic (i.e., oxidation) current is measured and integrated over the final 50 ms of the 1000 ms pulse (Fig. 1b; [45]).

2.1 Materials

2.1.1 Stereotaxic Equipment

1. Stereotaxic frame (Model 1504 with Model 1211 base plate; this stereotaxic frame allows for the use of 3–4 manipulator arms simultaneously), mouse or rat nose bar and rat ear bars (model 855) produced by David Kopf Instruments (Fig. 1d).
2. Stereotaxic drill (Model 1471 with Model 1469 stereotaxic holder) produced by David Kopf Instruments and trephine drill bits (Harvard Apparatus).
3. Stereotaxic cannula holder (Model 1776-P1) produced by David Kopf Instruments.

Fig. 1 (continued) brain area of interest (e.g., the nucleus accumbens), a combined reference and auxiliary electrode is placed into contact with the cortex (opposite hemisphere). A potentiostat, or electrometer, provides a circuit between the three electrodes and allows for the application of a voltage potential to the recording electrode via the auxiliary electrode, as well as the maintenance of a potential difference between the recording and reference electrode [45]. When a positive voltage is applied to the working electrode of sufficient amplitude it causes the oxidation of dopamine at its surface. (**d**) Anesthetized animal recordings should be performed in a standard Faraday cage. Place the stereotaxic apparatus inside the cage and place the electrometer outside of the cage. The stereotaxic apparatus shown here allows for the use of 3–4 manipulator arms simultaneously. At minimum, one carrier is needed to hold the stearate-modified graphite paste recording electrode and a second carrier is needed to hold the combined reference and auxiliary electrode. Additional carrier arms then allow for mounting stimulating electrodes (i.e., for PPT/LDT electrical stimulation) or injection cannulae (for intracranial microinjection of drugs or pharmacological agents). (**e**) Stearate-modified carbon paste electrode constructed from a Teflon-coated stainless steel wire. Extruding the Teflon coating ~1 mm beyond the tip of the stainless-steel wire creates a well that is filled with stearate-modified graphite paste. (**f**) Diagram showing a combined reference/auxiliary electrode

4. Various surgical instruments: scalpel, forceps, small animal clippers etc. (Fine Science Tools).
5. Hamilton syringes (Model 7002KH), injector tubing, guide cannulae (26 gauge), injector cannulae (33 gauge; Plastics One Inc.) and a micro-infusion pump (Harvard Apparatus).
6. Various surgical supplies: gauze, cotton tip applicators, betadine, latex gloves, etc.
7. Urethane for anesthesia (Sigma-Aldrich).
8. Temperature-regulated heating pad (e.g., TC-1000; CWE Inc., New York, NY).

2.1.2 Equipment for Construction of Stearate-Modified Graphite Paste Recording Electrodes and Reference/Auxiliary Electrodes

1. Glass mortar and pestle, a 25 ml beaker, a 250 ml beaker, appropriately sized magnetic stir bar, a glass capillary heat-sealed on one end, and a stirring hotplate.
2. Stearate (99.9 % purity), silicone oil, and graphite powder (particle size < 10 μm; all available from Sigma-Aldrich), 0.9 % saline solution.
3. Teflon-coated stainless-steel wire (0.008″ o.d.; Medwire, Mount Vernon, NY), silver wire (0.008″ o.d.; A&M Systems Sequim WA), stainless steel wire (0.008″ o.d.; A&M Systems Sequim WA).
4. #5 forceps, a glass plate, masking tape, #10 scalpel blade, Teflon tape, a compound microscope that will allow for viewing of the electrode tip, a 9 V battery, black wax, a needle tip, standard 1 ml pipette tip, epoxy.

2.1.3 Materials for In Vitro Calibration of Graphite-Paste Electrodes

1. 0.01 M phosphate-buffered saline solution, dopamine hydrochloride, perchloric acid (Sigma-Aldrich), and double distilled water.

2.1.4 Materials for In Vivo Electrochemical Recordings in Anesthetized Rats or Mice

1. Bipolar concentric stimulating electrode (e.g., SNE-100, Rhodes Medical Co., Woodland Hills, CA or CBARD75, FHC, Bowdoin, ME).
2. Programmable pulse generator (e.g., Master 9, AMPI, Jerusalem, Israel) connected to a stimulus isolator (e.g., ISO-Flex, AMPI, Jerusalem, Israel) to apply stimulation current pulses.
3. Electrometer (e.g., EChempro, GMA Technologies, Vancouver, Canada).

2.2 Methods

2.2.1 Preparation of Stearate-Modified Graphite Paste

Graphite paste electrodes are treated to enhance selectivity for dopamine by incorporating a fatty acid (stearic acid) into the graphite paste [46, 47]. The fatty acid provides an anionic recording surface that slows down the electron transfer kinetics of anions (i.e., ascorbic acid and dopamine metabolites, such as DOPAC [45]).

1. Using the 25 ml glass beaker and stir bar dissolve 75 mg of stearate in 1 ml silicone oil heated to 40 °C (Note 1).

2. Once the stearate crystals are fully dissolved remove from heat and transfer contents of beaker into glass mortar. Add 1 g of graphite powder and thoroughly mix using the glass pestle. The consistency of the mixture will start off paste-like and with additional mixing will become powdery.
3. Transfer stearate-modified graphite paste to a light-proof glass vial for storage.

2.2.2 Construction of Stearate-Modified Graphite Paste Recording Electrodes

1. Cut a length of Teflon-coated stainless-steel wire from the spool using wire cutters. Trim the length of wire to ~10 cm using a #10 scalpel blade. Straighten the cut length of wire carefully.
2. Strip ~1 cm of Teflon coating form one end of the wire. This will provide the top end of the implantable electrode where the electrometer can be connected.
3. Use a pair of #5 forceps to grip exposed 1 cm length of stainless-steel wire. Holding the electrode by the forceps, gently pass the coated portion between thumb and forefinger to loosen the Teflon coating. Be careful not to pull the Teflon coating off the stainless-steel wire.
4. Extrude the Teflon coating 2–3 mm past the tip of the stainless-steel wire on the opposite end.
5. While holding the uncoated portion of the wire between the thumb and forefinger of one hand, lay the wire flat on the glass plate, maintaining a loose grip on the wire. Use a thin-edged razor blade to cut the well's surface. Hold the blade perpendicular to the wire approximately 5 mm behind the end of the stainless-steel wire on the end where you previously extruded the Teflon coating. Gently lower the blade onto the wire and let blade rotate the wire across the glass surface for one full rotation. Remove the excess Teflon using forceps.
6. Hold the stainless-steel wire at the electrode contact end in one hand. Then with the thumb and forefinger of the other hand around the center of the electrode gently push the remaining Teflon coating so that the edge of the well extrudes ~1 mm beyond the tip of the wire (Fig. 1e). Examine the well surface under a microscope. The well should have a clean and unblemished concentric Teflon surface. And uneven cut will create Teflon spirals visible under the microscope. If this occurs cut another well as described above.
7. Grip and hold the wire firmly using #5 forceps at a distance from the well surface that is about 2–3 mm over the depth the electrode is to be implanted (~1 cm for a mouse). Loop the wire tightly around the tip of the forceps for two complete revolutions. Carefully slide out the tip of the forceps. Wrap a small (3–4 mm wide × 2 cm long) strip of masking tape around the loop.

8. Place a small quantity of the stearate-modified graphite paste mixture onto the clean surface of a glass plate. If needed clean the glass surface with ethanol. Holding the electrode around the contact end between thumb and forefinger gently push the well into the graphite paste mixture to fill the well. Avoid large particles of graphite powder and do not apply pressure as this will crush the fragile well.
9. Clean the excess paste from the well's side and surface by gently rubbing the tip across a suspended piece of Teflon tape.
10. Then, hold the electrode perpendicular to the glass surface (well facing down) and drop the electrode onto the glass surface to tightly pack the graphite paste into the Teflon well (Fig. 1e).
11. Examine the electrode's surface under a microscope. The graphite surface should appear shiny and smooth and the walls of the well's surface should be free of graphite. If cracks or grooves are observed additional packing and/or additional graphite paste may be required (Note 2).

2.2.3 Construction of Reference/Auxiliary Electrode (Fig. 1f)

1. Use gloves to handle silver wire. Wrap a length of silver wire (~10 cm) around the shaft of a stereotaxic drill bit to achieve about 10–15 closely spaced, but not touching, loops.
2. Place the coil in the 250 ml beaker filled with 0.9 % saline ensuring the entire coil is submerged. Attach the coil to the anode (+) of the 9 V battery. Attach a second short length of bare silver wire to the cathode (−) of the battery. The process of silver chloriding the wire will darken the surface of the coil. Set aside the coil in saline until ready for use.
3. Obtain a standard 1 ml pipette tip. Use a heated 18 g needle tip to burn a hole into the side of the pipette tip ~1 cm from the top rim. Next, melt a circular groove around the circumference of the pipette tip at the level of the hole. Melt a second circular groove just above where the pipette tip tapers.
4. Insert the silver/silver chloride coil into the pipet and feed the excess wire on one end through the hole. Cover the hole with black wax to secure the position of the coil.
5. Obtain a 30 cm piece of stainless-steel wire and tightly wrap it around the exposed piece of Ag/AgCl wire. Then wrap this wire around the pipette (starting in the groove) and a closely placed dowel to form a loop. This will provide a contact for the electrometer to connect to the Ag/AgCl reference electrode.
6. Obtain a second 30 cm piece of stainless-steel wire. Coil this wire around the groove created just above the tapered end of the pipette tip leaving plenty of access on one end. Start looping the wire downward to the tip creating a coil on the way down. Wind the wire around the groove at the tip several times

and begin looping upward again to the middle groove. Wrap around the middle groove several times and twist remainder around the other end of the wire left in place previously. Then wrap this wire around the pipette and a closely placed dowel to form a loop. This will provide a contact for the electrometer to connect to the auxiliary electrode. Cover the middle groove and wire with epoxy to ensure the position of the auxiliary electrode is secure.

7. Fill interior of pipette with 0.9 % saline while epoxy is drying. Insert an appropriately sized rubber stopper in the top end of the pipette tip (Fig. 1f). Set aside until ready to use (Note 3).

2.2.4 In Vitro Testing of Graphite Paste Electrodes

1. Place a glass container containing 15 ml of a 0.01 M phosphate-buffered saline solution on a battery-operated magnetic stirrer.
2. Submerge the graphite paste electrode and the combination reference/auxiliary in the solution.
3. Use the electrometer to obtain a linear sweep voltammogram by ramping the potential applied to the working electrode from −0.15 V to 0.5 V vs. the Ag/AgCl electrode at a rate of 10 mV/s.
4. Add discrete quantities of a 2 mM solution of dopamine (37.9 g of dopamine hydrochloride dissolved in 90 ml double distilled water and 10 ml of 0.1 M perchloric acid) to achieve a 1 μM concentration of dopamine in the phosphate-buffered saline solution. Stir the solution gently for a period of 5 s (Note 4).
5. Obtain another linear sweep voltammogram. Repeat the process for between 3 and 5 additions of 1 μM dopamine to confirm, first, that peak current was always obtained at the same potential and, second, that the relative increases in peak current were consistent across consecutive 1 μM additions of dopamine.

2.2.5 In Vivo Electrochemical Recordings in Anesthetized Rats or Mice

1. Anesthetized animal recordings should be performed in a standard Faraday cage. Place the stereotaxic apparatus inside the cage. Place the electrometer outside of the cage.
2. Anesthetize mice or rats with urethane (1.5 g/kg, i.p.; initial dose supplemented with 0.3 g/kg 30 min later).
3. Secure mouse or rat in a stereotaxic frame. Maintain body temperature at 37 ± 0.5 °C with a temperature-regulated heating pad.
4. Drill a hole into the skull above the brain area of interest (e.g., NAc or dorsolateral striatum) large enough to accommodate the graphite paste recording electrode. Drill a second larger hole to accommodate the tip of the combined reference/auxiliary electrode. This hole should be placed so that the

combined reference/auxiliary electrode is in contact with the opposite brain hemisphere at a distance that does not interfere with placement of the recording electrode.

5. Mount the combined reference/auxiliary electrode on a stereotaxic carrier and connect the electrometer. Ensure the reference/auxiliary electrode is completely filled with 0.9 % saline (air bubbles commonly occur at the tip). Place a drop of saline on to the hole and place the reference/auxiliary electrode into contact with the cortical surface.
6. Attach the graphite paste recording electrode to another stereotaxic carrier and lower into the brain area of interest (e.g., NAc) according to coordinates obtained from a stereotaxic brain atlas. Ensure that the tip of the electrode does not come into contact with blood.
7. Attach the electrometer to the recording electrode (Fig. 1d).
8. Perform repetitive chronoamperometric measurements of oxidation current by applying potential pulses from −0.15 to 0.30 V to the recording electrode (vs. reference electrode) for 1 s duration at 30 or 60 s intervals.

2.2.6 Using Muscarinic Knockout Mice to Study the Role of Muscarinic Receptor Signaling in Mesopontine Excitation of Mesolimbic or Nigrostriatal Dopamine Signaling

1. Obtain muscarinic receptor knockout mouse and wild-type controls (available from Taconic Bioscience, Inc.).
2. Secure anesthetized mouse in stereotaxic frame as described above.
3. Drill an additional hole into the skull above the LDT or PPT and lower a bipolar concentric stimulating electrode into either LDT or PPT. Use a programmable pulse generator connected to a stimulus isolator to apply current pulses with the following parameters: 1 s, 35 Hz train of 400 μA pulses (1 s intertrain interval) applied over a 60 s period (1050 pulses in total). These parameters are designed to mimic spontaneous firing patterns of PPT neurons in awake, naturally aroused animals [25].
4. Comparisons of electrically evoked nucleus accumbens or dorsal striatum dopamine efflux between wild-type and knockout mice will reveal the contribution of muscarinic receptors to each of the three phases of dopamine efflux [41].

2.2.7 Using In Vivo Pharmacology to Study the Role of Midbrain Muscarinic Receptor Signaling in Mesopontine Excitation of Mesolimbic or Nigrostriatal Dopamine Signaling

1. Secure anesthetized mouse or rat in stereotaxic frame as described above.
2. Drill an additional hole above the VTA (when studying LDT electrically evoked nucleus accumbens dopamine efflux) or the SNc (when studying PPT electrically evoked dorsal striatal dopamine efflux) to allow for insertion of a guide cannula. The guide cannula can be held by a stereotaxic carrier or can be chronically implanted by securing it to the skull surface using a combination of stainless-steel jeweler's screws and dental cement.

3. Apply electrical stimulation to the LDT or PPT and observe the triphasic response pattern [38, 40–42].
4. Infuse a muscarinic receptor antagonist, or vehicle, into the VTA or SNc (Note 5) and repeat electrical stimulation.
5. Compare the predrug, vehicle-treated, and antagonist-treated response patterns to determine the contributions of muscarinic receptors.

2.3 Notes

1. Be careful not to overheat the silicone oil. This is easily accomplished by removing the beaker from the heat source the moment the last few crystals of stearate are observed to dissolve into the silicon oil.
2. The microscope should allow for viewing the tip of the graphite paste electrode. Wells have to be recut or repacked if necessary.
3. The saline solution that fills the pipette tip should be replaced regularly. When not in use the tip of the reference/auxiliary electrode should be submerged in saline solution.
4. Prepare fresh dopamine solution for in vitro calibration.

3 Opioid Rewards and Mesopontine/M_5 Activation of Dopamine Neurons

M_5-KO mice show deficits in responding to morphine, either in conditioned place preference or in morphine-induced locomotion tasks, but not stimulant- or saline-induced locomotion tasks [48–51]. Similarly, PPT or LDT lesions significantly reduce morphine-induced dopamine release [52, 53]. Finally, M_5-KO mice lose their dopamine response to morphine entirely [42]. This suggests that morphine acts indirectly through the PPT/LDT to M_5 pathway to activate dopamine neurons, and for dopamine-dependent behavioral effects of morphine.

These effects of morphine depend on μ-opioid receptors in the ventral tegmental region that inhibit GABA neurons. The most critical GABA neurons for opioid rewards and for morphine-induced locomotion are located just caudal to VTA in the rostromedial tegmental nucleus (RMTg; [22, 54]). RMTg GABA neurons express together μ-opioid, nociceptin, $GABA_A$ and M_4 muscarinic receptors, all of which are normally inhibitory. We have proposed that together these receptors inhibit GABA neurons and thereby disinhibit cholinergic and dopaminergic output neurons needed for opioid reward and dopamine effects [55].

To excite VTA and RMTg neurons, we infused HSV-M_5-GFP bilaterally into M_5-KO and wild-type mice (Fig. 2) (See protocol below). Viral infections work best when infecting cells, not axons, due to the larger surface area of dendrites and somata which clearly

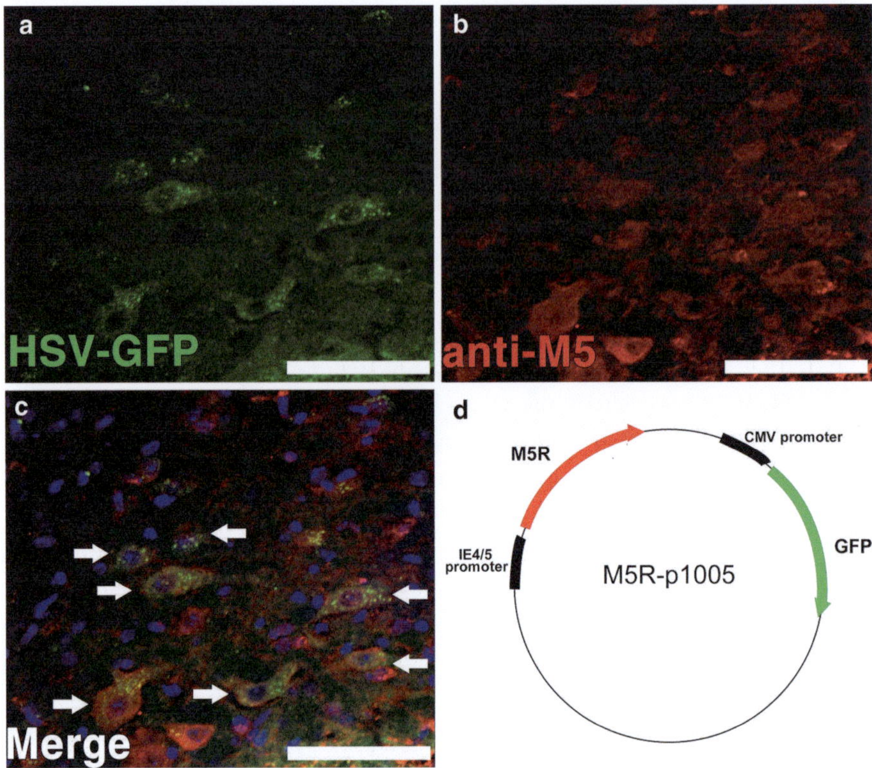

Fig. 2 (**a–c**) HSV-M_5-GFP Transfected Neurons and M_5 Expression. The HSV-M_5-GFP vector was infused into VTA sites in M_5-KO mice. Expression of viral GFP can be seen in (**a**). Sections were double-labeled with M_5 antibodies (*red*: **b**) and DAPI. (**c**) Merged image displaying triple-labeled cells (*purple nucleus*, with *yellow spots* in cytosol) of viral GFP, M_5-expression, and DAPI (20× magnification, bar represents 75 μm) All cells labeled with viral GFP were also M_5 positive (*white arrows* indicate triple-labeling). As the mouse was an M_5 knockout, all observed M_5 expression was virally mediated. (**d**) Schematic of the M_5R-p1005 Vector. Expression of M_5R was driven by a constitutive IE4/5 promoter while GFP expression was driven by a CMV promoter. Replication-deficient Herpes simplex virus (*HSV*)-derived particles were made from this vector as previously described ([60]; Figure modified from [22])

express GFP after viral transfections (Fig. 2a–c). This HSV-M_5 method relies on expression of extra M_5 receptors, so that endogenous activation of cholinergic neurons and endogenous ACh release is needed to activate the M_5 receptors.

Although these two sites are less than 1 mm apart, VTA infections strongly facilitated morphine-induced locomotion, while RMTg infections blocked morphine-induced locomotion (Fig. 3). Both VTA and RMTg sites receive direct projections from many LDT and caudal PPT cholinergic neurons [55]. Therefore, the same populations of PPT and LDT neurons that excite VTA and SNc neurons via M_5 muscarinic receptors can simultaneously inhibit RMTg GABA neurons via inhibitory muscarinic M_4 receptors. This suggests that morphine-mediated inhibitory inputs

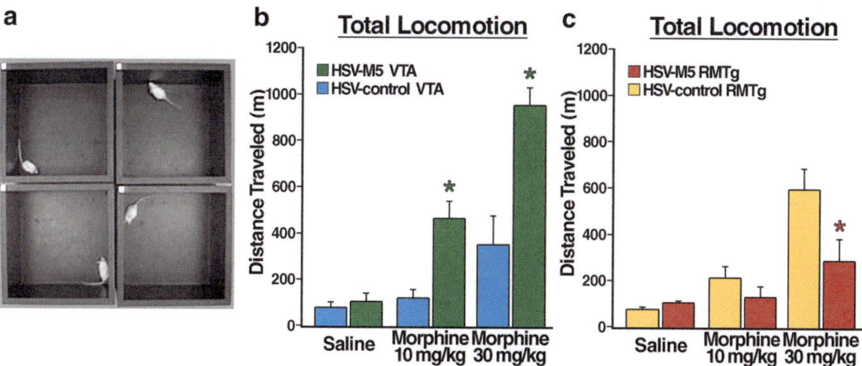

Fig. 3 (**a**) Open-field apparatus for locomotion. Mice were placed in the center of arenas while a video camera recorded their horizontal movements. A computerized tracking program, Noldus Ethovision, thereafter scored the video and calculated the distance traveled. (**b–c**) Total saline and morphine-induced locomotion following HSV transfections. (**b**) In VTA, HSV-M_5 infusions (*green*) significantly increased distance traveled at both morphine doses over HSV-control vector (*blue*) over the 2 h period. No differences were observed for saline. (**c**) In RMTg, HSV-M_5 infusions (*red*) significantly reduced total locomotion over HSV-control vector (*orange*) only at 30 mg/kg. (M_5-VTA, $n=7$; M_5-RMTg, $n=15$; Control-VTA, $n=6$, Control-RMTg, $n=7$.) (figure modified from [55])

to RMTg are facilitated by M_4 inhibition at the same time that cholinergic activation of dopamine neurons occurs. Then RMTg GABA connections to PPT and VTA release these neurons from inhibition to facilitate the arousing and rewarding outputs of cholinergic and dopaminergic systems. Therefore, the critical roles of RMTg inputs, and of caudal PPT, LDT and M_5 receptor outputs, to opioid reward and locomotion, appear to depend on close interconnections, supported by appropriate muscarinic receptor modulations.

These muscarinic effects can be studied by local site-specific viral transfections of neurons, or by neuron-specific viral transfections, using TH::Cre, ChAT::Cre or GAD2::Cre mouse lines. To identify which RMTg GABA neurons are critical for morphine-induced locomotion, we used AAV-M_3D and AAV-M_4D to activate or inhibit mCherry-labeled GABA RMTg neurons in GAD2::Cre mice. This method for activating genetically modified muscarinic receptors (i.e., M_3D for excitation like M_5, and M_4D for inhibition like M_4) relies on exogenously administered clozapine-N-oxide (CNO) [56]. AAV induces long-term infections of identified neurons, while systemic CNO control of these neurons occurs via muscarinic receptors with mutant binding sites. For future gene therapy applications, this is a powerful method for remote control of specifically targeted brain neurons (see protocol below).

4 Local Gene Control with HSV-M_5-GFP Viral Transfections Including Cell-Specific Control with AAV-M_3D, AAV-M_4D

4.1 Materials

4.1.1 Stereotaxic Equipment

1. Stereotaxic frame, mouse nose bar (Model 926) and rat ear bars (model 855) produced by David Kopf Instruments.
2. Stereotaxic drill and drill bits.
3. Stereotaxic cannula holder (Model 1776-P1) produced by David Kopf Instruments.
4. Various surgical instruments: scalpel, forceps, etc. (Fine Science Tools).
5. 10 μl Hamilton microsyringes (Model 7002KH) connected with Tygon tubing (0.02 mm i.d.) to injector cannula (33 gauge) produced my Plastics One Inc.
6. Small animal clippers.
7. Various surgical supplies: gauze, cotton tip applicators, betadine surgical scrub, nitrile gloves, etc.
8. Isoflurane for anesthesia (Sigma-Aldrich).
9. Temperature-regulated heating pad (e.g., TC-1000; CWE Inc., New York, NY).
10. Syringe pump.
11. Mineral oil.

4.1.2 Equipment Needed for Behavioral Testing of Open-Field Locomotion

1. Open-field arena(s) (measuring 31 × 31 × 31 cm).
2. Video camera suspended above open-field arena(s).
3. Video tracking software (e.g., Noldus Ethovision V7.0 (Groningen, Netherlands)).

4.1.3 Materials for HSV-M_5 Transfections in M_5-KO or WT Mice

1. M_5-KO mice and wild-type controls (CD1/129).
2. M_5-HSV-GFP and M_5-control-GFP (see Note 5 for details on HSV vectors).
3. Morphine sulfate pentahydrate (Sigma, St. Louis, MO) (Note 6).

4.1.4 Materials for AAV-M3D or M4D Transfections in GAD2::Cre Mice

1. GAD2::Cre mice (Jackson Laboratory, Gad2$^{tm2(cre)Zjh}$/J, 010802).
2. AAV-FLEX-M_3D-mCherry or AAV-FLEX-M_4D-mCherry (see Note 7 for details on AAV vectors and DREADDs).
3. Morphine sulfate pentahydrate (Sigma, St. Louis, MO) (Note 8).
4. Dimethyl sulfoxide (DMSO).
5. Clozapine-N-oxide (CNO) (Note 9).

4.2 Methods

4.2.1 Viral Transfections in Mice

1. Fill Hamilton syringe/Tygon tubing/injector cannula assembly with mineral oil (or water) (Note 10). Place filled Hamilton syringe into syringe-pump and eject 4–10 μl. Wipe tip of injector, and withdraw 0.5–1.0 μl to create an air bubble. Place injecNor cannula into virus solution, and slowly load vector (rate of 0.1–3.0 μl/min). To conserve virus, consider loading enough viral vector to perform several surgeries (Note 11).
2. Anesthetize mice based on an approved Institutional Animal Care and Use Committee (IACUC) protocol. Anesthesia should be induced at 3 % isoflurane and maintained at 1.5 % isoflurane delivered via a vaporizer at 0.7 l/min O_2.
3. Secure mouse in a stereotaxic frame. Maintain body temperature at 37 ± 0.5 °C with a temperature-regulated heating pad.
4. Drill burr holes into the skull above the brain area of interest large enough to accommodate an injector cannula. Slowly lower injectors into burr holes to desired D/V coordinates into site of interest.
5. Begin injecting virus slowly (0.1–3.0 μl/min) until the desired volume of the virus has been dispensed 0.1–1.0 μl of virus. Note: volume and rate of injection can depend on specific type of virus, titer, desired spread, and limitations of syringe pump.
6. Allow for the injector cannula to remain in place for an additional 5–10 min to ensure vector diffusion.
7. Following diffusion period, slowly raise the cannula out of the skull.
8. Depending on experimental design, repeat steps 5–7 for alternate hemisphere or additional sites.
9. Once all viral injections are complete, retract the cannula, suture the wound, and apply standard post-operative care.
10. Depending on biosafety protocol, segregate cages from colony under HEPA-filtered cage tops for 48 h.

4.2.2 Behavioral Testing: Open-Field Locomotion

Open-field locomotion is a reliable measure of drug effect in rodents. Drug-induced hyperactivity is easy to quantify and has been shown, in rodents, in response to several different drugs of abuse. This drug-elicited, goal-directed behavior is an adaptive survival mechanism to encourage exploration, and risk taking. Increases in drug-induced locomotion often reflect activation of the mesolimbic DA system. Thus, running can be taken as a behavioral index reflective of the functioning and integrity of the DA system [57].

While there are several other methods to measure open-field locomotion (automated photocell cages, scored by observation), here we will be discussing video-recorded trials and offline-scoring using Noldus Ethovision tracking software (Fig. 3a) [22, 50].

1. Behavioral experiments should always be carried out at the same time of the day taking into account that the animals show minimal motor activity around noon under a light regime with lights on from 6 a.m. to 6 p.m.
2. To ensure best results, perform open-field testing in a room with dim red lighting and with a 70 dB white-noise generator.
3. Assign mice to specific open-fields such that they are consistently tested in the same environment.
4. Before commencing tests, habituate mice to the open-field arenas. Place each mouse in the center of its open-field and allow it to freely explore the environment. It is recommended to habituate mice at least once before testing.
5. Begin test days with a brief habituation period (30–60 min). Place each mouse in the center of its open-field and video-record their behavior.
6. Briefly remove mice from open-fields and return them to individual cages. Using dry paper towel, lightly wipe down each open-field and remove any feces and urine. Do not use ethanol or any other cleaner as the odors may influence behavior.
7. Administer injections to each mouse, before returning them to the center of their open-field. Begin recording video for the duration of your test phase.
8. If mice are performing additional tests on that day, repeat steps 6–7. When mice have completed testing, return them to their home cages and housing room.
9. Analyze videos using software (e.g., Noldus Ethovision) and calculate the measures of interest (horizontal distance traveled, rotations, etc.).

4.2.3 HSV-M_5 Transfections

A Herpes simplex viral vector (HSV) was used to transfect M_5 DNA [58] in a small field, to precisely define neurons causing the behavioral changes [59, 60] (Fig. 2d). Due to the time course of the HSV transfections, an accelerated testing period was required as maximal gene expression for HSV vectors occurs 24–72 h following infusion [59, 61]. M_5-KO and WT mice were run in groups of 4–6 mice over a period of 5 days [22]. On day 1, surgery was performed and either the HSV-M_5-GFP or HSV-control-GFP vector was bilaterally infused (at a volume of 0.5 μl per hemisphere at 2.5 μl/min) into the VTA or RMTg (relative to bregma: VTA—AP: −3.40, ML: ±0.50, DV: −4.40; RMTg—AP: −3.90, ML: ±0.40, DV: −3.88 [62]. On day 2, mice received standard post-surgical treatment and were allowed the day to recover.

Behavioral testing occurred on days 3 and 4, and each day began with a 60-min habituation period. On day 3, following habituation, each mouse was briefly removed from its test box, received a saline injection and was immediately placed back in the center of its box. Video recording occurred over the next 2 h to track saline-induced locomotion. After the 2 h trial, mice were again briefly removed from their boxes, injected with 10 mg/kg morphine i.p., and were returned to the center of their respective boxes. Again, video acquisition occurred for 2 h. Following the morphine trial, mice were returned to their home cages and returned to the housing room. On day 4, following habituation, mice were injected with 30 mg/kg i.p. morphine. Again, locomotion was recorded for a 2 h period after which, the mice were returned to their home cages in the housing room. In order to minimize order effects over the 2 consecutive test days, all mice received increasing morphine doses (0, 10, 30 mg/kg) with each subsequent test. Video recordings were scored using Noldus Ethovision V7.0 (Groningen, Netherlands), which calculated the distance traveled (in m) for each 15-min period.

4.2.4 AAV-M_3D or M_4D Transfections of RMTg in GAD2::Cre Mice

To study the locomotor effects of muscarinic signaling on RMTg GABA neurons, we transfected GAD2::Cre mice with genes encoding the mutant muscarinic receptors M_3D or M_4D specifically into RMTg GABA neurons [56, 63] using Cre-responsive adeno-associated viral vectors (AAV) [55].

GAD2::Cre mice received bilateral microinjections of either AAV-M3D or AAV-M4D vectors (at a volume of 0.2 μl per hemisphere at a rate of 0.1 μl/min) into the RMTg (relative to bregma: RMTg—AP: −3.90, ML: ±0.40, DV: −3.88) [62].

Behavioral testing began 4–5 weeks after AAV infusions and occurred on alternate days for a total of 6 sessions. Testing sessions began with each mouse placed in the center of its test box for 1-h habituation, then briefly removed for an i.p. injection and returned to its box for a 3-h test. Test sessions 1 and 4, however, served as no-injection control trials, where, following habituation, each mouse was removed from its box, briefly handled, and then returned for a 3-h test without any injection. On test days 2 and 3, mice received either saline/vehicle alone, or CNO (counterbalanced). On test days 5 and 6, mice received either morphine alone, or morphine with CNO (counterbalanced). Following each testing session, mice were returned to their home cages, then to the housing room. All test sessions were video-recorded using overhead digital cameras. Recordings were analyzed using Noldus Ethovision V7.0 (Groningen, Netherlands) to calculate total horizontal distance traveled for each 30-min period.

4.3 Notes

5. A cDNA containing the full-length reading frame of mouse M_5 receptor DNA (M_5R) was subcloned into the bi-cistronic amplicon vector p1005 (Fig. 2d). The resulting vector, HSV-M_5-GFP, contained mM_5R after the HSV-derived IE4/5 promoter along with the cDNA for green fluorescent protein (GFP) after a CMV promoter. Replication-deficient Herpes

simplex virus-derived particles were made from this vector as previously described [60]. A control virus (p1005), HSV-control-GFP was also constructed; this was identical to the HSV-M_5 with the M_5R gene excluded. Both viruses (M_5R-p1005 and p1005) had titers $>1 \times 10^8$ infectious units/ml.

6. Morphine sulfate pentahydrate dissolved in sterile saline and injected i.p. at a volume of 10 ml/kg body weight and concentrations of 10 mg/kg and 30 mg/kg calculated according to free base weight.
7. To selectively excite or inhibit RMTg GABA neurons in mice, we used two mutant muscarinic receptors, M_3D and M_4D, also known as DREADDS: Designer Receptors Exclusively Activated by Designer Drugs [56]. We employed a Cre recombinase-dependent, AAV expression system called the AAV-FLEX approach. The AAV-FLEX vector carries a reversed and double-floxed effector gene to enable specific expression in transgenic mice expressing Cre recombinase under the control of the GAD2 promoter. The coding sequences of the M_3D or M_4D DREADDs were linked to the fluorescent protein, mCherry, and packaged into a Cre recombinase-dependent AAV vector to allow expression of either M_3D or M_4D and mCherry fusion proteins. In the absence of Cre recombinase, transgenes (M_3D-mCherry or M_4D-mCherry) are inverted with respect to the promoter between two pairs of heterotypic, antiparallel loxP sites, and thus transgene expression is off. When introduced to Cre-expressing cells, however, the transgene orientation is inverted by Cre-mediated excision, leading to the activation of the trans-gene expression. These mutant receptors are designed to be activated by CNO which can be systemically administered. Further, these receptors have been evolved such that they do not respond to endogenous ACh [64]. These receptors are identical to M_3 and M_4 muscarinic receptors except they both have the same 2 point mutations in the ACh binding site [65].
8. Morphine sulfate pentahydrate dissolved in sterile saline and injected i.p. at a volume of 10 ml/kg body weight and concentration of 10 mg/kg calculated according to free base weight.
9. CNO was dissolved in 0.5 % DMSO in sterile saline and injected i.p. at a volume of 10 ml/kg body weight for a final concentration of 2 mg/kg. CNO was generously donated through the National Institute of Mental Health's Chemical Synthesis and Drug Supply Program (Bethesda, MD).
10. Both mineral oil and water can be used to fill the assembly. Mineral oil is preferred, however, due to its greater viscosity and increased accuracy.

11. Depending on length of surgery, and properties of the viral vector, injectors may be loaded with enough vector for multiple sites and/or surgeries. If pursuing this option, ensure to load injector with excess vector. After infusing virus into first animal, wipe injector with ethanol, dry injector, and withdraw an additional air bubble (1.0 μl). Before beginning the following surgery, push out a volume greater than the withdrawn air bubble (e.g., 1.2 μl) to ensure that virus is correctly loaded at tip of injector.

5 Conclusions

M_2 and M_4 receptors directly inhibit LDT and PPT cholinergic neurons mediating arousal. Chronoamperometry studies in M_5 knockout mice show that activation of PPT/LDT cholinergic inputs to the midbrain results in sustained activation of dopamine neurons mainly via M_5 receptors. Behavioral studies in which M_5 receptors are downregulated (i.e., oligonucleotides or gene knockout) or upregulated (i.e., HSV M_5 transfection) show that M_5 receptors critically mediate sustained arousal and reward-seeking in response to brain-stimulation, food, and opioid rewards. These dopamine and cholinergic neuron populations are both inhibited by VTA and RMTg GABA neurons. RMTg GABA neurons express inhibitory M_4 and μ-opioid receptors that work together to disinhibit cholinergic and dopamine neurons in arousal states [55]. GFP-labeled transfections of these neurons with specific HSV- or AAV-muscarinic genes allows for studying the behavioral effects of inhibiting or activating these neurons using Cre lines and clozapine nitric oxide.

Acknowledgements

We thank our many collaborators and coauthors, including Gina Forster and Anthony Miller (electrochemistry), Haoran Wang, Sheena Josselyn, and Asim Rashid (HSV-M_5), Jun Chul Kim and Bryan Roth (AAV-M_3D and AAV-M_4D), and Junichi Takeuchi, Zheng-ping Jia, John Roder, and Juergen Wess (knockouts).

References

1. Bonner TI, Young AC, Brann MR, Buckley NJ (1988) Cloning and expression of the human and rat m5 muscarinic acetylcholine receptor genes. Neuron 1:403–410
2. Yeomans JS (2012) Muscarinic receptors in brain stem and mesopontine cholinergic arousal functions. Handb Exp Pharmacol 208:243–259
3. Yasuda RP, Ciesla W, Flores LR, Wall SJ, Li M, Satkus SA, Weisstein JS, Spagnola BV, Wolfe BB (1993) Development of antisera selective for m4 and m5 muscarinic cholinergic receptors: distribution of m4 and m5 receptors in rat brain. Mol Pharmacol 43: 149–157

4. Levey AI (1993) Immunological localization of m1-m5 muscarinic acetylcholine receptors in peripheral tissues and brain. Life Sci 52: 441–448
5. Vilaro MT, Palacios JM, Mengod G (1990) Localization of m5 muscarinic receptor mRNA in rat brain examined by in situ hybridization histochemistry. Neurosci Lett 114:154–159
6. Weiner DM, Brann MR (1989) The distribution of a dopamine D2 receptor mRNA in rat brain. FEBS Lett 253:207–213
7. Takeuchi J, Fulton J, Jia ZP, Abramov-Newerly W, Jamot L, Sud M, Coward D, Ralph M, Roder J, Yeomans J (2002) Increased drinking in mutant mice with truncated M5 muscarinic receptor genes. Pharmacol Biochem Behav 72:117–123
8. Yamada M, Lamping KG, Duttaroy A, Zhang W, Cui Y, Bymaster FP, McKinzie DL, Felder CC, Deng CX, Faraci FM, Wess J (2001) Cholinergic dilation of cerebral blood vessels is abolished in M(5) muscarinic acetylcholine receptor knockout mice. Proc Natl Acad Sci U S A 98:14096–14101
9. Redgrave P, Horrell RI (1976) Potentiation of central reward by localised perfusion of acetylcholine and 5-hydroxytryptamine. Nature 262: 305–307
10. Bauco P, Wise RA (1994) Potentiation of lateral hypothalamic and midline mesencephalic brain stimulation reinforcement by nicotine: examination of repeated treatment. J Pharmacol Exp Ther 271:294–301
11. Yeomans JS, Kofman O, McFarlane V (1985) Cholinergic involvement in lateral hypothalamic rewarding brain stimulation. Brain Res 329:19–26
12. Blaha CD, Allen LF, Das S, Inglis WL, Latimer MP, Vincent SR, Winn P (1996) Modulation of dopamine efflux in the nucleus accumbens after cholinergic stimulation of the ventral tegmental area in intact, pedunculopontine tegmental nucleus-lesioned, and laterodorsal tegmental nucleus-lesioned rats. J Neurosci 16:714–722
13. Yeomans JS, Baptista M (1997) Both nicotinic and muscarinic receptors in ventral tegmental area contribute to brain-stimulation reward. Pharmacol Biochem Behav 57:915–921
14. Yeomans JS, Takeuchi J, Baptista M, Flynn DD, Lepik K, Nobrega J, Fulton J, Ralph MR (2000) Brain-stimulation reward thresholds raised by an antisense oligonucleotide for the m5 muscarinic receptor infused near dopamine cells. J Neurosci 20:8861–8867
15. Sharf R, McKelvey J, Ranaldi R (2006) Blockade of muscarinic acetylcholine receptors in the ventral tegmental area prevents acquisition of food-rewarded operant responding in rats. Psychopharmacology (Berl) 186: 113–121
16. Rezayof A, Nazari-Serenjeh F, Zarrindast MR, Sepehri H, Delphi L (2007) Morphine-induced place preference: involvement of cholinergic receptors of the ventral tegmental area. Eur J Pharmacol 562:92–102
17. Rada PV, Mark GP, Yeomans JS, Hoebel BG (2000) Acetylcholine release in ventral tegmental area by hypothalamic self-stimulation, eating, and drinking. Pharmacol Biochem Behav 65:375–379
18. Mesulam MM, Mufson EJ, Wainer BH, Levey AI (1983) Central cholinergic pathways in the rat: an overview based on an alternative nomenclature (Ch1-Ch6). Neuroscience 10: 1185–1201
19. Oakman SA, Faris PL, Kerr PE, Cozzari C, Hartman BK (1995) Distribution of pontomesencephalic cholinergic neurons projecting to substantia nigra differs significantly from those projecting to ventral tegmental area. J Neurosci 15:5859–5869
20. Oakman SA, Faris PL, Cozzari C, Hartman BK (1999) Characterization of the extent of pontomesencephalic cholinergic neurons' projections to the thalamus: a comparison with projections to midbrain dopaminergic neurons. Neuroscience 94:529–547
21. Watabe-Uchida M, Zhu L, Ogawa SK, Vamanrao A, Uchida N (2012) Whole-brain mapping of direct inputs to midbrain dopamine neurons. Neuron 74:858–873
22. Wasserman DI, Wang HG, Rashid AJ, Josselyn SA, Yeomans JS (2013) Cholinergic control of morphine-induced locomotion in rostromedial tegmental nucleus versus ventral tegmental area sites. Eur J Neurosci 38:2774–2785
23. Isa T, Hall WC (2009) Exploring the superior colliculus in vitro. J Neurophysiol 102: 2581–2593
24. Dean P, Redgrave P, Westby GW (1989) Event or emergency? two response systems in the mammalian superior colliculus. Trends Neurosci 12:137–147
25. Steriade M, Datta S, Paré D, Oakson G, Curró Dossi RC (1990) Neuronal activities in brainstem cholinergic nuclei related to tonic activation processes in thalamocortical systems. J Neurosci 10:2541–2559
26. Paré D, Steriade M, Deschênes M, Bouhassira D (1990) Prolonged enhancement of anterior thalamic synaptic responsiveness by stimulation of a brain-stem cholinergic group. J Neurosci 10:20–33

27. McCormick DA (1992) Neurotransmitter actions in the thalamus and cerebral cortex and their role in neuromodulation of thalamocortical activity. Prog Neurobiol 39:337–388
28. Semba K, Fibiger HC (1992) Afferent connections of the laterodorsal and the pedunculopontine tegmental nuclei in the rat: a retro- and antero-grade transport and immunohistochemical study. J Comp Neurol 323:387–410
29. Manns ID, Alonso A, Jones BE (2000) Discharge properties of juxtacellularly labeled and immunohistochemically identified cholinergic basal forebrain neurons recorded in association with the electroencephalogram in anesthetized rats. J Neurosci 20:1505–1518
30. Dringenberg HC, Olmstead MC (2003) Integrated contributions of basal forebrain and thalamus to neocortical activation elicited by pedunculopontine tegmental stimulation in urethane-anesthetized rats. Neuroscience 119: 839–853
31. Lepore M, Franklin KB (1996) N-methyl-D-aspartate lesions of the pedunculopontine nucleus block acquisition and impair maintenance of responding reinforced with brain stimulation. Neuroscience 71:147–155
32. Bechara A, van der Kooy D (1989) The tegmental pedunculopontine nucleus: a brainstem output of the limbic system critical for the conditioned place preferences produced by morphine and amphetamine. J Neurosci 9: 3400–3409
33. Leonard CS, Llinás R (1994) Serotonergic and cholinergic inhibition of mesopontine cholinergic neurons controlling REM sleep: an in vitro electrophysiological study. Neuroscience 59:309–330
34. Kohlmeier KA, Ishibashi M, Wess J, Bickford ME, Leonard CS (2012) Knockouts reveal overlapping functions of M(2) and M(4) muscarinic receptors and evidence for a local glutamatergic circuit within the laterodorsal tegmental nucleus. J Neurophysiol 108: 2751–2766
35. Chapman CA, Yeomans JS, Blaha CD, Blackburn JR (1997) Increased striatal dopamine efflux follows scopolamine administered systemically or to the tegmental pedunculopontine nucleus. Neuroscience 76:177–186
36. Yeomans JS, Mathur A, Tampakeras M (1993) Rewarding brain stimulation: role of tegmental cholinergic neurons that activate dopamine neurons. Behav Neurosci 107:1077–1087
37. Mathur A, Shandarin A, LaViolette SR, Parker J, Yeomans JS (1997) Locomotion and stereotypy induced by scopolamine: contributions of muscarinic receptors near the pedunculopontine tegmental nucleus. Brain Res 775:144–155
38. Forster GL, Blaha CD (2000) Laterodorsal tegmental stimulation elicits dopamine efflux in the rat nucleus accumbens by activation of acetylcholine and glutamate receptors in the ventral tegmental area. Eur J Neurosci 12: 3596–3604
39. Tzavara ET, Bymaster FP, Davis RJ, Wade MR, Perry KW, Wess J, McKinzie DL, Felder C, Nomikos GG (2004) M4 muscarinic receptors regulate the dynamics of cholinergic and dopaminergic neurotransmission: relevance to the pathophysiology and treatment of related CNS pathologies. FASEB J 18:1410–1412
40. Forster GL, Blaha CD (2003) Pedunculopontine tegmental stimulation evokes striatal dopamine efflux by activation of acetylcholine and glutamate receptors in the midbrain and pons of the rat. Eur J Neurosci 17:751–762
41. Forster GL, Yeomans JS, Takeuchi J, Blaha CD (2001) M5 muscarinic receptors are required for prolonged accumbal dopamine release after electrical stimulation of the pons in mice. J Neurosci 22:RC190
42. Steidl S, Miller AD, Blaha CD, Yeomans JS (2011) M_5 muscarinic receptors mediate striatal dopamine activation by ventral tegmental morphine and pedunculopontine stimulation in mice. PLoS One 6:e27538
43. Ohno K, Hondo M, Sakurai T (2008) Cholinergic regulation of orexin/hypocretin neurons through M(3) muscarinic receptor in mice. J Pharmacol Sci 106:485–491
44. Michel FJ, Robillard JM, Trudeau LE (2004) Regulation of rat mesencephalic GABAergic neurones through muscarinic receptors. J Physiol 556:429–445
45. Blaha CD, Phillips AG (1996) A critical assessment of electrochemical procedures applied to the measurement of dopamine and its metabolites during drug-induced and species-typical behaviors. Behav Pharmacol 7:675–708
46. Kawagoe KT, Zimmerman JB, Wightman RM (1993) Principles of voltammetry and microelectrode surface states. J Neurosci Methods 48:225–240
47. Blaha CD, Lane RF (1983) Chemically modified electrode for in vivo monitoring of brain catecholamines. Brain Res Bull 10:861–864
48. Basile AS, Fedorova I, Zapata A, Liu X, Shippenberg T, Duttaroy A, Yamada M, Wess J (2002) Deletion of the M5 muscarinic acetylcholine receptor attenuates morphine reinforcement and withdrawal but not morphine analgesia. Proc Natl Acad Sci U S A 99: 11452–11457
49. Fink-Jensen A, Fedorova I, Wortwein G, Woldbye DP, Rasmussen T, Thomsen M, Bolwig TG, Knitowski KM, McKinzie DL,

Yamada M, Wess J, Basile A (2003) Role for M5 muscarinic acetylcholine receptors in cocaine addiction. J Neurosci Res 74:91–96

50. Steidl S, Yeomans JS (2009) M5 muscarinic receptor knockout mice show reduced morphine-induced locomotion but increased locomotion after cholinergic antagonism in the ventral tegmental area. J Pharmacol Exp Ther 328:263–275
51. Schmidt LS, Miller AD, Lester DB, Bay-Richter C, Schülein C, Frikke-Schmidt H, Wess J, Blaha CD, Woldbye DP, Fink-Jensen A, Wortwein G (2010) Increased amphetamine-induced locomotor activity, sensitization, and accumbal dopamine release in M5 muscarinic receptor knockout mice. Psychopharmacology (Berl) 207(4):547–558
52. Miller AD, Forster GL, Yeomans JS, Blaha CD (2005) Midbrain muscarinic receptors modulate morphine-induced accumbal and striatal dopamine efflux in the rat. Neuroscience 136:531–538
53. Forster GL, Falcon AJ, Miller AD, Heruc GA, Blaha CD (2002) Effects of laterodorsal tegmentum lesions on behavioral and dopamine responses evoked by morphine and d-amphetamine. Neuroscience 114:817–823
54. Jhou TC, Xu SP, Lee MR, Gallen CL, Ikemoto S (2012) Mapping of reinforcing and analgesic effects of the mu opioid agonist endomorphin-1 in the ventral midbrain of the rat. Psychopharmacology (Berl) 224:303–312
55. Wasserman DW, Tan JM, Kim J, Yeomans JS (2014) Muscarinic control of rostromedial tegmental nucleus GABA neurons and morphine-induced locomotion. Poster presented at Society for Neuroscience, Washington, D.C., 15–19 November 2014.
56. Rogan SC, Roth BL (2011) Remote control of neuronal signaling. Pharmacol Rev 63: 291–315
57. Tzschentke TM (2001) Pharmacology and behavioral pharmacology of the mesocortical dopamine system. Prog Neurobiol 63:241–320
58. Neve RL, Geller AI (1995) A defective herpes simplex virus vector system for gene delivery into the brain: comparison with alternative gene delivery systems and usefulness for gene therapy. Clin Neurosci 3:262–267
59. Carlezon WA, Boundy VA, Haile CN, Lane SB, Kalb RG, Neve RL, Nestler EJ (1997) Sensitization to morphine induced by viral-mediated gene transfer. Science 277:812–814
60. Han JH, Kushner SA, Yiu AP, Hsiang HLL, Buch T, Waisman A, Bontempi B, Neve RL, Frankland PW, Josselyn SA (2009) Selective erasure of a fear memory. Science 323: 1492–1496
61. Liu M, Thankachan S, Kaur S, Begum S, Blanco-Centurion C, Sakurai T, Yanagisawa M, Neve R, Shiromani PJ (2008) Orexin (hypocretin) gene transfer diminishes narcoleptic sleep behavior in mice. Eur J Neurosci 28: 1382–1393
62. Franklin KBJ, Paxinos G (2007) The mouse brain stereotaxic coordinates, 3rd edn. Academic, San Diego, CA
63. Taniguchi H, He M, Wu P, Kim S, Paik R, Sugino K, Kvitsiani D, Fu Y, Lu J, Lin Y, Miyoshi G, Shima Y, Fishell G, Nelson SB, Huang ZJ (2011) A resource of Cre driver lines for genetic targeting of GABAergic neurons in cerebral cortex. Neuron 71:995–1013
64. Armbruster BN, Li X, Pausch MH, Herlitze S, Roth BL (2007) Evolving the lock to fit the key to create a family of G protein-coupled receptors potently activated by an inert ligand. Proc Natl Acad Sci U S A 104:5163–5168
65. Wess J, Eglen RM, Gautam D (2007) Muscarinic acetylcholine receptors: mutant mice provide new insights for drug development. Nat Rev Drug Discov 6:721–733

INDEX

Jaromir Myslivecek and Jan Jakubik (eds.), *Muscarinic Receptor: From Structure to Animal Models*, Neuromethods, vol. 107, DOI 10.1007/978-1-4939-2858-3,

MIX
Papier aus verantwortungsvollen Quellen
Paper from responsible sources
FSC® C105338

If you have any concerns about our products,
you can contact us on
ProductSafety@springernature.com

In case Publisher is established outside the EU,
the EU authorized representative is:
Springer Nature Customer Service Center GmbH
Europaplatz 3, 69115 Heidelberg, Germany

Printed by Libri Plureos GmbH
in Hamburg, Germany